T0335625

VOLUME ONE HUNDRED AND NINE

Advances in
COMPUTERS

A Deep Dive into NoSQL Databases:
The Use Cases and Applications

VOLUME ONE HUNDRED AND NINE

Advances in
COMPUTERS

A Deep Dive into NoSQL Databases: The Use Cases and Applications

Edited by

PETHURU RAJ
Vice President
Site Reliability Engineering (SRE) Division,
Reliance Jio Infocomm. Ltd. (RJIL), Bangalore, India

GANESH CHANDRA DEKA
Deputy Director (International Cooperation & Technology),
Ministry of Skill Development & Entrepreneurship,
Government of India, New Delhi, India

ACADEMIC PRESS
An imprint of Elsevier

Academic Press is an imprint of Elsevier
50 Hampshire Street, 5th Floor, Cambridge, MA 02139, United States
525 B Street, Suite 1800, San Diego, CA 92101-4495, United States
The Boulevard, Langford Lane, Kidlington, Oxford OX5 1GB, United Kingdom
125 London Wall, London, EC2Y 5AS, United Kingdom

First edition 2018

Notices

Knowledge and best practice in this field are constantly changing. As new research and experience broaden our understanding, changes in research methods, professional practices, or medical treatment may become necessary.

Practitioners and researchers must always rely on their own experience and knowledge in evaluating and using any information, methods, compounds, or experiments described herein. In using such information or methods they should be mindful of their own safety and the safety of others, including parties for whom they have a professional responsibility.

To the fullest extent of the law, neither the Publisher nor the authors, contributors, or editors, assume any liability for any injury and/or damage to persons or property as a matter of products liability, negligence or otherwise, or from any use or operation of any methods, products, instructions, or ideas contained in the material herein.

ISBN: 978-0-12-813786-4
ISSN: 0065-2458

For information on all Academic Press publications
visit our website at https://www.elsevier.com/books-and-journals

  Working together to grow libraries in developing countries

www.elsevier.com • www.bookaid.org

Publisher: Zoe Kruze
Acquisition Editor: Zoe Kruze
Editorial Project Manager: Shellie Bryant
Production Project Manager: James Selvam
Cover Designer: Christian J. Bilbow

Typeset by SPi Global, India

CONTENTS

Preface   *ix*

**1. A Detailed Analysis of NoSQL and NewSQL Databases for Bigdata
   Analytics and Distributed Computing**   **1**
   Pethuru Raj

   1. The Emergence of the Digital Era   2
   2. Envisioning the Digital Intelligence-Inspired Transformations   4
   3. The Significance of Next-Generation Data Analytics   5
   4. The Technologies and Tools for BDA   7
   5. The Prominent Big Data Analytics Use Cases   13
   6. Real-Time Analytics   16
   7. Streaming Analytics   17
   8. IoT Data Analytics   18
   9. Cognitive Analytics   19
  10. Edge/Fog Device Analytics   20
  11. Machine Data Analytics by Splunk   22
  12. The Rapid Rise of the Cloud Paradigm for Next-Generation Data Analytics   24
  13. The Emergence of Data Analytics Platforms   25
  14. Cloud Databases   25
  15. The Rewarding Role of NoSQL Databases   27
  16. Why NoSQL Databases?   31
  17. The Classification of NoSQL Databases   32
  18. Key-Value Data Stores   32
  19. The Challenges   34
  20. The Prominent Use Cases   34
  21. The Best Practices   34
  22. The Key Characteristics   34
  23. Key-Value Database vs Cache   35
  24. Columnar Databases   36
  25. Document Databases   38
  26. Graph Databases   41
  27. When to Use NoSQL and SQL Databases   43
  28. How NoSQL Databases Differ?   44
  29. The Tangible Benefits of NoSQL   45
  30. Summary   46
    Further Reading   47
    About the Author   48

2. NewSQL Databases and Scalable In-Memory Analytics   49
Siddhartha Duggirala

1.	Introduction	50
2.	In-Memory Databases	52
3.	Use Cases	66
4.	In-Memory Databases	69
5.	Conclusion	73
	References	73
	About the Author	76

3. NoSQL Web Crawler Application   77
Ganesh Chandra Deka

1.	Introduction to Web Crawlers	78
2.	Features of Web Crawlers	82
3.	Types of Web Crawler	83
4.	APIs of Web Crawler	87
5.	Challenges of Web Crawler Design	88
6.	Application of NoSQL Databases in Web Crawling	89
7.	Conclusion	93
	Key Terms and Definitions	94
	References	95
	Further Reading	99
	About the Author	100

4. NoSQL Security   101
Neha Gupta and Rashmi Agrawal

1.	Fundamentals of NoSQL Security	102
2.	Security Features of Various NoSQL Databases	111
3.	Injection Attacks on NOSQL Database	116
4.	Challenges in Designing, Implementing, and Deploying NoSQL Databases	120
5.	NoSQL Security Reference Architecture	122
6.	Techniques to Mitigate the Attacks on NoSQL Databases	126
7.	Proposed Security and Privacy Solutions for NoSQL Data Stores	128
8.	Conclusion	129
	Glossary	129
	References	130
	Further Reading	130
	About the Authors	131

5. Comparative Study of Different In-Memory (No/New) SQL Databases   **133**

Krishnarajanagar G. Srinivasa and Srinidhi Hiriyannaiah

1. Introduction   134
2. Advanced Database Processing   134
3. In-Database Analytics   138
4. NewSQL Databases   141
5. Case Study on Alteryx (Demonstration/Installation, Creation of Database/Record)   151
6. Conclusions   153
References   153
About the Authors   155

6. NoSQL Hands On   **157**

Rebika Rai and Prashant Chettri

1. Introduction   158
2. Document-Oriented Databases   159
3. Graph-Based Databases   197
4. Key–Value Databases   226
5. Hybrid SQL–NoSQL Databases   258
Key Terminology and Definitions   274
References   275
Further Reading   276
About the Authors   277

7. The Hadoop Ecosystem Technologies and Tools   **279**

Pethuru Raj

1. Introduction   280
2. Demystifying the Big Data Charters   281
3. Describing the Big Data Paradigm   282
4. Big Data Analytics: The Evolving Challenges   283
5. The Other Technologies and Tools in the Hadoop Ecosystem   297
6. How Does It Work?   317
7. Conclusion   319
Further Reading   319
About the Author   320

8. Biological Big Data Analytics   321

Mohammad Samadi Gharajeh

1. Introduction   322
2. What Is Big Data Analytics   323
3. Machine Learning for Big Data Analytics in Plants   330
4. Big Data Analytics in Bioinformatics   335
5. Big Data Analytics in Healthcare   339
6. Healthcare Information System Use Cases   343
7. Conclusions and Future Works   350
Glossary   351
References   352
Further Reading   354
About the Author   355

9. NoSQL Polyglot Persistence   357

Ganesh Chandra Deka

1. Introduction   358
2. Programming Paradigm   360
3. Polyglot Persistence   365
4. Polyglot Persistence and NoSQL   366
5. Big Data and Polyglot Persistence   370
6. Polyglot Persistence in e-Commerce   374
7. Polyglot Persistence in Healthcare   378
8. Research Trends in Polyglot Persistence   380
9. Conclusion   382
Glossary   383
References   384
Further Reading   389
About the Author   389

PREFACE

In the IoT-driven Bigdata world, the traditional SQL database is incapable of accommodating multistructured data. As the data size becomes massive, the common interactive processing of data becomes difficult. Hence the concept of NoSQL databases is drawing a huge support from both practitioners and academicians these days. NoSQL databases are capable of tackling massive volumes of polystructured data. These specially crafted databases aka NoSQL databases are intrinsically designed to meet the huge data storage, processing, and analysis requirements of Bigdata. This edited book titled "A Deep Dive Into NoSQL Databases: The Use Cases and Applications" is a collection of highly curated chapters on diverse topics in NoSQL databases, contributed by accomplished academicians, researchers, and database professionals from IT companies.

The introductory chapter of the book titled "A Detailed Analysis of NoSQL and NewSQL Databases for Bigdata Analytics and Distributed Computing" is primarily prepared to tell all about the various NoSQL and NewSQL databases and how they come handy in augmenting, accelerating, and automating the highly complicated phenomenon of next-generation data analytics.

Chapter 2 titled "NewSQL Databases and Scalable In-Memory Analytics" deliberates upon the prospects of in-memory NewSQL databases in Bigdata analytics.

Indexing of a huge number of webpages requires a cluster with several petabytes of disk space. Since the NoSQL databases are highly scalable, use of NoSQL database for storing the web crawler data is increasing along with the surging popularity of NoSQL databases. Chapter 3 titled "NoSQL Web Crawler Application" discusses about the prospects and application of NoSQL databases in web crawler application to store and analyze data collected by the web crawler.

Chapter 4 titled "NoSQL Security" discusses the various security threats of NoSQL databases, security architecture of NoSQL databases, and the steps that can be taken to secure the NoSQL database.

Amazon, Facebook, Google, and other reputed IT corporate houses use the world wide web as a large distributed data repository. This large data repository on the web cannot be processed with traditional RDBMS systems. Chapter 5 titled "Comparative Study of Different In-Memory

(No/New) SQL Databases" discusses about the application of in-memory databases with advanced data processing techniques for distributed data repository.

Chapter 6 titled "NoSQL Hands On" explains how to *Download, Install,* and *Use* some of the mostly used open source NoSQL database.

Trillions of digitized objects and connected devices and millions of Polyglot-persistent software services are interacting with one another *locally* as well as over variety of *networks*. *Apache Hadoop Ecosystem* technologies and tools are the best way forward to squeeze out the relevant knowledge from these interconnected devices and software services. Chapter 7 titled "The Hadoop Ecosystem Technologies and Tools" details about the emerging technologies and platforms under the *Apache Hadoop Ecosystem*.

Chapter 8 titled "Biological Big Data Analytics" presents various Bigdata analytics tools for bioinformatics systems and some use cases of healthcare information system (HIS).

Chapter 9 titled "NoSQL Polyglot Persistence" is about the application of Polyglot persistence in e-Commerce and Healthcare. This chapter also discusses the prospects of NoSQL database on Polyglot persistent software development and the research trends in NoSQL Polyglot persistent.

The salient feature of the book also includes lots of references for advance reading for researchers, use cases for practitioners, and hands-on for students and practitioners. In a nutshell, this book is stuffed with a number of elegantly written chapters describing the ways and means of surmounting the above-mentioned challenges. This book will be useful for *researchers, PG* as well as *UG* students of Computer Science, and *practitioners*.

PETHURU RAJ PH.D
GANESH C. DEKA

CHAPTER ONE

A Detailed Analysis of NoSQL and NewSQL Databases for Bigdata Analytics and Distributed Computing

Pethuru Raj
Site Reliability Engineering (SRE) Division, Reliance Jio Infocomm. Ltd. (RJIL), Bangalore, India

Contents

1. The Emergence of the Digital Era	2
2. Envisioning the Digital Intelligence-Inspired Transformations	4
3. The Significance of Next-Generation Data Analytics	5
3.1 Big Data Analytics	6
4. The Technologies and Tools for BDA	7
4.1 The Rise of Big Data Appliances	9
4.2 Big Data Processes	10
4.3 Apache Hadoop	10
5. The Prominent Big Data Analytics Use Cases	13
6. Real-Time Analytics	16
7. Streaming Analytics	17
8. IoT Data Analytics	18
9. Cognitive Analytics	19
10. Edge/Fog Device Analytics	20
11. Machine Data Analytics by Splunk	22
11.1 IBM Accelerator for Machine Data Analytics	23
12. The Rapid Rise of the Cloud Paradigm for Next-Generation Data Analytics	24
13. The Emergence of Data Analytics Platforms	25
14. Cloud Databases	25
15. The Rewarding Role of NoSQL Databases	27
16. Why NoSQL Databases?	31
17. The Classification of NoSQL Databases	32
18. Key-Value Data Stores	32
19. The Challenges	34
20. The Prominent Use Cases	34
21. The Best Practices	34
22. The Key Characteristics	34

Advances in Computers, Volume 109
ISSN 0065-2458
https://doi.org/10.1016/bs.adcom.2018.01.002

1

23. Key-Value Database vs Cache   35
24. Columnar Databases   36
25. Document Databases   38
    25.1 Document Examples   39
26. Graph Databases   41
    26.1 Use Cases   42
27. When to Use NoSQL and SQL Databases   43
28. How NoSQL Databases Differ?   44
29. The Tangible Benefits of NoSQL   45
30. Summary   46
Further Reading   47
About the Author   48

Abstract

In the recent past, different database management systems are emerging and evolving fast in order to systematically and spontaneously tackle the growing varieties and vagaries of data structures, schemas, sizes, speeds, and scopes. That is, all kinds of data have to be carefully and consciously collected, cleansed, and crunched in order to squeeze out actionable and timely insights out of exponentially exploding data heaps. Data is turning out be a strategic asset for any growing and glowing organization across the world in order to devise perfect and precise strategies and roll out correct activities in time with all the clarity, continuity, and confidence. The noteworthy point here is that you cannot throw out any data (internal as well as external) as data can potentially emit viable and venerable information and insights when subjected to decisive and deeper investigations. Not only for data transformation, ingestion, mining, processing, and analytics but also for effective data engineering, management, governance, representation, exchange, persistence, and science, we need efficient technologies, tools, and tips. As we are heading toward the dreamt knowledge era, the role and responsibility of new-generation database systems are bound to escalate in the days ahead. This chapter is primarily prepared to tell all about the various NoSQL and NewSQL databases and how they come handy in augmenting, accelerating, and automating the highly complicated phenomenon of next-generation data analytics.

1. THE EMERGENCE OF THE DIGITAL ERA

The aspect of automation is productively pervasive and penetrative these days. With IT being recognized as the most prominent and dominant enabler of every business vertical, it is being made to consciously embark on the highly fruitful automation spree. For example, the cloud space is being continuously stuffed and sandwiched with a bevy of automation technologies and tools to make elastic, programmable, and workload-aware IT infrastructures. There are competent techniques and tools for realizing virtualized IT environments quickly and easily. Similarly, there are advanced software solutions for beneficially integrating different and distributed cloud environments. Cloud orchestration tools are very popular for setting up cloud environments in an automated and augmented manner. Cloud broker solutions are fast emerging and evolving in order to pinpoint perfect cloud solutions and services in the increasingly software-defined multicloud era. In addition, there are other abstractions and articulations for autonomic, self-servicing, and federated clouds.

Thus the technology space is quite upbeat these days. Definitely, the trends and the transitions happening in the hot field of information and communication technologies (ICT) space are noteworthy and even praiseworthy as they are being pronounced as the ones that are intrinsically and decisively capable of instigating and impacting every human being. Precisely speaking, the ICT field has been the principal and prominent enabler of business tasks thus far and now it is being projected and proclaimed as the most crucial component for amply empowering people in their everyday decisions, deals, and deeds. In this regard, we often hear, read, and even sometimes experience about a bunch of trend-setting technologies and tools. Let me summarize and categorize the delectable advancements in the ICT discipline into three trends.

1. *Tending toward the trillions of digitized elements/smart objects/sentient materials*: The surging popularity and pervasiveness of a litany of digitization and edge technologies such as diminutive sensors, actuators, and other implantables, application-specific system-on-a-chip (SoC) innovations, miniaturized RFID tags, disappearing, disposable yet indispensable codes, controllers, chips, stickers and labels, nanoscale specks, smart dust, and particles, illuminating LED lights, etc. come handy in readily and rewardingly enabling every kind of common, casual, and cheap things in our everyday environments to join in the mainstream computing. All kinds of ordinary and tangible things in our midst get succulently transitioned into purposefully contributing and participating articles to accomplish extraordinary people-centric tasks. This tactically as well as strategically beneficial transformation is performed through the well-intended and designed application of disposable and disappearing yet indispensable digitization technologies. In short, every concrete thing gets systematically connected with one another as well as the remotely held business and IT services, platforms, middleware, etc. This phenomenon is being aptly termed as the Internet of Things (IoT).

   In other words, every kind of physical, mechanical, and electrical systems and electronics in the ground level gets hooked to various software applications and data sources at the far away as well as nearby cyber/virtual environments. Resultantly, the emerging domain of cyber-physical systems (CPS) is gaining immense attention lately.

2. *Ticking toward the billions of connected devices*: A myriad of electronics, machines, instruments, wares, vehicles, equipment, drones, pumps, wearables, hearables, robots, smartphones, and other devices across industry verticals are intrinsically instrumented at the design and manufacturing stages in order to embed the connectivity capability. These devices are also being integrated with software services and data sources at cloud environments to be enabled accordingly. That is, all kinds of smartly instrumented and integrated devices are bound to exhibit intelligent behavior all the time.

3. *Envisioning millions of software services*: With the accelerated adoption of microservices architecture (MSA), enterprise-scale applications are being expressed and exposed as a dynamic collection of fine-grained, loosely coupled, network-accessible, publicly discoverable, API-enabled, composable, lightweight, and polyglot services. This arrangement is all set to lay down a stimulating and sustainable foundation for producing next-generation software applications and services through service assembly at runtime. The emergence of the scintillating concepts such as Docker containers, IT agility, and DevOps in conjunction with MSA clearly foretells that the days of software engineering is bound to flourish bigger and better. Even hardware resources and assets are being software-defined in order to incorporate the much-needed flexibility, maneuverability, and extensibility. That is, everything is to be delivered as a service for various users with appropriate SLAs.

In short, every tangible thing becomes smart, every device becomes smarter, and every human being tends to become the smartest. The disruptions and transformations brought in by the above-articulated advancements are really mesmerizing. The IT has touched every small or big entity decisively in order to produce context-aware, service-oriented, event-driven, knowledge-filled, people-centric, and cloud-hosted applications. Data-driven insights and insights-driven enterprises are indisputably the new normal.

2. ENVISIONING THE DIGITAL INTELLIGENCE-INSPIRED TRANSFORMATIONS

As articulated earlier, all kinds of digitized items, connected devices, and software services produce a lot of digital data when they individually as well as collectively interact and collaborate with one another. We now live in the age where the number of electronic, embedded, and networked devices has already overtaken the human population worldwide. The extreme and deeper connectivity and integration are the grandeur source for big, fast, streaming, and machine data. Besides the machine-to-machine (M2M) communication, the data generated by billions of users over billions of devices for human-to-machine (H2M) interaction is in the range of zettabytes. Then there are other sources of data. Apart from the mesmerizing number of mobiles, there are implantables, wearables, and portables in plenty in our everyday environments. The volumes of data getting generated and garnered are definitely massive. The velocity with which the data gets captured, transmitted, and processed is also varying fast. The data variety is on the rise with numerous data sources. The data veracity and viscosity too ought to be taken into consideration. Finally, the value hidden inside data volumes is literally game-changing. Precisely speaking the digital data is simply tremendous and trendsetting. For the ensuing digital era, we need viable mechanisms in place in order to seamlessly capture all sorts of digital data and do purposeful analysis programmatically in order to squeeze out actionable insights in time. The knowledge discovery and dissemination have to be made pervasive and penetrative in order not to lose any kind of usable and reusable information locked inside data heaps.

We, therefore, need pioneering data analytics platforms, databases, warehouses, and lakes, highly optimized IT infrastructures, data communication networks, synchronized processes, enabling architectures, data virtualization, and visualization toolsets to set smooth the complicated process of knowledge discovery and dissemination out of big data. Digital data has to be subjected to a variety of technology-supported investigations in order to extract the right and relevant insights.

3. THE SIGNIFICANCE OF NEXT-GENERATION DATA ANALYTICS

The data volume, variety, velocity, veracity, viscosity, and value are changing rapidly. We need versatile processes, path-breaking algorithms, agile methods, integrated platforms, architectural and design patterns, evaluation metrics, key guidelines, best practices, etc. to simplify and streamline the process of turning data into information and into knowledge. Powerful analytics is the need of the hour in order to leverage all kinds of data emanating from heterogeneous sources. The traditional analytical techniques are found to be inadequate as we are being bombarded with multistructured data in massive quantities. Having understood the emerging needs, researchers, and IT professionals across the globe have brought in forth a number of delectable advancements in the field of data analytics through a host of new-generation systems, solutions, and services. Primarily there is big, fast, streaming and IoT data analytics. There are batch, real-time, interactive, and iterative data processing. In the big data analytics and characterizing big data, we are to see each of them in detail.

3.1 Big Data Analytics

As widely known, with the uninhibited growth of data sources, the resulting data sizes are ranging from terabytes and petabytes to exabytes. This phenomenon is being termed as big data and the implications of big data are vast and varied. The principal activity is to do a variety of tool-based and mathematically sound analyses on big data for instantaneously gaining big insights out of big data. It is a well-known fact that any organization having the innate ability to swiftly and succinctly leverage the accumulating data assets is bound to be successful in what they are operating, providing, and aspiring. That is, besides instinctive decisions, informed decisions go a long way in shaping up and confidently steering organizations. Thus, just gathering data is no more useful, but IT-enabled extraction of actionable insights in time out of those data serves well for the betterment of businesses. Data analytics is the formal discipline in IT for methodically doing data collection, filtering, cleaning, translation, storage, processing, mining, and analysis with the aim of extracting hidden intelligence. With this renewed focus, big data analytics (BDA) is getting more market and mind shares across the world. There are fresh analytical types and capabilities being unearthed out of BDA. The various industry verticals including energy, utility, retail, supply chain, healthcare, etc. are consciously embracing the BDA field to get big insights out of big data.

Characterizing Big Data—Big data is the general term used to represent massive amounts of data that are not stored in the relational form in traditional enterprise-scale SQL databases. New-generation database systems are being unearthed in order to store, retrieve, aggregate, filter, mine, and analyze big data efficiently. The following are the general characteristics of big data.

- Data storage is defined in the order of petabytes, exabytes, etc. in volume to the current storage limits (gigabytes and terabytes).
- There can be multiple structures (structured, semistructured, and less-structured) for big data.

- Big data comprises multiple types of data sources (sensors, machines, mobiles, social sites, log files, operational data, etc.).
- Data is time-sensitive (near real time as well as real time). That means Big data consists of data collected with relevance to the time zones so that timely insight can be extracted.

In short, big data services, applications, platforms, appliances, and infrastructures need to be designed in a way to facilitate the usage and leverage of big data. There are business, technology, and use cases for BDA. Product vendors, independent software vendors (ISVs), system integrators (SIs), cloud service providers (CSPs), consulting organizations, worldwide governments, managed services providers (MSPs), and other agencies have been working overtime in order to make BDA pervasive, productive and persuasive.

This recent entrant of BDA into the continuously expanding technology landscape has generated a lot of interest among industry professionals as well as academicians. Big data has become an unavoidable trend and it has to be solidly and succinctly handled in order to derive workable insights. There is a dazzling array of tools, techniques, and tips evolving in order to quickly capture data from diverse distributed resources and process, analyze, and mine the data to extract actionable business insights to bring in technology-sponsored business transformation and sustenance. Hadoop-based analytical products are capable of processing and analyzing any data type and quantity across hundreds of commodity server clusters. In short, analytics is the thriving phenomenon in every sphere and segment today. Especially with the automated capture, persistence, and processing of the tremendous amount of multistructured data getting generated by men as well as machines, the analytical value, scope, and power of data are bound to blossom further in the days to unfold. Precisely speaking, data is a strategic asset for organizations to insightfully plan to sharply enhance their capabilities and competencies and to embark on the appropriate activities that decisively and drastically power up their short- as well as long-term offerings, outputs, and outlooks. Business innovations can happen in plenty and be sustained too when there is a seamless and spontaneous connectivity between data-driven and analytics-enabled business insights and business processes.

4. THE TECHNOLOGIES AND TOOLS FOR BDA

There is no doubt that consolidated and compact platforms accomplish a number of essential actions toward simplified BDA and knowledge discovery. However, they need to run in optimal, dynamic, and converged infrastructures to be effective in their operations. In the recent past, IT infrastructures went through a host of transformations such as optimization, rationalization, automation, and simplification. The cloud idea has captured the attention of infrastructure specialists these days as the cloud paradigm is being proclaimed as the most pragmatic approach for achieving the ideals of infrastructure optimization. Hence with the surging popularity of cloud computing, every kind of IT infrastructure (servers, storages, and network solutions) is being consciously subjected to a series of modernization tasks to empower them to be policy-based, software-defined, service-oriented, programmable, etc. That is, BDA is to be performed in centralized/federated, virtualized, automated, shared, and optimized cloud infrastructures (private, public, and hybrid). Workload-aware IT environments are being readied for the big data era. Application-aware networks are the most sought-after communication infrastructures for data transmission and processing. Fig. 1 vividly illustrates all the relevant and resourceful components for simplifying and streamlining BDA.

The success of any technology is to be squarely decided based on the number of mission-critical applications it could create and sustain. That is, the applicability or employability of the new paradigm to as many application domains as possible is the main deciding factor for its successful journey. As big insights are becoming mandatory for multiple industry segments, there is a bigger scope for big data applications.

It is an unassailable truth that an integrated IT environment is a minimum requirement for attaining the expected success out of big data. Deploying big data platforms in an IT environment that lack a unified architecture and does

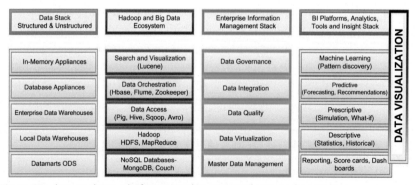

Fig. 1 Big data analytics platforms, appliances, products, and consoles.

not seamlessly and spontaneously integrate distributed and diverse data sources, metadata, and other essential resources would not produce the desired insights. Such deployments will quickly lead to a torrent of failed big data projects, and in a fragmented setup, achieving the desired results remains a pipe dream forever. Hence a unified and modular architecture is the need of the hour for taking forward the ideals of the big data discipline. Deploying big data applications in a synchronized enterprise or cloud IT environment makes analytics simpler, faster, cheaper, and accurate, while remarkably reducing deployment and operational costs.

In the ensuing era of big data, there could be multiple formats for data representation, transmission, and persistence. The related trend is that there are databases without any formal schema. SQL is the standard query language for traditional databases, whereas, in the big data era, there are NoSQL databases that do not support the SQL. Special file systems such as Hadoop Distributed File System (HDFS) are being produced in order to facilitate big data storage and access. Thus analytics in the big data period is quite different from the analytics on the traditional SQL databases. However, there is a firm place for SQL-based analytics and hence there is an insistence on converging both to fulfill the varying needs of business intelligence (BI). Tools and technologies that provide a native blending of classic and new data analytics techniques will have an inherent advantage.

4.1 The Rise of Big Data Appliances

Appliances (hardware and virtual) are being prescribed as a viable and value-adding approach for scores of business-critical application infrastructure solutions such as service integration middleware, messaging brokers, security gateways, data integrators, and load balancing. They are prefabricated with the full software stack so that their deployment and time to operation are quick and simple. There are XML and SOA appliances in plenty in the marketplace for eliminating all kinds of performance bottlenecks in business IT solutions. In the recent past, EMC Greenplum and SAP HANA appliances are stealing and securing the attention. SAP HANA is being projected as a game-changing and real-time platform for business analytics and applications. While simplifying the IT stack, it provides powerful features like significant processing speed, the ability to handle big data, predictive capabilities, and text mining capabilities. Thus the emergence and evolution of appliances represent a distinct trend as far as big data is concerned.

4.2 Big Data Processes

There are a number of noteworthy developments in the form of converged architecture, consolidated infrastructures, integrated platforms, versatile middleware, powerful databases, easy-to-use data lakes, etc. Still, we need highly synchronized processes in order to augment and accelerate big data analytics. Already analytics-attached processes are emerging and evolving consistently. That is, analytics has become such an essential activity to become tightly coupled with processes. Also, analytics as a service (AaaS) paradigm is on the verge of massive adaptation and hence analytics-enabling process integration, innovation, control, governance, and management aspects will gain more decisive prominence and dominance in the days to unfold.

The data is flowing endlessly from countless sources these days. The number of data sources is consistently on the climb. Innumerable sensors, varying in size, scope, structure, smartness, etc., are pouring data continuously. Financial institutions and stock markets are emitting a lot of data every second, e-commerce and business transactions generate a lot of data not to be casually treated, aircraft engines are generating a lot of operational data to be elegantly monetized, and the logs generated by hardware and software system are being received, stored, processed, analyzed, and acted upon ceaselessly. Monitoring agents are working tirelessly producing a lot of usable and useful data, business events are captured, knowledge discovery is initiated, information visualization is realized, etc. to empower enterprises' operations.

In this section, you can read more about the Hadoop technology. As elucidated before, big data analysis is not a simple affair and there are Hadoop-based software programming frameworks, platforms, and appliances emerging to tackle the growing complications. The Hadoop programming model has turned out to be the central and core method to propel the field of big data analysis. The Hadoop ecosystem is continuously spreading its wings wider and enabling modules are being incorporated freshly to make Hadoop-based big data analysis simpler, succinct, and supple.

4.3 Apache Hadoop

Apache Hadoop is an open source framework that allows for the distributed processing of large data sets across clusters of commodity computers and virtual machines using a simple programming model. Hadoop was originally designed to scale up from a single server to thousands of machines, each offering local computation and storage. Rather than rely on high-end hardware modules to deliver the much-needed scalability and availability, the

Hadoop software library is designed to detect and handle failures of IT resources. Hadoop is destined to deliver a highly available service on top of a cluster of cheap computers, each of which may be prone to failures. Hadoop is based out of the modular architecture and thereby any of its components can be swapped with competent alternatives if such a replacement brings praiseworthy advantages.

Because of the newness and the associated complexity of Hadoop, there are several areas wherein confusion reigns and restrains its full-fledged assimilation and adoption. The Apache Hadoop product family includes the Hadoop Distributed File System (HDFS), MapReduce, Hive, HBase, Pig, Zookeeper, Flume, Sqoop, Oozie, Hue, and so on. HDFS and MapReduce together constitute the core of Hadoop. For applications in BI, data warehousing (DW), and big data analytics, the core Hadoop is usually augmented with Hive and HBase and sometimes with Pig. The Hadoop file system excels with big data and it is file-based comprising multistructured (structured, semistructured, and nonstructured) data. HDFS is a distributed file system designed to run on clusters of commodity hardware. HDFS is highly fault tolerant because it automatically replicates file blocks across multiple machine nodes and is designed to be deployed on low-cost hardware. HDFS provides high-throughput access to application data and is suitable for applications that have large data sets. Because it is file-based, HDFS itself does not offer random access to data and has limited metadata capabilities when compared to a DBMS. Likewise, HDFS is strongly batch-oriented and hence has limited real-time data access functions. To overcome these challenges, you can layer HBase over HDFS to gain some of the mainstream DBMS capabilities. HBase is modeled after Google's Bigtable, and hence HBase, like Bigtable, excels with random and real-time access to very large tables containing billions of rows and millions of columns. Today HBase is limited to straightforward tables and records with little support for more complex data structures. The Hive meta-store gives Hadoop some DBMS-like metadata capabilities.

When HDFS and MapReduce are combined, Hadoop easily parses and indexes the full range of data types. Furthermore, as a distributed system, HDFS scales well and has a certain amount of fault-tolerance based on data replication even when deployed on commodity hardware. For these reasons, HDFS and MapReduce can complement existing BI/DW systems that focus on structured and relational data. MapReduce is a general-purpose execution engine that works with a variety of storage technologies including HDFS, other file systems, and some DBMSs.

As an execution engine, MapReduce and its underlying data platform handle the complexities of network communication, parallel programming, and fault-tolerance. In addition, MapReduce controls hand-coded programs and automatically provides multithreading processes, so they can execute in parallel for massive scalability. The controlled parallelization of MapReduce can apply to multiple types of distributed applications, not just analytic ones. In a nutshell, Hadoop MapReduce is a software programming framework for easily writing massively parallel applications which process massive amounts of data in parallel on large clusters (thousands of nodes) of commodity hardware in a reliable and fault-tolerant manner. A MapReduce job usually splits the input data set into a number of independent chunks, which are processed by the map tasks in a completely parallel manner. The framework sorts the outputs of the maps, which are then inputted to the reduce tasks which, in turn, assemble one or more result sets.

Hadoop is not just for new analytic applications, but it can revamp old ones too. For example, analytics for risk and fraud is based on statistical analysis or data mining. This analytics process benefits immensely from the much larger data samples, whose HDFS and MapReduce can wring from diverse data sources. Further on, most of the 360-degree customer views include hundreds of customer attributes. Hadoop has the inherent capability to include thousands of attributes and hence is touted as the best-in-class approach for next-generation precision-centric analytics. Hadoop is a promising and potential technology that allows large data volumes to be organized and processed while keeping the data on the original data storage cluster. For example, clickstreams and weblogs can be turned into browsing behavior (sessions) by running MapReduce programs (Hadoop) on the compute cluster and generating aggregated results on the same cluster. The attained results are then loaded into a relational DBMS system to be queried using structured query languages.

HBase is the mainstream Apache Hadoop database. It is an open source, nonrelational (column-oriented), scalable, and distributed database management system that supports structured data storage. Apache HBase is the right approach when you need random and real-time read/write access to your big data. This is for hosting of very large tables (billions of rows × millions of columns) on top of clusters of commodity hardware. Just as Google Bigtable leverages the distributed data storage provided by the Google File System, Apache HBase provides Bigtable-like capabilities on top of Hadoop and HDFS. HBase does support writing applications in Avro, REST, and Thrift. There is a separate chapter on the Hadoop ecosystem covering all about its origin, the widespread growth, and impacts besides some of the distinct use cases.

5. THE PROMINENT BIG DATA ANALYTICS USE CASES

Enterprises can understand and gain the value of big data analytics based on the number of value-add use cases and how some of the hitherto hard-to-solve problems can be easily tackled with the help of BDA technologies and tools. Every enterprise is mandated to grow with the help of powerful data analytics capability. As elucidated before, with big data, big analytics is the norm for businesses to take informed decisions proactively. Several domains are eagerly enhancing their IT capability to have the embedded analytics in place and there are several reports eulogizing the sophistication of BDA. The following are some of the prominent use cases.

Customer Satisfaction Analysis—This is the prime problem for most of the product organizations across the globe. There is no foolproof mechanism in place to understand the customers' feelings and feedbacks about their products. Gauging the feeling of people correctly and quickly goes a long way for enterprises to ring in proper rectifications and recommendations in product design, development, servicing, and support. And this has been a vital task for any product manufacturers to be relevant for their customers and product consumers. Thus customers' reviews regarding the product quality need to be carefully collected through various internal as well as external sources such as channel partners, distributors, sales and service professionals, retailers, and in the recent past, through social sites, microblogs, surveys, opinions, comments, call center details, etc. However, the issue is that the data being gleaned are extremely unstructured, repetitive, unfiltered, and unprocessed. The streamlined extraction of actionable insights out of the data volumes becomes a difficult affair here and hence leveraging big data analytics processes, platforms, and practices for a single, consolidated, and 360-degree view of customers (SVoC) helps enterprises to gain sufficient insights into the much-needed customer mindset and to solve their problems effectively and to overcome the identified concerns and complaints in their new product lines.

Market Sentiment Analysis—In today's competitive and knowledge-driven market economy, business executives and decision makers need to gauge the market environment deeply to be successful in their decisions, deals, and deeds. What are the products shining in the market, where the market is heading, who are the real competitors, what are their top-selling products, how they are doing in the market, what are the bright spots and prospects, and what are customers' preferences in the short- as well as long-term prospects through a deeper analysis legally and ethically. This information is available in a variety of websites, social media sites, and other public domains. BDA on this data can provide an organization with the much-needed information about Strength, Weakness, Opportunities, and Threats (SWOT) for their product lines.

Epidemic Analysis—Epidemics and seasonal diseases like flu start and spread with certainly noticeable patterns among the people and so it is pertinent to extract the hidden information to put a timely arrest on the outbreak of the infection. It is all about capturing all types of data originating from different sources, subjecting them to a series of investigations to extract actionable insights quickly, and contemplating the appropriate countermeasures.

In a similar line, with the leverage of the innumerable advancements being accomplished and articulated in the multifaceted discipline of big data analytics, various industry segments are jumping on the big data bandwagon in order to make themselves ready to acquire superior competencies especially in anticipation, ideation, implementation, and improvisation of premium and path-breaking services and solutions for the world market. BDA brings forth fresh ways for businesses and governments to analyze a vast amount of unstructured data (streaming as well as stored) to be highly significant and sagacious to their customers and constituencies.

Using Big Data Analytics in Healthcare—The healthcare industry has been a late adopter of ICT when compared to other industries such as banking, retail, and insurance. As per the McKinsey report on Big data from June 2011, if US health care could use big data creatively and effectively to drive efficiency and quality, it was estimated that the potential value from data in this sector could be more than $300 billion in value every year.

Reduce Hospital Readmission—One major cost in healthcare is the hospital readmission costs due to lack of sufficient follow-ups and proactive engagement with patients. These follow-up appointments and tests are often only documented as free text in patients' hospital discharge summaries and notes. This unstructured data can be mined using text analytics. If timely alerts were to be sent, appointments scheduled, or education materials dispatched, proactive engagement could potentially reduce readmission rates by over 30%.

Patient Monitoring: Inpatient, Outpatient, Emergency Visits, and ICU—Everything is becoming digitized. With rapid and rewarding progress in the technology space, diminutive sensors are embedded in weighing scales, blood glucose devices, wheelchairs, patient beds, X-ray machines, etc. Digitized devices generate large streams of data in real time that can provide insights into patient's health and behavior promptly. If this data is captured in time, it can be put to use to improve the accuracy of information and enable practitioners to better utilize the limited resources of service providers. It will also significantly enhance patient experience at a health care facility by providing proactive risk monitoring, improved quality of care and personalized attention. Big data can enable complex event processing (CEP) by providing real-time insights to doctors and nurses in the control room.

Preventive Care for ACO—One of the key accountable care (ACO) goals is to provide preventive care. Disease identification and risk stratification will be very crucial to business function. Managing real-time feeds coming in from HIE, pharmacists, providers, and payers will deliver key information to apply risk stratification and predictive modeling techniques. In the past, companies were limited to historical claims and HRA/survey data, but with HIE, the whole dynamic to data availability for health analytics has changed dramatically. Big data tools can sharply enhance the speed of processing and data mining.

Epidemiology—Through HIE, most of the providers, payers, and pharmacists will be connected through networks in the near future. These networks will facilitate the sharing of data to better enable hospitals and health agencies to track disease outbreaks, patterns, and trends in health issues across a geographic region or across the world allowing determination of the source and containment plans.

Patient Care Quality and Program Analysis—With the exponential growth of data and the need to gain insight from information comes to the challenge to process the voluminous variety of information to produce metrics and key performance indicators (KPIs) that can improve patient care quality and medical aid programs. The enlarging domain of big data analytics provides the reference architecture, tools, and techniques that will allow processing terabytes and petabytes of data to provide deeper analytic capabilities to its stakeholders.

It is conventional to do batch processing of big data, but the real challenge is to do real-time analytics on big data. There are fresh use cases expecting such a turnaround in the computing arena.

6. REAL-TIME ANALYTICS

Not only big data but also we come across fast and real-time data. That is, the volume/size of data is comparatively on the lower side, but the data velocity is accelerating. That is, the speed at which data gets generated has to be equally matched by the data capture, transmission, and processing speeds in order to make sense out of it. It is all about real-time gathering and processing of data in order to emit real-time insights that in turn lead to faster decision making and action. You cannot create batches out of fast data to do lazy and leisure processing. Whenever there is some data hitting the analytics platform, it has to be crunched quickly to produce useful knowledge to be disseminated to the concerned systems, devices, and people swiftly to ponder about and proceed with the next course of actions.

In short, data has to be captured, transmitted, translated, and analyzed at the frequently varying velocity in order to have trustworthy and timely insights. We have legions of sensors, sensors-attached operational systems, and actuators leading to a lot of perceivable and noteworthy events. And these event messages comprising crucial data, documents, and details ought to be procured fast and subjected to the appropriate analytics processes to generate actionable knowledge.

In the recent past, the aspect of real-time analytics has gained greater weight and several product vendors have been flooding the market with a number of highly synchronized and state-of-the-art solutions (software as well as hardware) for facilitating on-demand, ad hoc, and real-time analysis of online as well as offline data. There are a number of advancements in this field due to its huge and hitherto untapped potentials. Especially worldwide companies leverage real-time analytics of their transaction, operational, performance, security, and log data toward considerably reducing operational expenditures while increasing the productivity. Stream computing drives continuous and cognitive analysis of massive volumes of streaming data with submillisecond response times. There are enterprise data warehouses, analytical platforms, in-memory appliances, etc. Enterprise-grade data warehousing delivers deep operational insights with advanced in-database analytics. The EMC Greenplum Data Computing Appliance (DCA) is an integrated analytics platform that accelerates analysis of big data assets within a single integrated appliance. IBM PureData System for Analytics architecturally integrates database, server, and storage into a single, purpose-built, and easy-to-manage system to speed up the analytics activity. Then SAP

HANA is an exemplary platform for real-time analytics of real time and big data. Platform vendors are conveniently tied up with infrastructure vendors especially CSPs to take analytics to the cloud so that the goal of data analytics as a service (DAaaS) sees a neat and nice reality sooner than later. There are multiple startups with innovative product offerings to speed up and simplify the complex part of next-generation data analysis.

7. STREAMING ANALYTICS

Streaming analytics is typically the systematic analysis of large and in-motion data. The data is continuously streamed from different and distributed sources. These data streams primarily pour in because of multiple sequential as well as parallel events that occur as the result of an action or set of actions. There are events in the form of threshold breakups, triggers, state changes, actions, decisions, etc. Simple events are easy to handle, whereas business events in the extremely connected world are a bit complicated as they are generally made out of multiple discrete, basic, and atomic events.

Event streams are generally continuous and represent the boundary-less flow. As articulated earlier, due to the growing multiplicity of connected devices, digitized entities, software-defined infrastructures, software applications, social websites, and other data-generating sources, the event-processing capability is steadily incorporated into the corporate IT infrastructure and has gained an immense adoption in the recent past. With the unprecedented growth of the IoT adoption, the unique contributions of real-time analytics technologies and tools for making sense out of event streams are bound to grow further. Precisely speaking, as the data volumes see exponential growth, the potential benefits of next-generation data analytics are to reach greater heights in the days ahead.

We often hear about simple event processing and CEP techniques and tools. These integrated tools come handy in squeezing out actionable insights from event streams in real time through data manipulation, normalization, cleansing, advanced analytics, and pattern-of-interest detection. Streaming analytics represents the integrated set of powerful algorithms, technologies, and tips to facilitate the act of knowledge discovery out of event streams.

Traditional approaches rely on batch processing, which is done as per agreed schedules. The batch analytics is appropriate for polystructured big data and inherently reactive. The timeliness of data is immaterial here and hence businesses could majorly react to past events or conditions.

The historical information batched together and subjected to a variety of investigations. There are platforms and cloud infrastructures for doing batch processing. Traditional architectures are sorely limited because they have real difficulty in efficiently managing and tracking the consumption of event streams. The streaming analytics could analyze millions of events per second, enabling submillisecond response times and instant decision making and actuation with all the clarity and confidence.

By contrast, the systematic analytics of event streams can capture events, assess them, make decisions, and share the outputs all within specific time windows. The operational efficiency is to go up, customer satisfaction is on the rise, and productivity is bound to grow with the successful and smart streaming analytics processes, platforms, patterns, and practices in place. Employing continuous queries is the way forward for stream processing. Streaming analytics connects to external data sources, enabling applications to integrate certain data into the application flow or to update an external database with processed information. Streaming analytics is the ability to constantly calculate statistical analytics while moving within the stream of data. Streaming analytics allows management, monitoring, and real-time analytics of live streaming data.

Stream analytics provides deeper and decisive insights through the real-time analytics of streaming data and knowledge visualization tools such as dashboards, report-generation tools, and others. Business activity and performance monitoring are the key benefits being accrued out of streaming analytics. The knowledge discovered and disseminated can substantially improve sales, reduce costs, identify errors, and provide information to react faster to risks to mitigate them. Streaming analytics accelerates decision making and provides access to business metrics and reporting.

8. IoT DATA ANALYTICS

We all know that the IoT domain is very strategic and futuristic. Everything in our everyday environments gets systematically digitized and readied for the era of ambient intelligence (AmI). Devices are getting succulently instrumented to be interconnected and intelligent in their behavior. The faster maturity and stability of edge technologies (sensors, actuators, tags, chips, codes, controllers, etc.) have led to the realization of CPS, connected devices, smart objects, sentient materials, etc.

When these empowered entities interact and collaborate for fulfilling various business processes in an automated fashion, the data getting generated, captured, garnered, stocked, and processed are extremely large in size. The speed, structure, and scope change a lot with the accumulation of IoT data. There are several industry use cases for putting IoT data analytics capability in place. There are corporates collecting and analyzing data from sensors on manufacturing equipment, pipelines, weather stations, smart meters, delivery trucks, and other types of machinery in order to extract usable and reusable information.

IoT analytics supplies competent information that help decision makers, executives, and other stakeholders in steering their organizations with all kinds of strategic planning and execution. Enterprise-grade software applications are being empowered by the insights by IoT data to be adaptive and people-centric. The IoT data analytics capability ultimately leads to a variety of hitherto unknown and insights-driven applications are to be designed, developed, deployed, and delivered. The principal advantages of IoT include the substantial reduction in maintenance costs, avoiding equipment failures, customer delight, and improving business operations. Retailers, restaurant chains, and makers of consumer goods can use data from mobiles, wearables, implantables, portables, fixed systems, and nomadic devices to do targeted marketing and promotions.

IoT data can be big, fast, and streaming and hence there are highly synchronized platforms to capture, cleanse, and crunch IoT data to emit out venerable insights. The volume, variety, velocity, viscosity, virtuosity, and value of IoT data are literally unprecedented. The problem is that while all this data is potentially useful, its value goes down quickly if the rich context required making sense of the data and act intelligently on it is not there. There are IoT data analytics platforms from both the open source community and the commercial-grade vendors. Further on, there are IoT application enablement platforms (AEPs) in plenty in order to work with analytics platform to extract and use the unearthed insights to produce sophisticated applications. Thus the analytics and application development go hand in hand. There are several challenges such as data virtualization, security, and knowledge visualization. The real-time analytics of the huge amount of IoT is another brewing challenge for algorithm developers, software engineers, data scientists, and infrastructure providers.

9. COGNITIVE ANALYTICS

There is a special significance for cognitive analytics. The cognitive analytics platform is capable of self-learning, training, modeling, and understanding all about the incoming data. This platform is also capable of bringing forth newer hypotheses through evidence. The key machine and deep learning algorithms are optimally implemented and firmed up inside the platform in order to automate the process of analytics. Model creation, training the model created, and testing the refined model in real-world environments and examinations are the key process in machine learning. That is, empowering machines to learn through algorithms and models is the game-changing phenomenon for the unprecedented popularity of cognitive analytics. Traditional analytics approaches are mandating users to feed the data to be examined at first and to write the appropriate processing logic/queries using any one of the programming, script, and query languages. That is, data and logic are the key ingredients needed for proceeding with analytical steps. But for cognitive analytics, data is enough and the cognitive analytics platform is to work on the supplied data to create and harden pioneering predictive models. Thus cognitive analytics is touted as the automated analytics. IBM Watson is the first and foremost cognitive analytics platform to work with all kinds of data. Watson is capable of answering users' questions by smartly leveraging the embedded NLP feature and Watson is blessed with enough power to interact with humans and machines in their native communication languages. Many industry verticals are leaning toward cognitive analytics in order to be ahead of their competitors.

10. EDGE/FOG DEVICE ANALYTICS

With the faster proliferation of IoT devices (resource-constrained as well as resource-intensive) in our daily environments, the volume of IoT devices data is growing exponentially as they are seamlessly and spontaneously interacting with one another for fulfilling different personal, professional, and social needs. The off-premise and on-premise clouds are being categorized as the most appropriate IT infrastructure for collecting and crunching IoT data to extract actionable insights. However, there is a new twist now. That is, for enabling real-time analytics, it is being recommended to create ad hoc IoT device clouds locally to capture, stock, partition, and analyze all sorts of IoT data. This new phenomenon is being termed as fog or edge device analytics. The following is a sample and simple list indicating why IoT data has to be captured and processed in time.
• Data that helps cities predict accidents and crimes

- Data that gives doctors real-time insight into information from pace-makers or biochips
- Data that optimize productivity across industries through predictive maintenance on equipment and machinery
- Data that creates truly smart homes with connected appliances
- Data that provides critical communication between self-driving cars

The other prominent reasons being bandied for choosing edge analytics include:

1. In the case of faraway and nearby clouds, a lot of precious and expensive network bandwidth gets wasted. With edge analytics, there is a signifi-cant reduction in the usage of network bandwidth.
2. All kinds of IoT data can be captured and processed quickly. The resulting knowledge can be disseminated to the right actuators to start the identified action in time.
3. A lot of irrelevant, repetitive, and redundant data can be filtered out at the source itself.
4. The IoT data security is fully ensured as the data need not pass through the open, public, and penetrable networks.

With the availability of billions of connected devices, the realization of fog or edge device clouds is getting speeded up in order to accomplish real-time analytics and actions. The picture below vividly illustrates how fog or edge computing is gaining immense popularity in the recent past. Interestingly, industrial IoT is the greatest beneficiary.

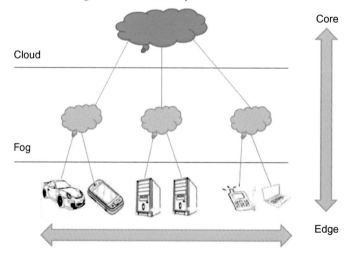

There are edge appliances and solutions in order to simplify and stream-line edge data collection, storage, processing, and knowledge dissemination.

11. MACHINE DATA ANALYTICS BY SPLUNK

All your IT applications, platforms, and infrastructures generate data every millisecond. The software and hardware systems generate a lot of data to be consciously captured and analyzed in time. Similarly, all our everyday machines, devices, robots, drones, cars, instruments, consumer electronics, aircraft engines, micro grids, ships, etc. are also generating a lot of usable and useful data to be picked and processed. A variety of actionable and timely insights emanate out of users' transactions, customer behavior, sensors and actuators networks, machines' operations, security threats, fraudulent activity, and more. Machine data hold critical insights useful across the enterprise.

- Monitor end-to-end transactions for online businesses providing 24 × 7 operations
- Understand customer experience, behavior, and usage of services in real time
- Fulfill internal SLAs and monitor service provider agreements
- Identify spot trends and sentiment analysis on social platforms
- Map and visualize threat scenario behavior patterns to improve security posture
- Any anomalies can be detected and acted upon

Making sense out of machine data has been challenging. The traditional database management solutions are found wanting as far as the act of extracting knowledge out of humongous machine data. Machine data originates from a multitude of disparate, distributed, and decentralized machines. The data is massive in volume, polystructured, and difficult to fit into a predefined schema. Machine data is time-series based, requiring new approaches for capture, stocking, management, and analysis. The most valuable insights from machine data are often needed in real time. Traditional business intelligence, data warehouse, or IT analytics solutions are simply not engineered for this class of high-volume, dynamic, time-critical, low-latency, and unstructured data.

As indicated in the beginning, machine-generated data is more voluminous than man-generated data. The applications, servers, network devices, storage and security appliances, sensors, browsers, compute machines, cameras, and various other systems deployed to support business operations are continuously generating information relating to their status, interactions, and activities. Machine data can be found in a variety of formats such as application log files, call detail records, user profiles, KPIs, and clickstream data associated with user web interactions, data files, system configuration files, alerts, and tickets. Machine data are generated by both machine-to-machine (M2M) and human-to-machine (H2M) interactions. Outside of the traditional IT infrastructure, every processor-based system including HVAC controllers, smart meters, GPS devices, actuators and robots, manufacturing systems, and RFID tags and consumer-oriented systems such as medical instruments, personal gadgets and gizmos, aircraft, scientific experiments, and automobiles that contain embedded devices is continuously generating machine data. The list is constantly growing. Machine data can be structured or unstructured. The growth of machine data has accelerated in recent times with the trends in IT consumerization and industrialization. That is, the IT infrastructure complexity has gone up remarkably driven by the adoption of portable devices, virtual machines, bring your own devices (BYODs) containers, and cloud-based services.

The goal here is to aggregate, parse, and visualize these data to spot trends and act accordingly. By monitoring and analyzing data emitted by a deluge of diverse, distributed and decentralized data, there are opportunities galore. Someone wrote that sensors are the eyes and ears of future applications. Environmental monitoring sensors in remote and rough places bring forth the right and relevant knowledge about their operating environments in real time. Sensor data fusion leads to develop context and situation-aware applications. With machine data analytics in place, any kind of performance degradation of machines can be spotted in real-time and corrective actions can be initiated with full knowledge and confidence. Security and surveillance cameras pump in still images and video data that in turn help analysts and security experts to preemptively stop any kind of undesirable intrusions. Firefighting can become smarter with the utilization of machine data analytics.

11.1 IBM Accelerator for Machine Data Analytics

Machines produce huge amounts of data that contain valuable and actionable information. However, accessing and working with that information requires large-scale import, extraction, transformation, and statistical analysis. IBM Accelerator for Machine Data Analytics, a set of end-to-end applications, helps you import, extract, index, transform, and analyze your data to

- Search within and across multiple log entries based on a text search, faceted search, or a timeline-based search to find events
- Enrich the context of log data by adding and extracting log types into the existing repository

- Link and correlate events across systems
- Uncover patterns.

These analytical capabilities are capable of producing diagnostic, prognostic, predictive, prescriptive, and personalized insights. Thus every bit of data getting generated and garnered is being transformed into information and knowledge through a bevy of path-breaking analytical platforms. The techniques for transitioning data into knowledge are matured and stabilized fast. The awareness of the advanced analytics has led to the heightened adoption. Not only businesses but also individuals are excited about the emerging and evolving analytical competencies. The insights unearthed are to act as a key inducer for realizing and sustaining highly adaptive, knowledge-filled, service-oriented, event-driven, cloud-hosted, people-centric, and situation-aware applications.

12. THE RAPID RISE OF THE CLOUD PARADIGM FOR NEXT-GENERATION DATA ANALYTICS

The cloud movement is expediently thriving and trend-setting a host of delectable novelties. A number of tectonic transformations on the business front are being activated and accentuated with faster and easier adaptability of the cloud IT principles. The cloud concepts have opened up a deluge of fresh opportunities for innovators, individuals, and institutions. IT professionals could conceive and concretize new-generation business services and solutions through the advancements happening in the cloud space. Without an iota of doubt, a dazzling array of path-breaking and mission-critical business augmentation models and mechanisms have emerged and they are consistently evolving toward the much-needed perfection as the cloud technology grows relentlessly in conjunction with other enterprise-class technologies. There are many books depicting and describing the steady growth of the cloud paradigm. One of the key applications leveraging the cloud concepts is the data analytics. As cloud infrastructures are inherently elastic enabling the realization and running of scalable applications, the complicated and data-intensive analytical applications are being easily fit in for cloud environments. There are public, private, and hybrid clouds for data analytics. Even the edge or fog devices are capable of forming ad hoc clouds in order to do real-time and local data analytics. Thus clouds (far away, nearby and localized) are being proclaimed as the most appropriate IT environments for doing comprehensive, contextual, and cognitive data analytics.

13. THE EMERGENCE OF DATA ANALYTICS PLATFORMS

Integrated platforms are essential in order to automate several tasks enshrined in the data capture, analysis, and knowledge discovery processes. A converged platform comes out with a reliable workbench to empower developers to facilitate application development and other related tasks such as data security, virtualization, integration, transformation, mining, visualization, and dissemination. Special consoles, engines, and runtimes are being attached in new-generation platforms for performing the important activities such as management, governance, enhancement, orchestration, and brokerage. Analytical platforms are being designed and developed to work upon any kind of data. Platforms guarantee batch, real-time, stream, interactive, and iterative processing. There are hundreds of connectors, adapters, and drivers for integrating and interacting with different and distributed data sources, connected devices, social, mobile, embedded, and enterprise-scale applications, data management systems, knowledge visualization tools, etc. In-memory and in-database analytics are gaining momentum for high-performance and real-time analytics. Thus platforms need to be fitted with new features, functionalities, and facilities in order to provide next-generation insights out of data volumes. As accentuated earlier, these platforms are primarily destined to run on cloud infrastructures. IBM Bluemix, being run on IBM cloud centers, is the key data analytics platform. GE Predix is another popular data analytics platform running on different cloud environments (public, private, and edge clouds).

Thus we have seen how different kinds of data formats and forms mandate multiple data analytics approaches. The continued rise of cloud IT brings the adaptivity, affordability, and automation while speeding up the new-generation data analytics.

14. CLOUD DATABASES

RDBMSs are an integral and indispensable component in enterprise IT and their importance is all set to grow and not to diminish. However, with the advent of cloud infrastructures, the opportunity to offer any DBMS as an offloaded and outsourced service is gaining momentum. Carlo Curino and his team members have introduced a new transactional "database as a service" (DBaaS) offering. A DBaaS promises to move the operational burden of provisioning, configuration, scaling, performance tuning, backup, privacy, and access control from the database users to the service operator, effectively offering lower overall costs to users. The DBaaS being provided by leading CSPs does not address three important challenges: efficient multitenancy, elastic scalability, and database privacy. The authors argue that before outsourcing database software and management into cloud environments, these three challenges need to be suppressed and surmounted.

The key technical features of this DBaaS: (1) a workload-aware approach to multitenancy that identifies the workloads that can be colocated on a database server achieving higher consolidation and better performance over existing approaches, (2) the use of a graph-based data partitioning algorithm to achieve near-linear elastic scale-out even for complex transactional workloads, and (3) an adjustable security scheme that enables SQL queries to run over encrypted data including ordering operations, aggregates, and joins. An underlying theme in the design of the components of DBaaS is the notion of workload awareness; by monitoring query patterns and data accesses, the system obtains information useful for various optimization and security functions, reducing the configuration effort for users and operators. By centralizing and automating many database management tasks, a DBaaS can substantially reduce operational costs and perform well. There are myriad advantages of using cloud databases, some of which are as follows:

- Fast and automated recovery from failures to ensure business continuity
- Either built-in to a larger package with nothing to configure, or comes with a straightforward GUI-based configuration
- Cheap backups, archival & restoration
- Automated on-the-go scaling with the ability to simply define the scaling rules or manually adjust
- Potentially lower cost, device independence, and better performance
- Scalability & automatic failover/high availability
- Anytime, anywhere, any device, any media, any network discoverable, accessible, and usable
- Less capital expenditure and usage-based payment
- Automated provisioning of physical as well as virtual servers in the cloud.

Some of the disadvantages include

- Security and privacy issues
- Constant Internet connection (bandwidth costs!) requirement

- Loss of controllability over resources
- Loss of visibility on database transactions
- Vendor Lock-in

Thus newer realities such as NoSQL and NewSQL database solutions are fast arriving and being adopted eagerly. On the other hand, the traditional database management systems are being accordingly modernized and migrated to cloud environments to substantiate the era of providing everything as a service. Data as a service, insights as a service, etc. are bound to grow considerably in the days to come as their realization technologies are fast maturing and stabilizing.

15. THE REWARDING ROLE OF NoSQL DATABASES

Our software applications are becoming steadily web and cloud-enabled. They are web-scale in their capability and capacity. The traditional online transaction processing (OLAP) systems need to be augmented accordingly. The cloud paradigm, as accentuated earlier, is the new one-stop IT solution and hence databases, data warehouses, data marts and cubes, data lakes, and their management software solutions are modernized artistically and migrated to cloud environments. The other noteworthy factor is that distributed computing is penetrating into every industry vertical and becoming pervasive and persuasive. The SQL databases are hugely limited by the lack of horizontal scalability, flexibility, availability, discoverability, etc. In short, for next-generation web-scale and cloud-based applications, the conventional SQL databases are found wanting in many aspects. There are research works initiated for upgrading SQL databases in the form of parallel and clustered SQL databases. Thus the use of NoSQL databases has mushroomed in the recent past.

The CAP theorem is another key foundation for the enormous success of NoSQL databases. This theorem (consistency, availability, and partition tolerance) states that it is impossible for any distributed system to simultaneously provide all the three capabilities. That is, any distributed system guarantees any two of the three stated later.

- Consistency (all nodes see the same data at the same time)
- Availability (a guarantee that every request receives a response about whether it was successful or failed)
- Partition Tolerance (the system continues to operate despite arbitrary message loss or failure of a part of the system).

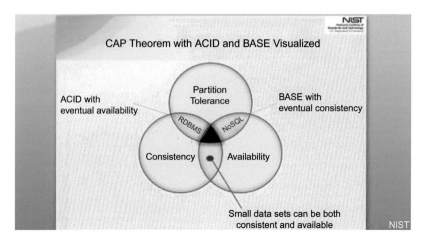

CAP Theorem with ACID and BASE Visualized

ACID with eventual availability

Partition Tolerance

BASE with eventual consistency

RDBMS   NoSQL

Consistency   Availability

Small data sets can be both consistent and available

ACID vs BASE—SQL databases fulfill ACID (atomicity, consistency, isolation, and durability) properties and accommodate only structured data. Another factor is that when the amount of data becomes humungous, the interactive interactions with SQL databases are bound to face enormous difficulties. SQL databases are very famous for transactional applications. RDBMSs were designed to manage structured data in manageable fields, rows, and columns such as dates, social security numbers, corresponding addresses, and transaction amounts. Conversely, most NoSQL databases tout their schema-less capability and this ostensibly allows for the ingestion of multistructured data. This works especially well for documents and metadata associated with a variety of unstructured data types. BASE (basically available, soft state, and eventually consistent) implies the database will, at some point, classify and index the content to improve the findability/discoverability of data or information contained in the text or the object.

We all know that HDFS is being positioned as the one-stop distributed file system solution for Hadoop-based data analytics. However, for interactive processing, the contributions of NoSQL databases are vast and varied. This section is specially crafted for telling all about the humble origin and the exciting journey of NoSQL databases. The next-generation databases are mandated to be nonrelational, distributed, open source, and horizontally scalable. The original inspiration is the modern web-scale databases. Additional characteristics such as schema-free, easy replication support, simple API, eventually consistent/BASE (not ACID), etc. are also being demanded. The traditional relational database management systems (RDBMSs) use structured query language (SQL) for accessing and manipulating data that reside in structured columns of relational tables. However, the forthcoming era is leaning toward unstructured data, which is typically stored in key-value pairs in a data store. These data stores, therefore, cannot be manipulated and managed using SQL. Such data are stored in NoSQL data stores and are accessed via different commands.

- *Flexible Data Model*—Relational and NoSQL data models are very different. The relational model takes data and separates it into many inter-related tables that contain rows and columns. Tables reference each other through foreign keys, which are stored in columns as well. When looking up data, the desired information needs to be collected from many tables and combined before it can be provided to the application or the users. Similarly, when writing data, the write needs to be coordinated and performed on many tables.

  NoSQL databases follow a very different model. For example, a document-oriented NoSQL database takes the data you want to store and aggregates it into documents using the JSON format. Each JSON document can be thought of as an object to be used by your application. A JSON document might, for example, take all the data stored in a row that spans 20 tables of a relational database and aggregate it into a single document/object. The resulting data model is flexible and easy to distribute the resulting documents. Another major difference is that relational technologies have rigid schemas while NoSQL models are schema-less. Changing the schema once data is inserted is a big deal, extremely disruptive and frequently avoided. However, the exact opposite of the behavior is desired in the big data era. Application developers need to constantly and rapidly incorporate new types of data to enrich their applications.

- *High Performance and Scalability*—To deal with the increase in the number of concurrent users (big users) and the amount of data (big data), applications and their underlying databases need to scale using one of two choices: scale up or scale out. Scaling up implies a centralized approach that relies on bigger and bigger servers with more quantity of resources such as memory, processing power, storage, and I/O capacity. Scaling out implies a distributed approach that leverages many commodity physical or virtual servers to tackle more user as well as data loads. Prior to the now famous NoSQL databases, the default scaling approach at the database tier was to scale up. This was dictated by the fundamentally centralized and shared-everything architecture of relational database technology.

  NoSQL databases were developed from the ground up to be distributed and scale-out databases. They use a cluster of physical or virtual servers to store big data and support all the standard database operations. To scale out, additional servers are joined to the cluster at runtime. The data and the various database operations are holistically spread across the larger cluster. Since commodity servers are expected to fail frequently, NoSQL databases are built to inherently tolerate and recover from such failures making them highly resilient. NoSQL databases provide a much easier and linear approach to database scaling. If 10,000 new users start using your application, simply add another database server to your cluster. There is no need to modify the application as you scale since the application always sees a single (distributed) database. NoSQL databases share some characteristics with respect to scaling and performance.

o *Auto-Sharding*—A NoSQL database automatically spreads data across servers without requiring applications to participate. Servers can be added or removed from the data layer without the much-worried application downtime. Most NoSQL databases support data replication storing multiple copies of same data across the cluster and even across data centers to ensure high availability (HA) and to support disaster recovery (DR). A properly managed NoSQL database system should never need to be taken offline, for any reason. Thus NoSQL databases are scalable and available.

o *Distributed Query Support*—Sharing a relational database can reduce or eliminate in certain cases the ability to perform complex data queries. NoSQL database systems retain their full query expressive power even when distributed across hundreds of servers.

o *Integrated Caching*—To reduce latency and increase sustained data throughput, advanced NoSQL database technologies transparently cache data in system memory. This behavior is transparent to the application developer and the operations team, compared to relational technology where a caching tier is usually a separate infrastructure tier that must be developed to and deployed on separate servers and explicitly managed by the operations team.

The business needs to leverage complex and connected data is driving the adoption of scalable and high-performance NoSQL databases. This new technology is to sharply enhance the data management capabilities of various businesses. Several variants of NoSQL databases have emerged over the past decade in order to handsomely handle the terabytes, petabytes, and even exabytes of data generated by networked embedded systems, enterprise applications, and services. They are specifically capable of stocking and processing multiple data types and massive quantities of data. That is,

NoSQL databases could accommodate data such as text, audio, video, social network feeds, weblogs, and much more that are not being handled efficiently by traditional databases. These data are highly complex and deeply interrelated. Therefore the brewing need is to unravel the truth hidden behind these huge yet diverse data assets. The unwrapped insights can be then used to enable business productivity and customer delight.

16. WHY NoSQL DATABASES?

B2C e-commerce and B2B e-business applications are highly transactional and the leading enterprise application frameworks and platforms such as Java Enterprise Edition (JEE) directly and distinctly support a number of transaction types (simple, distributed, nested, etc.). For a trivial example, flight reservation application has to be rigidly transactional; otherwise everything is bound to collapse. As enterprise systems are increasingly distributed, the need for transaction feature is being pronounced as a mandatory one.

In the recent past, web–scale social applications have grown fast and especially youth is totally fascinated by a stream of social computing sites, which are seeing an astronomical growth. It is no secret that the popularity, ubiquity, and utility of Facebook, LinkedIn, Twitter, Google +, and other social, professional, media sharing, musing, and blogging sites are surging incessantly. There is a steady synchronization between enterprise and social applications with the idea of adequately empowering enterprise applications with additional data-driven insights. For example, online sellers understand and utilize customers' choices, leanings, historical transactions, feedbacks, feelings, etc. in order to do more intensive and intimate business. That is, businesses are increasingly interactive, open, and inclined toward customers' participations to garner and glean their views toward bigger and better business opportunities. This intellectually inspiring integration is to result in richer enterprise applications (REAs). There are specialized protocols and web 2.0 technologies (Atom, RSS, AJAX, mash-up, etc.) to programmatically tag information about people, places, phenomena, and proclivity to dynamically conceive, conceptualize, and concretize people-centric and premium services.

The point to be conveyed here is that the current database technology has to evolve faster in order to accomplish this new-generation IT abilities. With the modern data being innately complicated, the NoSQL databases need to have the innate strength to cogently handle the multistructured and massive data. NoSQL databases should enable formulating complex queries on the data. Users should be able to ask questions such as "Who are all my contacts in Europe?" and "Which of my contacts ordered from this catalog?" The role and responsibility of NoSQL databases are on the climb. There are many business problems requiring the elegant services of NoSQL databases. There are a few important limitations of SQL databases and they could be eliminated with the adoption of NoSQL databases. Many enterprise-grade, web-scale, and cloud-based applications are increasingly depending on the unique capabilities of NoSQL database management systems in order to be distinct in their operations and outputs.

17. THE CLASSIFICATION OF NoSQL DATABASES

There are four major categories of NoSQL databases available today: key-value stores, column family databases, document databases, and graph databases. Each was designed to accommodate the huge volumes of data as well as to have room for future data types. The choice of NoSQL database depends on the type of data you need to store, its size, and complexity.

18. KEY-VALUE DATA STORES

These databases are the simplest of all the NoSQL databases: The basic data structure is a dictionary or map. You can store a value, such as an integer, a string, a JSON structure, or an array, along with a key used to reference that value. For example, a simple key-value database might have a value such as "Douglas Adams." This value is then assigned an ID, such as cust1237. Using a JSON structure adds complexity to the database. For example, the database could store a full mailing address in addition to a person's name. In the previous example, key cust1237 could point to the following information:

```
{name: "Douglas Adams",
 street: "782 Southwest St.",
 city: "Austin",
 state: "TX"}
```

Because key-value databases have no SQL-style query language to describe which values to fetch, keys are the primary mechanism used to reference data. Some key-value databases compensate for the lack of a query language by incorporating search capabilities. Instead of searching for the key, users can search values for particular patterns, for example, all values with a certain string in the name, such as "Douglas."

A key-value data model is quite simple. It stores data in key and value pairs where every key maps to a value. It can scale across many machines but cannot support other data types. Key-value data stores use a data model similar to the popular Memcached distributed in-memory cache, with a single key-value index for all the data. Unlike Memcached, these systems generally provide a persistence mechanism and additional functionality as well: replication, versioning, locking, transactions, sorting, and/or other features. The client interface provides inserts, deletes, and index lookups. Like Memcached, none of these systems offer secondary indices or keys. A key-value store is ideal for applications that require massive amounts of simple data like sensor data or for rapidly changing data such as stock quotes. Key-value stores support massive data sets of very primitive data. Amazon's Dynamo was built as a key-value store. Redis, Riak, and Oracle NoSQL database are examples of key-value databases.

The database uses a hash table to store unique key and pointers to each data value which in turn are stored in a schema-less way.

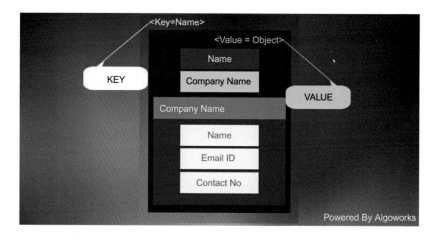

From the perspective of the database, "value" is just meaningless data that database just stores, without caring or knowing what is inside. For the database, it is the responsibility of the application to understand what was stored. There are no complex relations and hence the implementation is easy and since these databases only use a single primary key for access, these databases give great performance and can be very easily scaled.

19. THE CHALLENGES

- Since there are no column-type relations in databases, updating a part of value or querying the database for a specific information is not easy and fast as compared to SQL databases.
- The database model is not designed to provide consistency when multiple transactions are executed. It is the responsibility of the application developer to make sure its consistency.
- Also as the volume of data increases, maintaining millions of unique primary key becomes a difficult affair. You would need a well-designed complex character string generation algorithm.

20. THE PROMINENT USE CASES

- Storing user session data
- Maintaining schema-less user profiles
- Storing user preferences
- Storing shopping cart data

21. THE BEST PRACTICES

We need to avoid if
- we have to query the database by specific data value,
- we need relationships between data values
- we need to operate on multiple unique keys

22. THE KEY CHARACTERISTICS

Flexible data modeling—Because a key-value store does not enforce any structure on the data, it offers tremendous flexibility for modeling data to match the requirements of the application.

High performance—Key-value architecture can be more performant than relational databases in many scenarios because there is no need to perform the lock, join, union, or other operations when working with objects. Unlike traditional relational databases, a key-value store doesn't need to search through columns or tables to find an object. Knowing the key will enable very fast location of an object.

Scalability and reliability—Key-value stores scale out by implementing partitioning (storing data on more than one node), replication, and auto recovery. They can scale up by maintaining the database in RAM and minimize the effects of ACID guarantees (a guarantee that committed transactions persist somewhere) by avoiding locks, latches, and low-overhead server calls. Most key-value databases make it easy to scale out on-demand using commodity hardware. They can grow to virtually any scale without significant redesign of the database.

High availability—Key-value databases may make it easier and less complex to provide high availability than can be achieved with a relational database. Some key-value databases use a masterless and distributed architecture that eliminates single points of failure to maximize resiliency.

Portability and lower operational costs—Key-value stores are portable because they do not require a complex query language. You can move an application from one system to another without rewriting code or constructing new architecture. Companies can expand their product offerings on new operating systems, without affecting their core technology.

Operational simplicity— Some key-value databases are specifically designed to simplify operations by ensuring that it is as easy as possible to add and remove capacity as needed and that any hardware or network failures within the environment do not create downtime.

Please visit http://basho.com/resources/key-value-databases/ for more information.

23. KEY-VALUE DATABASE vs CACHE

Cache is sometimes likened to a key-value store because of its ability to return a value given a specific key. A cache transparently stores a pool of read data so that future requests for the data can be quickly accessed at a later time to improve performance. Caches and key-value store do have some differences. Where key-value stores can be used as a database to persist data, caches are used in conjunction with a database when there is a need to increase read performance. Caches are not used to enhance write or update performance yet key-value stores are very effective. Where key-value stores can be resilient to server failure, caches are stored in RAM and cannot provide you with transactional guarantees if the server crashes.

24. COLUMNAR DATABASES

Traditional databases store data by each row. The fields for each record are sequentially stored. Let's say you have a table like this:

```
+--+-----------+-------------------+----------+------------+------------+----------+
|ID| name      | address           | zip code | phone      | city       | country  |
+--+-----------+-------------------+----------+------------+------------+----------+
| 1| Benny Smith| 23 Workhaven Lane | 52683   | 14033335568| Lethbridge | Canada   |
| 2| Keith Page | 1411 Lillydale Drive| 18529 | 16172235589| Woodridge  | Australia|
| 3| John Doe   | 1936 Paper Blvd.  | 92512   | 14082384788| Santa Clara| USA      |
+--+-----------+-------------------+----------+------------+------------+----------+
```

This two-dimensional table would be stored in a row-oriented database like this:

```
1,Benny Smith,23 Workhaven Lane,52683,14033335568,Lethbridge,
Canada;2,Keith Page,1411 Lillydale Drive,18529,16172235589,
Woodridge,Australia;3,John Doe,1936 Paper
Blvd.,92512,14082384788,Santa Clara,USA;
```

As you can see, a record's fields are stored one by one, then the next record's fields are stored, then the next, and on and on. However, a columnar database would store this data:

```
1,2,3;Benny Smith,Keith Page,John Doe;23 Workhaven Lane,1411
Lillydale Drive,1936 Paper Blvd.;52683,18529,92512;14033335578,
16172235589,14082384788;Lethbridge,Woodridge,Santa Clara;
Canada,Australia,USA;
```

Each field is stored *by the column* so that each "id" is stored, then the "name" column, then the "zip codes," etc.

The advantages of columnar databases—The primary benefit is that some of your queries could become really fast. Imagine, for example, that you wanted to know the average age of all of your users. Instead of looking up the age for each record row by row (row-oriented database), you can simply jump to the area where the "age" data is stored and read just the data you need. In the above example, because of space constraints, I removed the age column. So when querying, columnar storage lets you skip over all the nonrelevant data very quickly. Hence, aggregation queries (queries where you only need to lookup subsets of your total data) could become really fast compared to row-oriented databases. Further, since the data type for each column is similar, you get better compression when running compression algorithms on each column (which would make queries even faster). And this is accentuated as your data set becomes larger and larger.

Disadvantages—There are many cases where you actually do need multiple fields from each row. And columnar databases are generally not great for these types of queries. That is, if there are many fields to be read, then the columnar storage is inefficient. In fact, if your queries are for looking up user-specific values only, row-oriented databases usually perform those queries much faster. Writing new data could take more time in columnar storage. If you are inserting a new record into a row-oriented database, you can simply write that in one operation. But if you are inserting a new record to a columnar database, you need to write to each column one by one. Resultantly, loading new data or updating many values in a columnar database could take much more time.

That's why you need to start with a row-oriented database as the back-end component of your application. Once the application becomes big, then, think about switching over to the columnar database to enable business analytics. Analytics queries are typically aggregation queries. When you have volume and variety of data, you might want to use a columnar database. It is very easy to add columns, and they may be added row by row, offering great flexibility, performance, and scalability. It is very adaptable. In short, columnar storage is good for queries that involve only a few columns and not good for online transaction processing.

Apache HBase is a columnar database and designed like Google's BigTable. HBase is highly scalable, sparse, distributed, persistent multi-dimensional sorted maps. The map is indexed by a row key, a column key, and a timestamp. Each value in the map is an uninterpreted array of bytes. When your big data implementation requires random, real-time read/write data access, HBase is a very good solution. It is often used to store results for later analytical processing.

The key characteristics

- *Consistency*—This is not a pure ACID implementation. However, HBase offers strongly consistent reads and writes. This is not based on the eventually consistent model.

- *Sharding*—Because the data is distributed by the supporting file system, HBase offers transparent, automatic splitting, and redistribution of its content.
- *High availability*—Through the implementation of region servers, HBase supports LAN and WAN failover and recovery. At the core, there is a master server responsible for monitoring the region servers and all meta-data for the cluster.

HBase implementations are best suited for

- High-volume, incremental data gathering, and processing
- Real-time information exchange (for example, messaging)
- Frequently changing content serving

A column family database can handle semistructured data because, in theory, every row can have its own schema. It has few mandatory attributes and few optional attributes. It is a powerful way to capture semistructured data but often sacrifices consistency for ensuring the availability attribute. Column family databases can accommodate huge amounts of data and the key differentiator is it helps to sift through the data very fast. Writes are really faster than reads so one natural niche is real-time data analysis. Logging real-time events is a perfect use case and another one is random and real-time read/write access to the big data. Apache Cassandra, the Facebook database, is a columnar database and capable of storing billions of columns per row. However, it is unable to support unstructured data types or end-to-end query transactions.

25. DOCUMENT DATABASES

A document database contains a collection of key-value pairs stored in documents. The document databases support more complex data than the key-value stores. While it is good at storing documents, it was not designed with enterprise-strength transactions and durability in mind. Document databases are the most flexible of the key-value style databases and are perfect for storing a large collection of unrelated and discrete documents. Unlike the key-value databases, these systems generally support secondary indexes and multiple types of documents (objects) per database and nested documents or lists. A good application would be a product catalog, which can display individual items, but not related items. You can see what is available for purchase, but you cannot connect it to what other products similar customers have bought after they viewed it.

Imagine a key-value database in which instead of storing "value," documents are stored. These documents are mainly stored in common notation formats like XML, JSON, BSON, etc.; therefore they give greater flexibility in querying. For example in key-value databases, you can only query on the basis of the primary key. In document databases, you can search by primary key and the value in documents. This is because document database embeds attribute metadata related to the stored content with each document which provides a node to query the database based on the content of the document.

25.1 Document Examples

Both examples use the same data and they are just written in different markup languages. Here is the first example, written in XML.

```
<artist>
  <artistname>Iron Maiden</<artistname>
  <albums>
    <album>
      <albumname>The Book of Souls</albumname>
      <datereleased>2015</datereleased>
      <genre>Hard Rock</genre>
    </album>
    <album>
      <albumname>Killers</albumname>
      <datereleased>1981</datereleased>
      <genre>Hard Rock</genre>
    </album>
    <album>
      <albumname>Powerslave</albumname>
      <datereleased>1984</datereleased>
      <genre>Hard Rock</genre>
    </album>
    <album>
      <albumname>Somewhere in Time</albumname>
      <datereleased>1986</datereleased>
      <genre>Hard Rock</genre>
    </album>
  </albums>
</artist>
```

And here is the same example, but this is written in JSON.

```
{
    '_id' : 1,
    'artistName' : { 'Iron Maiden' },
    'albums' : [
        {
            'albumname' : 'The Book of Souls',
            'datereleased' : 2015,
            'genre' : 'Hard Rock'
        }, {
            'albumname' : 'Killers',
            'datereleased' : 1981,
            'genre' : 'Hard Rock'
        }, {
            'albumname' : 'Powerslave',
            'datereleased' : 1984,
            'genre' : 'Hard Rock'
        }, {
            'albumname' : 'Somewhere in Time',
            'datereleased' : 1986,
            'genre' : 'Hard Rock'
        }
    ]
}
```

A document database is used for storing, retrieving, and managing semistructured data. Unlike traditional relational databases, the data model in a document database is not structured in a table format of rows and columns. The schema can provide far more flexibility for data modeling than the relational databases. A document database uses documents as the primary data structure for storage and queries. In this case, the term "document" may refer to a Microsoft Word or PDF document but is commonly a block of XML or JSON. Instead of columns with names and data types, a document contains a description of the data type and the value for that description. Each document can have the same or different structure. To add additional types of data to a document database, there is no need to modify the entire database schema as we do with a relational database. Data can simply be added by adding objects to the database. Documents are grouped into "collections," which serve a similar purpose to a relational table. A document database provides a query mechanism to search collections for documents with particular attributes. Document databases offer important advantages when specific characteristics are required, including:

- *Flexible data modeling*: There are several web-scale applications (cloud, IoT, analytics, operational, mobile, social, etc.) emerging and evolving. The traditional relational data models find it difficult to do justice to these applications and their data models. However, the newly incorporated document databases can support appropriate application data models.
- *Fast write performance*: Document databases prioritize write availability over strict data consistency. This ensures that writes will always be fast even if a failure in one portion of the hardware or network results in a small delay in data replication and consistency across the environment.
- Fast query performance: Document databases are blessed with powerful query engines and indexing features that provide fast and efficient querying capabilities.

Document databases are useful when there is a need to implement the following systems:

- Content management systems
- Blogging platforms
- Analytics platforms
- e-commerce platforms

Application developers avoid document databases if they have to run complex search queries or if their application requires complex multiple operation transactions. Document databases store and retrieve documents and the basic atomic stored unit is a document. As per the requirements, a smart document model has to be designed and used to attain the expected business success.

26. GRAPH DATABASES

We live in a connected world. Data are increasingly open and linked. Only a database that embraces relationships as a core aspect of its data model is able to store, process, and query connections efficiently. While other databases compute relationships expensively at query time, a graph database stores connections as first-class citizens, readily available for any "join-like" navigation operation. Accessing those already persistent connections is an efficient and constant-time operation and allows you to quickly traverse millions of connections per second. Even when the data size grows exponentially, graph databases excel at managing highly connected data and complex queries. Armed only with a pattern and a set of starting points, graph databases explore the larger neighborhood around the initial starting points. It can collect and aggregate information from millions of nodes and relationships leaving the billions outside the search perimeter untouched.

A graph is composed of two elements: a node and a relationship. Each node represents an entity (a person, place, thing, category or another piece of data), and each relationship represents how two nodes are associated. For example, the two nodes "cake" and "dessert" would have the relationship "is a type of" pointing from "cake" to "dessert." Relationships provide directed and named semantically relevant connections between two node entities. A relationship always has a direction, a type, a start node, and an end node. Like nodes, relationships can have any properties. In most cases, relationships have quantitative properties, such as weights, costs, distances, ratings, time intervals, or strengths. As relationships are stored efficiently, two nodes can share any number or type of relationships without sacrificing performance. Although they are directed, relationships can always be navigated regardless of direction.

A graph database uses nodes, relationships between nodes and key-value properties instead of tables to represent information. This model is typically substantially faster for associative data sets and uses a schema-less and bottom-up model that is ideal for capturing ad hoc and rapidly changing data. Today's interlinked data can be easily stored in a graph database as the data relationships can be elegantly captured and represented in graph databases. Any purposeful query can be easily answered via graph databases. A graph database accesses data using traversals. A traversal is how you query a graph, navigating from starting nodes to related nodes according to an algorithm, finding answers to questions like "what music do my friends like, that I don't yet own?" or "if this power supply goes down, what services are affected?" Using traversals, you can easily conduct end-to-end transactions that represent real user actions.

26.1 Use Cases

Today's enterprise uses graph databases in a diversity of ways:
- Fraud detection
- Real-time recommendation engines
- Master data management (MDM)
- Network and IT operations
- Identity and access management (IAM)

Graph databases are enormously useful in applications that have massively connected data, such as social and sensor networks and extremely useful in analytic applications which require predictions, recommendations, and consequence analysis engines. Graph databases ensure the following capabilities.

Performance—The data volume of any growing organization is bound to increase in the days ahead. This increment correspondingly enhances the data relationships. The conventional databases are not equipped to deal with this situation and hence there is a rush toward graph databases, which are intrinsically efficient in managing associated data at scale.

Flexibility—Business strategies and directions are changing with the rise in the number of data sources and complexity. The data structure, schema, and scope are varying and hence there is a need for database management solutions that are highly modifiable and sustainable. Graph databases fulfill this requirement at ease.

Agility—Application development gets speeded up with the graph databases that are perfectly aligned and associated with today's agile programming model.

For the connected applications, the role and responsibility of graph databases are growing drastically. All kinds of linked data are to be aggregated and sliced to extract actionable insights with the assistance of graph databases.

27. WHEN TO USE NoSQL AND SQL DATABASES

The image below clearly delineates the core aspect of both SQL and NoSQL databases.

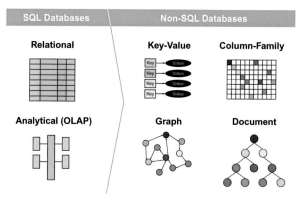

SQL Databases	NoSQL Databases
Centralized applications (ERP, CRM, SCM, etc.)	Decentralized applications (web, mobile, IoT, etc.)
Structured data and moderate data volumes	Multistructured data and high data volumes
High availability	Continuous availability
Moderate velocity data	High velocity data from sensors, machines, phones, etc.
Limited number of data sources	Multiple data sources from many distributed locations
Complex and nested transactions	Simple transactions
Vertical scalability	Horizontal scalability for high performance
Static schema	Dynamic schema

28. HOW NoSQL DATABASES DIFFER?

There are certain differences between SQL and NoSQL databases. Further on, there are some critical deviations between NoSQL database types as accentuated later.

Architecture—Some NoSQL databases like MongoDB are architected in a master/slave model, whereas NoSQL databases such as Cassandra are designed in a "masterless" fashion. That is, every node in a database cluster has the same capability and functionality. The architecture of a NoSQL database greatly impacts how well the database system supports various functional as well as nonfunctional requirements such as uptime, multigeography data replication, predictable performance, and more.

Data model—If the data you need to store can be represented in a simple table structure, RDBMS is sufficient. But when you have complex data with multiple levels of nesting, it cannot be modeled into relational tables. For example, multilevel nesting can easily be represented in a JSON format and many NoSQL databases. NoSQL databases are often classified by the data model they support. Some support a wide-row tabular store, while others sport a model that is either document-oriented, key-value, or graph. All aspects of the data model are not known at design time. Therefore, if the schema has to be modified, then NoSQL databases are the only option.

Data distribution model—Because of their architecture differences, NoSQL databases differ on how they support the reading, writing, and distribution of data. Some NoSQL platforms like Cassandra support writes and reads on every node in a cluster and can replicate/synchronize data between many data centers and cloud providers.

Development model: NoSQL databases differ on their development APIs with some supporting SQL-like languages (e.g., Cassandra's CQL).

High performance guaranteed—When database grows, the performance of SQL databases is bound to go down. But we are heading toward web-scale and high-performance applications. There are application domains and industry verticals generating massive volumes of data. It is not only collecting and storing data but also subjecting them to a variety of purposeful investigations to make sense out of data. Thus for big data analytics, the contributions of high-performing NoSQL databases are becoming crucial. As widely accepted, distributed computing is the way forward. NoSQL databases are scalable and distributed database systems.

29. THE TANGIBLE BENEFITS OF NoSQL

- *Continuously available*—This is available all the time even in the face of the most devastating infrastructure outages.
- *Geographically distributed*—Database instances are distributed and available everywhere you need it.
- *Operationally low latency*—The response time is very less and is suitable for intense operational applications.
- *Linearly scalable*—Database servers could be added in order to tackle higher user as well as date nodes. The system performance grows linearly with the addition of new servers in the cluster.
- *Lower total cost of ownership*—The leverage of commodity servers and open source software for NoSQL database management systems results in lower TCO and higher RoI.
- *Multiple types*—There are specific NoSQL types for specific application requirements and data models.

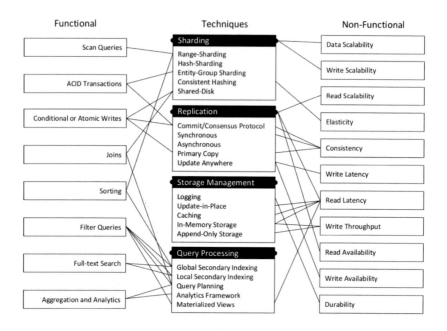

The above diagram taken from https://medium.baqend.com/nosql-databases-a-survey-and-decision-guidance-ea7823a822d depicts the *NoSQL Toolbox* where each technique is connected to the functional and nonfunctional capabilities.

Traditional relational database management systems (RDBMSs) provide powerful mechanisms to store and query structured data. They provide strong consistency and transaction guarantees. RDBMSs are matured and stabilized ensuring higher reliability, stability, and support. Yet, the data becomes big data and there is an insistence that big insights originate from big data. NoSQL databases are being touted as the right and relevant systems for the ensuing big data era.

30. SUMMARY

Big data analytics is now moving beyond the realm of intellectual curiosity and propensity to make tangible and trendsetting impacts on business operations, offerings, and outlooks. It is no longer hype or a buzzword and is all set to become a core requirement for every sort of business enterprise to be relevant and rightful to their stakeholders and end–users. Being an emerging and evolving technology, it needs a careful and cognitive analysis before its adoption and adaption.

Enterprises squarely and solely depend on a variety of data for their day-to-day functioning. Both historical and operational data have to be religiously gleaned from different and disparate sources, then cleaned, synchronized, and analyzed in totality to derive actionable insights that in turn empower enterprises to be ahead of their competitors. In the recent past, social computing applications throw a cornucopia of people data. The brewing need is seamlessly and spontaneously linking enterprise data with social data in order to enable organizations to be more proactive, preemptive, and people-centric in their decisions, discretions, and dealings. Data stores, bases, warehouses, marts, cubes, etc. are flourishing in order to congregate and compactly store different data. There are several standardized and simplified tools and platforms for accomplishing data analysis needs. Then there are dashboards, visual report generators, business activity monitoring (BAM), and performance management modules to deliver the requested information and knowledge of the authorized persons.

In this chapter, we have described how the various types of NoSQL databases come handy in storing, accessing, and mining big data to squeeze out trustworthy knowledge. NoSQL is a database technology designed to support the requirements of cloud applications and architected to overcome the scale, performance, data model, and data distribution limitations of relational databases.

FURTHER READING

[1] Oracle, "Big Data for the Enterprise" a White Paper, http://www.oracle.com/us/products/database/big-data-for-enterprise-519135.pdf, 2011.
[2] McKinsey Global Institute, Big Data: The Next Frontier for Innovation, Competition, and Productivity, http://www.mckinsey.com/insights/business_technology/big_data_the_next_frontier_for_innovation, 2011.
[3] Neo Technology, "NoSQL for the Enterprise", a Whitepaper, http://www.neotechnology.com/tag/nosql/, 2011.
[4] R. Bloor, "Enabling the Agile Business With an Information Oriented Architecture", the Bloor Group, 2011).
[5] http://bigdata-madesimple.com/a-deep-dive-into-nosql-a-complete-list-of-nosql-databases/.
[6] http://nosql-database.org.
[7] https://www.datastax.com/nosql-databases.
[8] P. Raj, Cloud Enterprise Architecture, CRC Press, 2012. http://www.crcpress.com/product/isbn/9781466502321.
[9] www.peterindia.net.

ABOUT THE AUTHOR

Pethuru Raj have been working in the Site Reliability Engineering (SRE) Center of Excellence, Reliance Jio Infocomm Ltd. (RJIL), Bangalore. Previously worked as a cloud infrastructure architect in the IBM Global Cloud Center of Excellence (CoE), IBM India Bangalore for 4 years. Prior to that, I had a long stint as TOGAF-certified enterprise architecture (EA) consultant in Wipro Consulting Services (WCS) Division. Also worked as a lead architect in the corporate research (CR) division of Robert Bosch, Bangalore. In total, I have gained more than 18 years of IT industry experience and 8 years of research experience. He finished the CSIR-sponsored PhD degree at Anna University, Chennai and continued with the UGC-sponsored postdoctoral research in the Department of Computer Science and Automation, Indian Institute of Science, Bangalore. Thereafter, I was granted a couple of international research fellowships (JSPS and JST) to work as a research scientist for 3.5 years in two leading Japanese universities. I have published more than 30 research papers in peer-reviewed journals such as IEEE, ACM, Springer-Verlag, Inderscience, etc. I have authored eight books thus far and focus on some of the emerging technologies such as IoT, Cognitive Analytics, Blockchain, Digital Twin, Docker Containerization, Data Science, Microservices Architecture, fog/edge computing, etc. I have contributed 30 book chapters thus far for various technology books edited by highly acclaimed and accomplished professors and professionals. For more details, please visit the Linkedin page https://www.linkedin.com/in/peterindia/.

CHAPTER TWO

NewSQL Databases and Scalable In-Memory Analytics

Siddhartha Duggirala
Bharat Petroleum Corporation Ltd., Mumbai, India

Contents

1. Introduction   50
2. In-Memory Databases   52
   2.1 SQL Databases   52
   2.2 NoSQL Databases   55
   2.3 NewSQL Databases   56
   2.4 What Are In-Memory Database Systems?   58
   2.5 How Are IMDBs Different?   61
3. Use Cases   66
4. In-Memory Databases   69
   4.1 Redis   69
   4.2 VoltDB   70
   4.3 Microsoft Hekaton   71
   4.4 SAP HANA   71
5. Conclusion   73
References   73
About the Author   76

Abstract

The velocity of the data generation is accelerated and the value associated with it diminishes with time. If not processed in real time businesses have to face lost opportunities with financial implications and sometimes, especially in industrial IoT scenarios, can result in fatal situations.

The question we will be trying to find an answer for is how do we organize scalable stored data with low-latency read and writes and process it in real time? SQL is the industry standard language that stands for Structured Query Language. This is the language used by relational databases to query and manage data. These databases form the bedrock for the world's transactions. These databases are strongly ACID compliant. NoSQL is Not Only SQL database which is a group of data management systems with common characteristic of high availability and massive scaling. NewSQL databases are massively scalable ACID-compliant databases. SAP HANA, NuoDB, and VoltDB

Advances in Computers, Volume 109
ISSN 0065-2458
https://doi.org/10.1016/bs.adcom.2018.01.004

are few NewSQL databases. In-memory databases are group of databases for which the primary data storage is in memory instead of disk storage. This provides huge performance improvements due to sheer access latency difference between memory and disk. In this chapter we will study the SQL databases and NoSQL databases in the purview of the IMDBs.

1. INTRODUCTION

Going back few decades when the first personal computers are ridiculed to be of any use to general public, enterprises and research institutions are the only ones to have computers. The central mainframes are the processing hubs, and different divisions of the enterprise or the research institute use to share the mainframe computers in mostly time–duplex manner. The self–fulfilling prophecy stated as a law, Moore's law, dictated that the number of processing chips will double, while the price of the same will fall by half every few years. As the world marched on, personal computers became a necessity from being an exotic piece of hardware for ultrarich. The earlier computers had a RAM of about 2 MB and storage capacity in orders of 64 MB. Nowadays not only the computers became smaller, but pack several GBs of RAM in low–end systems to 100s to 1000s GB for high–performance computers.

The first computers that took humans to moon has lesser processing power than your smart door bell.

Gershenfeld [1]

The openness of the Internet and the freedom of expression so much valued by the innovators of the Internet age led to a rich human expression through emails, blogs, and video logs, interconnecting the whole world in an essence making it a small village. The social networks help in connecting people from different regions, cultures, and languages, and with Internet as their tool make world listen to their voice and their stories.

Through our increased consumption of digital content as a source of entertainment, work, and simplifying our lives through online shopping, we develop digital communities with our own identities. As the networks grew larger, their immediate circles grew bigger, and people start connecting the physical world to the digital world. Converging their identities and ideas. The smart devices like security cameras which analyze burglary and inform relevant authorities or smart traffic systems which detect congestion and rerouting traffic, the industrial sensors deployed in harsh climates. The air pollution monitors, weather monitors, crop monitors, and our smart watches, everything is interconnected.

This spawned up new business models and new use cases that were never envisioned before. The world is generating humungous amounts of data. In the past data alone we have generated more than 1 TB/day. These cultural shifts along with our technological advances ask us a simple and rather important question: How to make use of this data? A longer version of the same question is how to make sense of this fast and huge data? How to take spot decisions on the data whose value decreases with time? Or how do we spot trends at macrolevel?

The question we will be to try and find an answer for, albeit in technical terms, how do we organize scalable stored data with low-latency read and writes and process it in real time?

A few specific examples for real-time processing requirements are in financial world placing a bid on the exchange when there are subsecond level differences, placing a relevant add on the webpage while user is browsing, finding out the users aborting the items placed in the cart of an commerce website, fraud detection on credit card usage, industrial air quality monitor, industrial fleet tracking. Data shows us what is currently happening in businesses, in real-time basis. With IoT as a compelling data source across industries, processing and taking advantage of this real-time insights is more than ever. However traditional systems cannot process this data within real-time constraints. Newer technologies like in-memory analytics and massively parallel processing data can process data fast enough to gain insights in real time. Hadoop popularized massively parallel processing but it is fit for batch processing, and Kafka/Spark streaming is used for streaming analytics. We will study some of these use cases in the succeeding sections.

The velocity of the data generation is accelerated, and the value associated with it diminishes with time. If not processed in real time businesses have to face lost opportunities with financial implications and sometimes, especially in industrial IoT scenarios, can result in fatal situations. Fog computing can be helpful in discerning the local situations in IoT scenarios but would require a central data store to discern the global patterns. As the number of Internet users and digital first services across the world increases, many of the major systems were hitting the upper limit on number of concurrent users. Online shopping, connected devices securities trading, session management in very large services like Facebook, Amazon, Salesforce, or banking institutions are prime examples for this. Fast and consistent transactions are desirable in all business lines; they however are decisive for financial trading, call connection in Telcos, payment processing where every second of wait time erodes the number of completed transactions, and decreasing user experience affecting the businesses bottom line. Users (or bots or smart devices) cannot and will not wait for slow transaction to finish. These transactional applications require high-performance data management system [2]. In-memory databases provide the required high concurrent read and writes. VoltDB, for example, can process about $10 \times$ more writes per second than other RDBMS [3]. Telcos and financials, unsurprisingly, use high-performance mainframes and cutting-edge memory-based data stores to provide the requirements.

The attention span for most users is in order of few seconds. If a website or application takes more time to load, then most users just give up and move on to other alternatives. Session tracking and storing session-level details for quick recall if any network or some other issues arises are regularly expected from most of the applications. In-memory data stores like Memcached provide a high-performance cache on top of data stores, enabling quick access of static as well as dynamic data. Redis [4] is a high-performance in-memory key-value data store used for session management.

We divide the chapter into three sections: in-memory databases, use cases, and brief introduction to different in-memory databases.

2. IN-MEMORY DATABASES

Any study of in-memory databases will be incomplete without learning the history of database systems that came before and understanding how in-memory databases are different from those database systems. What are current breed of solutions available? Relational databases, sharding, partitioning, distributed databases, grid processing, caching, and NoSQL databases.

2.1 SQL Databases

SQL is the industry standard language that stands for Structured Query Language. This is the language used by relational databases to query and manage data. These databases form the bedrock for the world's transactions. Established business invested millions of dollars in relational database through decades. The reasons for this range from standardized ecosystems to familiarity to simplicity. These databases are strongly ACID compliant.

ACID is an acronym for Atomicity, Consistency, Isolation, and Durability. Atomicity property requires every transaction to be "all or nothing." If any one part of the transaction fails, the whole transaction fails. Consistency ensures the data is in consistent state after each transaction. Isolation requires the concurrent transactions to result in system state like that of serial execution of transactions. Durability states the committed transactions to be reflected in database amid failures and errors [5].

Relational databases (hereafter we will refer them as DBMSs) were introduced in mid-1960s with a basic idea that application program should be separated from data. Due to the separation of concern, the developers spend more time on application logic rather than data access. IBM's System R and University of California's INGRES are the first relational databases. Oracle released first version of their DBMS around that time. Two of the today's major open-source DBMS projects are released in mid-1990s: MySQL and PostgreSQL (Fig. 1).

In the 2000s we saw the rise of web-scale applications like Think Google, E-Bay, and similar applications. These applications had more performance requirements than their predecessors. They need to support large number of concurrent users and required to be online all the time 24×7. But the relational databases were consistently overloaded. The databases and hardware fell short on the resource demands and performance requirements. The conventional databases are not designed to deal this scale. The obvious remedy many tried is to scale up: migrate the database to a bigger, meaner machine. This however becomes prohibitively costly after a level. The migration itself is a complex processing requiring significant downtime. Some companies created their customer middleware to shard single-node database over a cluster of commodity servers. The application deals with a single logical database. The middleware coordinates with the servers in

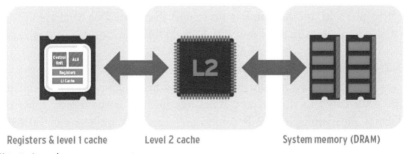

Registers & level 1 cache   Level 2 cache   System memory (DRAM)

Fig. 1 A cache memory system.

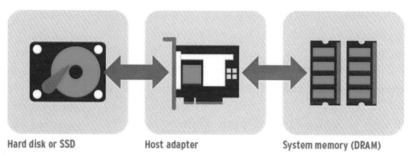

Hard disk or SSD   Host adapter   System memory (DRAM)

Fig. 2 A disk cache memory system.

the cluster for query execution. Middleware collates the responses from different nodes into a single response. Notable examples are Google's MySQL-based cluster and Facebook's MySQL cluster. This works well for simple operations like reading and updating single record. Complex queries and join operations are not supported by these middlewares. It required the developers to implement application-level joins (Fig. 2).

2.1.1 Scaling Databases

Any data store can be scaled in two dimensions:

- Scale UP: In this case you migrate the data store to a more powerful machine. You continue doing this to achieve ever-increasing performance requirements. For example, initially your database is hosted on a 1 GB RAM, Intel Duo Core system. You move this to Core I7 with 8 GB RAM and then you migrate the database to a powerful SMP with 128 GB RAM and multiple CPUs. Newer SMPs and high-end systems are based on NUMA architecture. Each CPU core has its local memory, and the access times of data depend on the location of the memory.
- Scale Out: While scaling out, you add an additional node to the cluster of databases nodes. For example, initially your database is hosted on a 1 GB RAM, Duo Core node. And if you want to increase the performance, you just add another similar node and distribute the database among these two. NoSQL and NewSQL DBs do this distribution automatically.

2.1.2 Transparent Sharding Middleware

Instead of writing your own database sharding middleware, you can use options available in the market. This middleware transparently splits the databases into multiple shards which are stored in single-node DBMSs. These nodes run same DBMS, have only a portion of overall database, and cannot be accessed independently. Middleware coordinates the queries and the transaction, and manages data placement, partitioning across the nodes, thereby presenting a single logical database to the application without any other changes. There is a drop-in replacement for the currently used database. Most of the middleware systems are based on MySQL supporting MySQL wire protocol. One particular downside is that this incurs redundant query planning and optimization on sharded nodes for complex queries. Another drawback is that these middleware do not take advantage of the aggregate memory capacities. Apache Ignite is an open-source in-memory grid which works with the individual shards and gives them in-memory processing capabilities [6].

Eventually some of the companies developed their own distributed DBMSs. The motivating factors for this move are that relational model is not one fit for all solution. Web-based applications focus on availability and concurrent users. The relational databases on the flip-side favor consistency at the expense of availability. It was also thought that SQL for simple queries is simply an overkill.

2.2 NoSQL Databases

Few distinguishing NoSQL features are high-performance writes, massive scalability, and write schema free. In NoSQL class of DBs ACID compliance is less fashionable, and some databases loosened the data consistency and data freshness in favor of scale and resilience. These databases support Basically Available Soft State Eventually Consistent (BASE). Basic Availability implies that the database appears to work most of the time. Soft state implies that data stores do not have to be write-consistent. Eventual consistency implies that the data store becomes consistent eventually at all replicas. The BASE consistency model provides less strict assurance that ACID that the data will be consistent in the future [7].

Not all NoSQL databases are built with goals in mind. Some NoSQLs are developed with emphasis on availability like DynamoDB and Cassandra. While some other NoSQL DBs focus on the flexibility of schema or data models. Key-value stores like Redis and document stores like MongoDB follow this design guideline. NoSQL DBs like Neo4j and Trinity focus on storing data according to their relationships, not rows or columns [8].

NoSQL gained attention from web-scale companies [5]. These companies were dealing with vast amount of unstructured data from myriad sources. The most important consideration was scalability, scale out massively on commodity servers [9]. This is what the NoSQL databases brought to the community. Applications requiring massive scale with nonrelational schemas benefit from NoSQLs.

NoSQL systems were architected for availability. CAP stands for Consistency, Availability, and Partition. Brewer's theorem (aka the CAP theorem) states that you can have two, but not all three of these properties. This makes NoSQLs poor fit for applications requiring strong consistency. Example for these use cases can be financial transaction, call go through in Telcos.

At the same time the database is always responsive, even in case of network partitions. It is ok to display temporarily inconsistent data rather than showing nothing, for example, it is ok for the comment on a Facebook post to be displayed late [10]. The NoSQLs provide high write throughput but eventually consistent. High-performance consistent workloads have little fit in NoSQL databases. These events occur millions of message a day, even millions of message per hour (or even second) all over the world.

Many applications cannot give up strong transaction consistency requirements. Developers using NoSQLs spend too much time writing code to handle inconsistent data. Having a transactionally consistent databases provides a useful abstraction, thus making them more productive. Yet, as we have seen earlier the only option available is either to scale up or to write custom sharding middleware, both prohibitively expensive for many. This has given rise to new breed of databases, NewSQL databases. These are best of the both worlds of SQL and NoSQL databases.

2.3 NewSQL Databases

The term NewSQL is coined by 451 group analyst to denote new class of databases providing transactional consistency and massive scalability. These are a class of modern relational DBs that intend to provide similar scalable performance of NoSQLs for OLTP workloads while maintaining ACID guarantees. In other terms, these databases provide transaction support of SQL databases and NoSQL massive scalability. Applications will use SQL to run high number of parallel transactions to ingest and modify data. Developers no longer have to deal with eventually consistent updates. Few guarantee stronger consistency, while others (MemSQL) provide tuneable consistency. These are cloud-first databases which can be massively scaled across geographic regions [5,7].

NewSQL systems employ lock-free concurrency control mechanism to support huge parallel transactions. They use shared-nothing distributed (or massively parallel) architecture for high scalability.

SAP HANA, NuoDB [11], MemSQL [12], VoltDB [13], Google Spanner [14], CockRoachDB, and Hyper [15] are few NewSQL databases. Among these Spanner and CockRoachDB are not memory-resident databases. Hyper is a research database not commercially available. Both SAP HANA and Hyper support multiple storage formats, i.e., row and columnar format.

NewSQL systems are built from scratch, rather than extending a database Hekaton for Microsoft SQL server [3]. They do not have the baggage of legacy systems. All these databases are based on shared–nothing distributed architectures. These databases have components for multinode concurrency, fault tolerance, and distributed query processing. Data is automatically partitioned and distributed among the nodes. While partitioning data colocation, value skews and geographical distribution are also considered. Different nodes send intraquery data directly without any central node, although one of the nodes takes the transaction coordinator job [16].

The NewSQL DBs manage their own primary storage. This means they are responsible for distributing the database across nodes, instead of relying on data grids (Pivotal Gemfire) and distributed file systems (HDFS). This helps in sending the queries to the data directly instead of moving data to query processing node. This results in significantly less network traffic. By employing sophisticated replication schemes NewSQLs achieve better performance and fault tolerance than other systems.

Few downsides in using NewSQL systems are that the organization will not be enthusiast changing. This inertia is the reason for slow adoption of NoSQL by enterprises. As a ripple effect there will be less people familiar with these NewSQL databases. The enterprises might also lose the investments in current administration tools. MemSQL and Clustrix [17] are maintaining compatibility with MySQL wire protocol. This means enterprises can use their existing investment and tools seamlessly with NewSQL DBs.

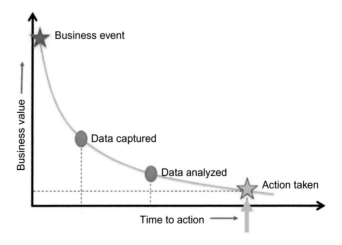

Scaling a legacy relational database has significant costs, whereas the NewSQL databases are designed to scale out using a shared nothing cluster all while maintaining interactivity and transactionality of legacy databases. These are designed for low-latency high-performance read and writes. VoltDB, for example, scales out almost linearly on inexpensive commodity hardware [18].

These systems are geographically distributed, and data is partitioned and replicated across nodes. This helps in achieving availability and fault tolerance. NewSQL favors consistency over availability. A NewSQL system will return exact same response to all clients and ensures no write conflicts. In case of network partitions, we cannot always be 100% consistent and 100% available. By choosing consistency these systems provide CP from CAP law.

2.4 What Are In-Memory Database Systems?

Speed of processing and customer insights have become deal breaker for all businesses, be it a small startup trying to sell a niche product or global conglomerate trying to sell a packaged solution. As we have seen earlier, legacy databases could not cope up with the high-performance demands of modern applications [19]. These databases store the data primarily on the block-addressable durable storage device like HDD [20]. Accessing and manipulating data requires costly I/O operations. The commercial databases came a long way with sophisticated data caches in memory. They also provide advanced indexing mechanisms to speed up the data access, although maintaining these indexes and caches has huge cost associated with it [10].

One idea that had somewhat of renaissance is the main memory-based databases. Unlike other databases, the main memory (RAM) is the primary storage of data in these systems. From now on let us refer to these as in-memory databases (shortly IMDBs) [21]. This eliminates the need for complex disk IO operations. The processors can directly access the memory in the RAM. Due to this the data in main memory is more vulnerable to software errors than in disks. DBMS no longer assumes that a transaction is accessing data not in memory and will have to wait till the data is fetched. The IMDBs can get better systems and can perform better because buffer pool manager, indexes, locking, latching, and heavy weight concurrency control schemes are no longer needed. Several NewSQL and NoSQL databases are based on main memory storage architecture, including commercial (Redis, Memcached, SAP HANA, VoltDB) and academic (Hyper/SyPer).

This idea of IMDBs is first studied in 1980s and first commercial systems appeared in 1990s. Altibase [22] and Oracle's TimesTen [23] are the early IMDBs. The primary idea behind these databases is that RAM access times are tremendously faster than disk I/Os. As an extension to that idea, database storing data in purely memory should be faster than disk-based DBs [21]. However the high cost of memory in the past decades prevented in fully loading multigigabyte databases in memory. They remained a niche solution. The price of a GB of RAM in 1990s is about $100K, as of today (2017) the same is priced at $5 dollars. And the prices are falling fast.

One downside is that the price of RAM is expensive than compared to disks. This however is drastically changing as the memory prices are coming down as a whole. Cloud and database-as-a-service provider allow the firms to affordably benefit from IMDBs. With almost nil capital expenses, the firms can take advantage of the faster processing times and real-time insights. IMDBs make it possible to do real-time analytics on the live data with high-throughput writes. This is referred to Hybrid Transaction/Analytical Processing (HTAP) [24]. OLAP data warehouses are generally segregated from OLTP databases. The business intelligence report is generally run on the data warehouses. OLTP or the transaction databases are optimized high-throughput short-lived reads and writes, whereas the OLAP data warehouses are optimized for long running batch processing. The process of moving data from OLTP to OLAP system involves a complex ETL process. Often this process runs once in a day or so when the peak workload on OLTP systems is heavy. This means the data in OLAP systems is older than the data in OLTP. Some OLAP processes take days, if not weeks, to complete. One of the reasons for this long execution time is that they manipulate large amounts of data on slow disks. If we shift the data from disk to RAM, these processes can gain $10–100 \times$ speedup [19].

Caching is one of the mechanisms used by disk-based databases to provide to high-performance demands [18]. Concept of caching is as old as early microprocessors. CPUs were faster than the memory they were accessing. As CPUs became faster, RAM became more of a bottleneck. This caused the hardware developers to place small amounts of very fast memory close to microprocessors. These caches are smaller than the traditional memory. For example, current generation of Intel Kaby Lake processors have three levels of caches L1, L2, and L3 with 8 MB shared among processors. CPUs use sophisticated algorithms like LRU and RR to manage these caches. Following the same basic ideas many of the modern databases provide cache technology. This allows them to load a part of database into RAM, which keeps majority of data on the disk. This approach gains the performance by limiting the heavy I/Os for keeping hot records in RAM. Algorithms much more sophisticated than LRU are used by modern databases.

The one drawback in using RAM as the main data storage is with the volatility of the memory. It does not provide persistent storage that disk storage provides, although most main memory databases often remedy this by persisting data by copying it to disk asynchronously. Hyper, for example, uses memory snapshots and log to maintain persistency of the data. In a more a stringent mechanisms to log the transaction to the disk first and then commit the transactions. This mechanism is called write-ahead logging. This however levies a performance penalty on writes [21].

Some of the incumbent database vendors advocate that their databases have in-memory capabilities due to their cache systems. It is possible to load entire database in a cache system. But this brings to us the question: Is an IMDB just a database with large enough cache? [18].

The short answer to that question is no. CPUs access data on disk different than that residing in memory. Disk read requires complex I/O operations which require 1000s of CPU cycles, whereas that of in-memory requires fewer than 100 CPU cycles. This translates to thousands of additional CPU cycles in disk I/O. Some databases optimize this by checking the cache system before issuing disk I/O. However if data is not in cache, CPU incurs long wait. In case the whole database is loaded into cache, this check can be avoided. The memory bus which CPU uses to access RAM is designed for much more speed than peripheral bus used by disk. If, though, this system performs well, it is not taking full advantage of the memory. Index structures will be designed for disk access, and access for data is through buffer manager [25].

The advent of SSDs provides an alternative to in-memory solutions. The Solid-State Drives (SSDs) which are persistent memory chip-based drive are in order of 10–100 times faster than hard drives. Hard drives are based on electromechanical systems which require moving disks to read data. The SSDs are backward compatible with hard drives. This means that the SSDs can be used as a drop-in replacement to hard drive and achieve huge speedups. Not surprisingly, traditional database vendors are pushing to use SSDs for speed boosts. The advantage of this approach is that the whole ecosystem of tools works as they were earlier, with speedup. Does replacing the mechanical disk by SSD make the existing database to an IMDB? The clue lies in the CPU I/O operation. CPU still requires complex and time-consuming I/O operations to access data. To make distinction simpler, these are not considered to be pure IMDBs [18]. Owing to the volatile nature of the main memory it is prudent to take back the database regularly. The hardware memory errors or power failure might cause the database to be transaction inconsistent. To remedy this undo and redo logs are written to persistent storage as soon as transactions are committed. In recovery process the latest backup is copied to memory, and the redo and undo log operations are executed from that checkpoint. VoltDB [13], for example, uses virtual memory snapshots and redo log to maintain the durability of the database.

As we have gathered till now, the data is being generated at unprecedented rates. In this scenario can we safely assume that the entire database fits in the main memory? For certain applications like customer profile or other master data maintenance, we can safely assume the data to fit in the main memory. However this is not the case with other applications. Not all records in the database are accessed frequently. Only a subset of the records is accessed. We call these records as hot data. Modern NewSQL databases push subset of data to persistent storage, thereby reducing the memory footprint. This allows the IMDBs to support data larger than available main memory without switching back to disk-based databases. H-Store's anticaching mechanism [26] evicts the cold records to disk and place a tombstone [27]. If any transaction tries to access this memory location with tombstone, then it is aborted. And the corresponding record is asynchronously read into main memory. Microsoft Hekaton [3] maintains a Bloom filter per index to reduce the in-memory storage overhead for tracking evicted tuples. In MemSQL administrator can manually specify which tables to be stored in columnar format. No metadata is stored for the records evicted to disk. The disk-resident data is stored in log structure format to minimize update overhead.

2.5 How Are IMDBs Different?

Different functional components of an IMDB should be treated different than DBMSs. Main memory is volatile and prone to software errors and power failures. This necessitates an efficient logging and backup mechanisms to ensure durability. Concurrency between several concurrent transactions is required to ensure the consistency. Owing to cloud-first, scale-out architecture of most NewSQL databases partitioning and replication mechanisms are to be efficient to ensure high availability and high performance [21,25,28–30].

2.5.1 Partitioning/Sharding

Most of the distributed IMDBs scale out horizontally by splitting the database into multiple disjoint sets. These subsets are called partitions or shards.

Distributed transaction processing on these shards is not entirely a new idea. The database's tables are partitioned horizontally based on one or more table's columns. These columns are referred to as partitioning attributes. The IMDB then assigns each of these partitions to different nodes. While assigning different tables partitions to different nodes, table relationships are also considered. This approach is called colocation of related data. Join performance through this colocation increases as the interquery network transfer is minimum. Each node takes responsibility for executing the queries on its database partition. Modern distributed IMDBs support native partitioning. Internode queries need to be distributed to different nodes and the results should be collated to form final response, though the collocation of data helps in reducing the multinode queries.

While splitting the database among several nodes, IMDBs take two distinct routes. For example, MemSQL uses heterogenous architecture [12], while VoltDB [13] uses homogenous architectures. In MemSQL nodes comprise execution-only aggregator nodes and leaf nodes which store actual data. In heterogenous architecture not all nodes are same. Storage node can also execute part of the transactions, thereby reducing the data sent to aggregator nodes. NuoDB is another database which uses similar heterogenous architecture. Transaction Engines are execute-only nodes and Storage Engines are data nodes. The data is not prepartitioned but rather the load balancing schemes are exposed. This ensures that the data used together generally resides on same TE node.

Modern IMDBs also support live migration of partitions. This helps in load balancing, either increasing or decreasing database capacity without any interruptions. This is slightly similar to rebalancing in NoSQL databases. ACID compliancy in NewSQL IMDBs complicates the matter a bit. Two approaches are used to achieve this. One is to use virtual or logical partitions and move these partitions. This makes it easier to create replicas as well. The other is to use a fine-grained approach to redistribute the individual tuples or tuple group.

2.5.2 Concurrency Control

The data access latency of memory is very low compared to disk-based systems. By avoiding disk I/O IMDBs are able to provide low latency and high transaction throughput. As a result we can expect transactions to complete faster in main memory systems. Traditionally the RDBMS uses lock-based concurrency models. The databases interleave the transactions, while one is waiting for data from disk. Since the waiting times in IMDBs are very less, the locks are held for shorter duration. Certain IMDBs lock the entire database before executing update transactions. This amounts to serial execution of transactions. Serial execution of transactions almost completely eliminates the cost of maintaining locks. Deadlocks cannot happen in IMDBs. However in multicore and distributed databases locking the whole database causes delays. Some IMDBs have multiple readers and a single writer. In this scheme multiple readers can concurrently execute read transactions, whereas the writer serially executes transactions. Concurrency protocol provides the atomicity and isolation guarantees required by the relational databases.

Distributed databases either use decentralized or centralized transaction coordination scheme. In decentralized scheme, different nodes maintain transactions on their nodes. Each node coordinates with others to determine concurrent transaction conflicts [21,25]. In centralized a central coordinator decides on which transactions to execute and when. Few of the concurrency protocols used in majority of the databases are as follows:

- Two-phase locking schemes (2PL): In this scheme transactions take lock on the data required by transaction. Once the transaction is committed and all nodes give an acknowledgment, final commit is logged.
- Timestamp ordering: Database assumes that the transactions are executed in the order of their timestamps. They are never interleaved when serializable ordering is violated. This protocol requires all the distributed nodes to have highly synchronized clocks.
- Multiversion concurrency control protocol (MVCC): Database creates a new version of tuple while updating it in a transaction. Multiple versions of the tuple are maintained. This allows for the read transactions to be completed without affecting write transactions and vice versa [31].

Most of the IMDBs eschew 2PL protocol due to the cost of locking and latching. Most of the modern IMDBs like MemSQL, Hyper, SAP HANA [32], and CockroachDB use decentralized MVCC [33]. Systems like NuoDB and Google's Spanner use a combination of 2PL and MVCC [14]. Using this approach the transactions need to acquire locks. And once a transaction updates a record, a version of tuple is created. Read-only queries do not need to acquire locks. VoltDB uses time order concurrency control. Instead of interleaving transactions VoltDB executes the transactions one at a time. Single-partition transactions are executed in decentralized manner. For multipartition transaction a centralized node coordinates with other partitions. VoltDB orders these transactions based on the logical timestamps. When a transaction executes at a partition, it has complete access to the data on the partition. Fine-grained locks are not required. Google Spanner uses hardware GPS clocks to keep ensure high precision synchronization.

2.5.3 Commit Processing

To guard against any software failures or power failures, it is essential to backup database. This ensures the durability and consistency guarantees. As the backup is done regularly, it is prudent to keep a log of transaction activity. Owing to volatility of the memory, the log must be stored on stable storage. Logging of data to persistent storage impacts the response times. Each transaction has to wait till the log is committed to storage. And the throughput is also affected if log becomes bottleneck. This is the only disk operation required by the IMDBs [34,35]. The following are few solutions that have been suggested and some of them are used by major IMDBs:

- Use stable storage for logging: A stable storage is used to hold portion of the log. As a transaction commits, the log information is written to stable storage. A dedicated processor asynchronous writes the data to log disks. Response time improves even though log bottleneck is still present. A modern variation of this approach is to use nonvolatile RAM storage. This device offers lesser read/write than disk storages.
- Precommitting transactions: Transaction locks are released as soon as its log record is placed in the log. This reduces the blocking delays for other concurrent transactions.
- Group commits: A log record is not sent to disk as soon as transaction commits. Instead the log records of several transactions are accumulated in memory. All these records are flushed to log disk in a single–disk operations. This reduces the number of disk operations. The log bottleneck and blocking wait are also eliminated. The downside is that in case of software failures some transactions might be lost.
- Asynchronous commits: The log records are flushed to database asynchronously. The transactions need not wait for log to be committed to disk. As with the group commit, certain transactions might be lost in case of failures.

2.5.4 Indexing

Indexing mechanisms help in reducing the response times of the queries. Traditionally indexing structures like B–Trees are optimized for disk access. As IMDBs have all the data in memory, these indexing mechanisms do not fare well. Indexing techniques like ART and optimized B + trees are used in

IMDBs. Supporting secondary indexing mechanisms on a distributed database is nontrivial. Secondary indexes contain attributes which may not be the partition parameters. Many of the IMDBs use partitioned indexes. Each node will contain indexes for its own data. In this way each transaction is issued at every node. Updating indexing is simplified as each node needs to update index on its partition only. The other way is to replicate the indexes at all nodes. Although this reduces the transaction time, maintaining indexes is complicated [21,34].

2.5.5 Data Representation

Databases can choose the storage representation of the data. Row-based storage representations are highly useful for transaction processing systems, while columnar representations give higher performance in analytical queries. Also the columnar compression algorithms reduce the memory footprint.

SAP HANA and Hyper both support hybrid storage representations [36]. In Oracle's in-memory tables are stored only in columnar format [23].

2.5.6 Recovery

Backups and checkpoints are essential in an IMDB to ensure the database is sufficiently fault tolerant [37]. Backups of the databases are to be maintained in stable storage devices like SSD or HDDs. Two things a database has to ensure are that the backups are up-to-date and the recovery times should be negligible. Log records stored during transaction committing are used to make DB consistent. The database is checkpointing at regular interval to eliminate the need of old logs.

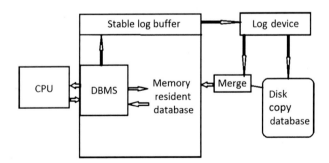

After a failure, IMDBs restore the database from latest checkpoint and redo the log to bring data to up-to-date. When loading the whole database into memory takes a lot of time, we can use lazy loading. Load records when they are demanded. In single-node systems, when a DB comes back online after a failure, the last checkpoint data is loaded from disk. Then the write-ahead log is replayed till it returns to the state at the moment of the crash. In distributed systems, however, multiple replicas for the same data are present. After the master node crash, some other node will be elected as a master and the transactions might have processed. So the failed node when it comes back online cannot directly use its last checkpoint and replay its log. It instead can take the current state of the new master and continue as a normal node. The other mechanism is to first restore the database as in single-node scenario. And then query the changes by importing the log from other nodes and replaying those logs. This will converge if the processing speed of log replay is faster than the rate of transactions.

2.5.7 Replication
To ensure high availability databases are replicated. All modern databases, including IMDBs, support replications in one form or the other. In strongly consistent databases the transaction must be executed at all the nodes to consider committed. All the replicas are synchronized. This causes significant delays due to slow networks or network partitions. NoSQL on the other hand supports eventual consistency model. In this model, not all replicas are needed to acknowledge committed transactions.

Two different replica models are present. One is active–active replication, in which a query is executed in all replicas in parallel in which query is processed at one node and the resultant state is propagated to the replica. VoltDB supports active–active replication configuration, in which all the transactions are guaranteed to be executed in same order, whereas most IMDBs cannot guarantee the order of execution of queries in replicas. This is because the execution order depends on external factors like network delays.

3. USE CASES
Session Caching
Whenever a user logs in to an application, user is authenticated and a session ID/session access token is generated. Traditionally this information is stored at the application server and session token is sent to user client (web browser or mobile app). These tokens are configured to be valid for a few minutes to few days, depending upon on the application. Banking apps traditionally set the validity to be few minutes. Applications like Gmail and Facebook set the session cookie validity for more than a few days.

The problem with this approach is that due to high concurrent usage of the applications, the app server will be overloaded for maintaining the session-level data. Another issue is that application server can crash or become unavailable along with it session data is lost. To cater to the high number of users, multiple application servers can be used. Each application server maintains certain sessions. This may cause certain application servers to be flooded with request, while the others are relatively idle.

Instead of storing session data at the application server level, if we just push the data to a global database we can resolve these issues. Since the session data is stored at the database, any application server can act on a user request. The load balancer can dynamically route the user queries to the free app servers. Whenever an application server fails, the session data is no longer lost and any other app server can cater the user requests. Moreover application server no longer needs to store the session data for long running sessions. This would free up app server resources considerably. The session data is accessed frequently. Any changes in the session state should be reflected in the cache with minimum delays. Using an in-memory database would be the best approach as the access times and performance are considerably better. Although adding a layer adds a little latency, this is tolerable because of the other benefits it provides us with. This user experience gains through this are multifold. Redis and Memcached are two in-memory databases which are used to cache user sessions to improve UX [8].

Caching frequently accessed posts or blog in an in-memory database is widely used in the industry to increase the responsiveness of the applications.

Retail Inventory Management

E-commerce platforms allow thousands of suppliers to sell their merchandise to millions of users across the globe. These suppliers range from individual designers to supermarkets, who may have both online and offline point of sales. Moreover users expect seamless integration of online shopping experience with offline presence. E-commerce players offer buy online and pick up at a store option to increase user experience. The UX flatters if user places an order for a particular item on one platform and when he reaches the store to pick it up, the stock is out of sale. Suppliers with multiple stores may find it useful to transfer items from store to other based on the demand for the merchandise.

To facilitate these demands a real-time inventory of all the items at the places needs to be tracked. Many of the major E-commerce players have 1000s of suppliers and processed about 10,000s of orders per minute.

This inventory should be consistent at all items, implying the database should be ACID compliant.

A globally distributed in-memory database stores the product, items, and supplier details, which is updated whenever there is any sale either at the store or on the E-commerce platforms. The end consumers can see items with available stocks. The suppliers can replenish the stocks dynamically to cater to their customers. E-commerce platforms can use this order data to display best sellers and trending items.

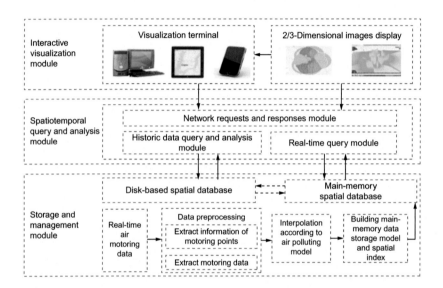

Real-Time Air Quality Monitoring

Across the world, thousands of people die every day due to pollution-related diseases. The air quality is deteriorating in all the major cities to dangerous levels. This global threat has unified the countries to stand together and curb the emissions. The first step is monitoring the air quality in real-time basis.

Air quality sensors are distributed across a geographical region. The sensors monitor the CO_2, NO, and CO ppm in the air and push the data to centralized storage repository. Then different reporting and insights are derived from this data on regular basis. The risk in using slower disk based is that the sudden air quality fluctuations cannot be detected in time. And this may cause unforeseen damages. Industries specifically need to monitor the air quality in their manufacturing plants, which otherwise will lead to safety hazards.

In-memory databases allow users to run real-time queries on the real-time data and receive data within milliseconds. The centralized data store supports complex event processing, or stream processing would be able to detect these fluctuations instantly and issue notifications to users and relevant authorities as well. This will potentially save thousands of lives. The old sensor data then can be pushed to cold storage devices like hard disk for historic analysis and pattern recognition [35].

In the following section we will briefly study different in-memory databases available in the market. Research databases by HyPer and H-Store are intentionally left out as we would like to discuss solutions available for usage instead of experimental builds. As we have seen till now, in-memory databases are not a new phenomenon. As the IMDBs gain popularity, the traditional database vendors like Oracle, Microsoft, and IBM started adopting in-memory features into their databases. Here we will study three of the databases built from scratch to take advantage of memory. Redis is a high-performance key-value store, which is commonly referred to as a NoSQL database. VoltDB is an in-memory NewSQL database. SAP HANA is an appliance from SAP which allows users to store data in different storage formats (Rows, Columns, Graphs, Json, and Texts).

4. IN-MEMORY DATABASES

4.1 Redis

Redis is an open-source high-performance in-memory key-value data store. The fundamental datatype is string. String can be either text or binary with maximum size of 512 MB. Redis also provides data structures for list, hash map, and sorted list. Redis is written in Java and C. It has over 100 open-source clients supporting Java, Python, C, R, Go, etc. Redis is an acronym for Remote Dictionary Server [4].

Redis supports active–passive asynchronous replication where data is replicated to multiple slave servers. This increases the read performance as well as the fault tolerance. Redis supports snapshots and creating append-only files to persist each change.

Redis supports PUB/SUB standard which makes it a very good applicant for chat rooms and event notification. Its sorted set data structure can be used in real-time leather board applications. Redis list data structure makes it easy to implement a central, lightweight queue.

Redis is single threaded, but the process requests asynchronously. It maintains a Hash-table structure for all key-value objects. Redis supports server-side scripting through Lua language. It allows applications to perform user-defined function inside the server, reducing the round trip. Since Redis is single threaded, a batch process can block all other requests.

Redis cluster [4] is a decentralized distributed data store. It supports automatic data sharding, master–slave replication, and live reorganization. Each Redis server keeps all the metadata information and uses gossip protocol for updating replicas. Redis cluster uses a hash slot partition strategy to assign a subset of hash slots to each server node. Each node is responsible for the key-value objects it has been assigned. A client can send request to any of the nodes; the node responds with the respective redirection. The partitioning of data is currently manual.

4.2 VoltDB

VoltDB is an in-memory, SQL-based, ACID-compliant, Cloud-first database. This is the database for high throughput and huge amounts of data-generating applications. VoltDB consistency model is based on serial execution of transactions at its own partitions. For multinode partitions central coordinators execute the transactions in serial order. The transactions are ordered based on the logical timestamps. This ensures that the transactions are executed in the same order on the replicas as on primary node. For this VoltDB supports both active–active and active–passive replication mechanisms. VoltDB is built on Java and C++. The queries are issued through SQL. User-defined functions or stored procedures can be written in Java [13].

VoltDB has Hadoop connectors for ingesting data into Hadoop for historical analysis. VoltDB is best suited for extremely high read and write throughput requirements, point of sale analytics, and real-time air quality monitoring. VoltDB is horizontally scalable on cluster of commodity servers. The entire database is divided into partitions, and each partition is attached to a single CPU core with a single thread. Applications cannot lock records in the VoltDB. Each transaction is isolated from one another due to serial execution and no transaction interleaving. Interactive queries are not supported. To support ACID properties on complex transaction, developers can use stored procedures. Each stored procedure executes atomically. As we have seen earlier, VoltDB supports cross-datacenter active–active replication. The VoltDB cluster can be configured to have cross-datacenter active–active synchronization. The user can be catered by the datacenter nearby him. In this case conflict might arise and these need to be taken care of.

When a node fails, VoltDB will wait for 1–3 s to confirm that the node is actually gone, not just a network delay, and then continue its work on surviving partitions. VoltDB supports C#, C++, Erlang, Go, Java, JDBC, JSON + HTTP (REST), Node.js, PHP, Python, and Ruby.

4.3 Microsoft Hekaton

Hekaton is a memory optimized OLTP engine integrated in MS SQL server [3]. Hekaton tables are created by the user explicitly. Users can combine normal table and Hekaton tables in the same SQL query. Hekaton engine is designed for high concurrency OLTP. This engine uses lock-free and latch-free data structures and optimistic MVCC protocols. It uses a sophisticated framework called Siberia to evict cold data to disk devices, thereby effectively managing larger-than-memory big data use cases. Hekaton adopts compile once execute multiple times strategy by precompiling SQL code into C code, which intern is translated to machine-level code at run time. Durability of the data is ensured by using incremental checkpoints and transaction logs. Group commits and asynchronous commits are also supported by Hekaton tables. Replicas are used to ensure high availability of the database.

Hekaton uses Siberia to perform offline classification of hot and cold data. Tuple accesses are logged first. These logs are analyzed offline to predict the top-k hot tuples with highest access frequencies. This incurs less overhead than LRU due to offline analysis. No additional information is stored in memory related to evicted records. Instead Bloom filter and adaptive range filters are used to filter access to disk.

4.4 SAP HANA

SAP HANA is a distributed in-memory database featured for Hybrid Transaction Analytical Processing. It has several database engines to support row/column engines for structured data, graph engine for semistructured data, and text engine for unstructured data. All the data is memory resident. If the database does not fit the memory, a portion of it is evicted to disk and reloaded upon demand. To reduce the inefficiency of insert or update in the columnar data layout, two delta structures are added. It supports multiple query language interfaces like SQL, MDX, WIPE, and R. It also has a business function library for user-defined functions. Transaction isolation is achieved through MVCC and optimized 2PL for commit processing. Periodic checkpointing and logging are implemented to provide necessary fault tolerance. HANA supports temporal queries based on Timeline Index [32].

In HANA, tables or partitions are configured manually to be either in row store or in the column store. It provides advisory tool which suggests best layout for a table based on data and query characteristics. SQL queries are executed similar to other databases, while other queries are transformed into calculation graph model. The leaf nodes denote the actual data and inner nodes denote the logical operations. HANA supports R scripting for ad-hoc analytics. There R functions can be used as logical operators in above calculation graph. Data transfer to R data frames is done through shared memory. It is possible to combine columnar and row tables in the same SQL query [38].

SAP HANA is tightly integrated with SAP application servers. The communications between SAP application server and HANA are done through shared memory. It worthwhile to note that HANA is a proprietary system developed to augment and enrich the SAP offerings. Although it can technically be used with non-SAP applications as data store, HANA is sold as an appliance.

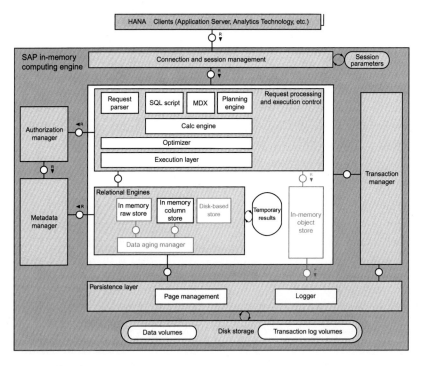

The above figure shows the high-level architecture of SAP HANA [32]. Authorization manager checks if the user has the authorization to issue particular set of queries. The metadata manager is the global repository of all the table structures, indexes, foreign relationships, and partition details. The relational engine are the in-memory data stores where the actual data is present. The persistent layer takes periodic checkpoints and logs committed transactions for durability and fault tolerance. The transaction manger coordinates the transactions issuing locks and deadlock prevention operations. The calculation and execution engines optimize the queries and translate them to calculation graph models. And the session manager manages the client sessions.

5. CONCLUSION

Data is being generated at massive scale at massive speeds. If we can analyze this data within time, we can unlock insights which can help business gain and retain customers. We have seen the evolution of databases from relational databases like MySQL, IBM DB2 to NoSQL databases which are massively scalable with minimum support to ACID properties. We touched mainly on the NewSQL database which forms the major chunk of modern high-performance low-latency read–write access utilizing in-memory data store. Making main memory the primary data storage involves removing or optimizing the bulky and complex functional components of a database. Multiversion concurrency control is being used by major NewSQL databases like Hyper and HANA. 2PL + MVCC is being used in Google Spanner and NuoDB. Logging is the only disk access which is optimized through precommitting log or group commits.

Few use cases for in-memory databases are real-time air quality monitoring, user session caching, and real-time inventory tracking, all of them require highly concurrent reads and writes. We touched upon few in-memory databases in the market like SAP HANA which is hybrid transaction processing system; VoltDB for high-performance writes/reads; Redis, an in-memory key-value store; and Microsoft Hekaton which is an in-memory engine in Microsoft SQL server. The future of the analytics is moving toward hybrid processing where the analytics queries will be run on the fresh transactions.

REFERENCES

[1] N. Gershenfeld, Physics of the future: how science will shape human destiny and our daily lives by the year 2100, Phys. Today 64 (10) (2011) 56.
[2] P. Raj, A. Raman, D. Nagaraj, S. Duggirala, High-performance big-data analytics, in: Computing Systems and Approaches, first ed., Springer, 2015.

[3] C. Diaconu, C. Freedman, E. Ismert, P.A. Larson, P. Mittal, R. Stonecipher, M. Zwilling, Hekaton: SQL server's memory-optimized OLTP engine, Proceedings of the 2013 ACM SIGMOD International Conference on Management of Data, ACM, 2013, pp. 1243–1254.

[4] Redis, from https://redis.io/documentation, 2018. Retrieved August 01, 2017.

[5] R. Cattell, Scalable sql and nosql data stores, ACM SIGMOD Rec. 39 (2011) 12–27.

[6] Apache Ignite, Open Source In-Memory Computing Platform—Apache Ignite, from https://ignite.apache.org/, 2018. Retrieved August 01, 2017.

[7] M. Aslett, How Will the Database Incumbents Respond to NoSQL and NewSQL? The 451 Group, 2011.

[8] K. Kovács, (2013). Cassandra vs MongoDB vs CouchDB vs Redis vs Riak vs HBase vs Couchbase vs Neo4j vs Hypertable vs ElasticSearch vs Accumulo vs VoltDB vs Scalaris comparison. Available at: http://kkovacs.eu/cassandra-vs-mongodb vs-couchdb-vs-redis.

[9] S. Binani, A. Gutti, S. Upadhyay, SQL vs. NoSQL vs. NewSQL—a comparative study, Database 6 (1) (2016) 43–46.

[10] P. Raj (Ed.), Handbook of Research on Cloud Infrastructures for Big Data Analytics, IGI Global, 2014.

[11] NuoDB, from http://www.nuodb.com/, 2018. Retrieved August 01, 2017.

[12] MemSQL, The Real-Time Data Warehouse, from http://www.memsql.com/, 2018. Accessed 8 January 2017.

[13] VoltDB In-Memory Database, from http://www.voltdb.com/, 2018. Retrieved August 01, 2017.

[14] S. Proctor, Google Cloud Spanner & NuoDB: A Comparison of Distributed, ACID-Compliant Database, NuoDB whitepaper, available from https://www.nuodb. com/techblog/google-cloud-spanner-nuodb-comparison-distributed-acid-compliant-databases, 2017.

[15] A. Kemper, T. Neumann, HyPer—Hybrid OLTP&OLAP High Performance Database System, Technical Report, TUM-I1010, May 19, 2010, from http://hyper-db.de/, 2018. Retrieved August 01, 2017.

[16] A. Pavlo, M. Aslett, What's really new with NewSQL? ACM SIGMOD Rec. 45 (2) (2016) 45–55.

[17] Clustrix Scale-out RDBMS, from http://www.clustrix.com/, 2018. Retrieved August 01, 2017.

[18] H. Zhang, G. Chen, B.C. Ooi, K.L. Tan, M. Zhang, In-memory big data management and processing: a survey, IEEE Trans. Knowl. Data Eng. 27 (7) (2015) 1920–1948.

[19] Chao, P., He, D., Sadiq, S., Zheng, K., & Zhou, X. (2017). A performance study on large-scale data analytics using disk-based and in-memory database systems, in: IEEE International Conference on Big Data and Smart Computing (BigComp), February 2017, IEEE 2017, pp. 247–254.

[20] J. Gray, In: Tape is dead, disk is tape, flash is disk, RAM locality is king, Gong Show Presentation at CIDR, 2007, pp. 231–242.

[21] H. Garcia-Molina, K. Salem, Main memory database systems: an overview, IEEE Trans. Knowl. Data Eng. 4 (6) (1992) 509–516.

[22] K. Jung, K. Lee, H. Bae, Implementation of storage manager in main memory database system altibase TM, in: Lecture Notes in Computer Science, Proceedings of the 10th International Conference on Real-Time and Embedded Computing Systems, 2004.

[23] T. Lahiri, M.A. Neimat, S. Folkman, Oracle TimesTen: an in-memory database for enterprise applications, IEEE Data Eng. Bull. 36 (2) (2013) 6–13.

[24] H. Plattner, A Common Database Approach for OLTP and OLAP Using an In-Memory Column Database, 2009.

[25] S. Harizopoulos, D.J. Abadi, S. Madden, M. Stonebraker, In: OLTP through the looking glass, and what we found there, Proceedings of the 2008 ACM SIGMOD International Conference on Management of Data, ACM, 2008, pp. 981–992.

[26] J. DeBrabant, A. Pavlo, S. Tu, M. Stonebraker, S.B. Zdonik, Anti-caching: a new approach to database management system architecture, PVLDB 6 (14) (2013) 1942–1953.

[27] H-Store, Next Generation OLTP Database Research, from http://hstore.cs.brown.edu/, 2018. Retrieved August 01, 2017.

[28] K. Alfons, T. Neumann, In: Main-memory database systems, 2014 IEEE 30th International Conference on Data Engineering (ICDE), IEEE, 2014, p. 1310.

[29] D.J. DeWitt, R.H. Katz, F. Olken, L.D. Shapiro, M.R. Stonebraker, D. Wood, Implementation techniques for main memory database systems, ACM SIGMOD Rec. 14 (2) (1984) 1–8.

[30] Giles, E., Doshi, K., & Varman, P. (2016). Persisting in-memory databases using SCM. In Big Data (Big Data), 2016 IEEE International Conference on (pp. 2981-2990). IEEE.

[31] T. Neumann, T. Mühlbauer, A. Kemper, Fast serializable multi-version concurrency control for main-memory database systems, SIGMOD, 2015, pp. 677–689.

[32] F. Färber, N. May, W. Lehner, P. Große, I. Müller, H. Rauhe, J. Dees, The SAP HANA database—an architecture overview, IEEE Data Eng. Bull. 35 (1) (2012) 28–33.

[33] CockroachDB, from https://www.cockroachlabs.com/, 2018. Retrieved August 01, 2017.

[34] Lehman, T. J., & Carey, M. J. (1986). Query processing in main memory database management systems, SIGMOD '86 Proceedings of the 1986 ACM SIGMOD International Conference on Management of Data, vol. 15, No. 2, pp. 239-250. ACM.

[35] Tang, Z., Xiong, W., Chen, L., & Jing, N. (2016). A real-time system for air quality monitoring based on main-memory database. In 2016 24th International Conference on Geoinformatics on (pp. 1-4). IEEE.

[36] A. Kemper, T. Neumann, HyPer: A Hybrid OLTP&OLAP Main Memory Database System Based on Virtual Memory Snapshots, ICDE, 2011, pp. 195–206.

[37] L. Camargos, F. Pedone, R. Schmidt, In: A primary-backup protocol for in-memory database replication, NCA 2006: Fifth IEEE International Symposium on Network Computing and Applications, 2006, IEEE, 2006, pp. 204–211.

[38] F. Färber, S.K. Cha, J. Primsch, C. Bornhövd, S. Sigg, W. Lehner, SAP HANA database: data management for modern business applications, ACM SIGMOD Rec. 40 (4) (2012) 45–51.

ABOUT THE AUTHOR

 **Siddhartha Duggirala** has graduated from IIT Indore. He has been working in Bharat Petroleum Corporation Ltd ever since. He has written multiple research chapters and conducted few training programs. His research interests include Data Analytics, IoT, Storage Infrastructures, Machine Learning.

CHAPTER THREE

NoSQL Web Crawler Application

Ganesh Chandra Deka

Directorate General of Training, Ministry of Skill Development & Entrepreneurship, New Delhi, India

Contents

1. Introduction to Web Crawlers	78
1.1 Search Engine Back-End Database Generation	81
2. Features of Web Crawlers	82
3. Types of Web Crawler	83
3.1 Focused Web Crawler	83
3.2 Incremental Crawler	84
3.3 Distributed Crawler	84
3.4 Parallel Crawler	84
3.5 Traditional Web Crawlers	85
3.6 Open Source Web Crawlers	85
3.7 IoT Web Crawlers	87
4. APIs of Web Crawler	87
5. Challenges of Web Crawler Design	88
6. Application of NoSQL Databases in Web Crawling	89
7. Conclusion	93
Key Terms and Definitions	94
References	95
Further Reading	99
About the Author	100

Abstract

With the advent of Web technology, the Web is full of unstructured data called Big Data. However, these data are not easy to collect, access, and process at large scale. Web Crawling is an optimization problem. Site-specific crawling of various social media platforms, e-Commerce websites, Blogs, News websites, and Forums is a requirement for various business organizations to answer a search quarry from webpages. Indexing of huge number of webpage requires a cluster with several petabytes of usable disk. Since the NoSQL databases are highly scalable, use of NoSQL database for storing the Crawler data is increasing along with the growing popularity of NoSQL databases. This chapter discusses about the application of NoSQL database in Web Crawler application to store the data collected by the Web Crawler.

Advances in Computers, Volume 109
ISSN 0065-2458
https://doi.org/10.1016/bs.adcom.2017.08.001

1. INTRODUCTION TO WEB CRAWLERS

With the advent of Web technology, data has exploded to a considerable amount. The Web is full of unstructured data called Big Data. However, these data are not easy to collect, access, and process at large scale. Web Crawler is an indispensable part of Web for accessing Web data. Web Crawler is a computer program for traversing through the hyperlinks, indexes them and index them. Web Crawler (aka Spiders, Robots, Wanderers) can be used for clustering of websites [1].

Crawlers have many purposes such as Web analysis, email address gathering, etc., although their main one is the discovery and indexing of pages for Web search engines [2]. Major utilities of Web Crawler include:

- Performing data mining. Web Crawling is the first step in the Web data mining process
- Gathering pages from the Web, automatically download documents from a Web server
- Analyze documents retrieved from a Web server and send data back to a search engine database
- Supporting a search engine
- Search engines, Data analysis, Automated Web interactions, Mirroring, and HTML/link validation

When a Web Crawler visits a Web page, it reads the visible text, the hyperlinks, and the content of the various tags used in the site, such as keyword-rich Meta tags. For using the information gathered from the crawler, a search engine determines what the site is about and indexes the information. Finally, all the text and metadata specifying the Web documents scanned by the crawler are stored in the search engine's Database.

The basic data structure of Crawler is the Uniform Resource Locator (URL) List. Web Crawlers surf from website to website, via embedded links in documents, copying everything it comes across. Crawler system main data sets are:

- Domains
- Links
- Pages
- HyperText Markup Language (HTML), Graphics Interchange Format (GIF), Joint Photographic Experts Group (JPG), American Standard Code for Information Interchange (ASCII) files
- Postscript and Active Server Pages (ASP)

The download time of a website is divided into the following four components [3]:

1. DNS resolution time
2. TCP latency
3. HTTP latency
4. Page transfer time

For instance the j-soup Java library gets the webpage in the form of HTML. The j-soup then parses the HTML from the present webpage and finally gets the text of full webpage. This library helps in extraction and manipulation of data in the real-world Web pages. The text appears in the text area present in the front interface of the Java application [4].

All the crawlers generally have the following components:

- a URL Fetcher
- Parser (Extractor)
- Multithreaded processes
- Crawler Manager
- Queue structure

Web Crawlers use a graph algorithm such as Breadth First Search (BFS) for navigation from page to page. Another option may be using a connectivity-based ordering metric such as Google's **PageRank**. The PageRank will be a better choice than using breadth first search; however, the challenge is computing the PageRank values. Computing the PageRank values for millions of pages is an extremely expensive computation.

A Web Crawler works in the following steps:

1. Remove URL from the unvisited URL list
2. Determine the IP address of its host name
3. Download the corresponding document
4. Extract any links contained in it.
5. If the URL is new, add it to the list of unvisited URLs
6. Process the downloaded document
7. Back to step 1

Fig. 1 illustrates the above mentioned steps of a typical Web Crawler.

Crawler first resolves the server **hostname** into an **IP** address to contact it using the **Internet Protocol**. The mapping from Domain Name to IP address is done by mapping with Domain Name Server (DNS) database. Web Crawlers identify **IP** address to a Web server by using the **User-agent** field in a Hypertext Transfer Protocol (HTTP) request, and each crawler has their own unique identifier [5].

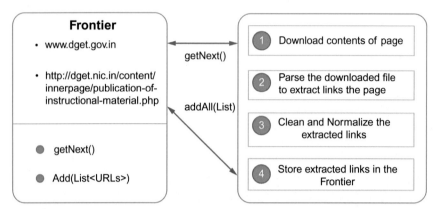

Fig. 1 Crawling Web Pages.

The following SQL query demonstrates how Web Crawler builds a search engine database. The query *inserts* the crawled *url* (mostly the HTML or RDF (Resource Description Framework)) in the "*Record*" table of the "*Crawler*" database [6].

> *sql = "INSERT INTO 'Crawler'.'Record' " + "('URL') VALUES " + "(?);";*

The authors Alessandro [6] developed and run the code in the following test environment:

- Operating System (OS): Ubuntu
- Library: JSoup core Java library (*http://jsoup.org/download*)
- JDBC driver for MySQL: MySQL Connector/J (*https://dev.mysql.com/downloads/connector/j/*)
- phpMyAdmin: GUI interface for using MySQL (WampServer, a Windows Web development environment, *http://www.wampserver.com/en/*)
- Scripting language: Hypertext Preprocessor (PHP)

Similarly, the Scrapy, an application framework for crawling websites and extracting data, uses Scrapy library (v1.3 is latest) along with *PyMongo* (v3.0.3) required for storing the data in *MongoDB* [7]. MongoDB and PyMongo are to be installed on the system to interconnect MongoDB and Python [8].

PyMongo is the Python distribution containing tools working with MongoDB [9]. Similarly, the raw data collected by the Web Crawlers can be stored in distributed column-based NoSQL database such as HBase.

1.1 Search Engine Back-End Database Generation

Web Crawler client program connects to a DNS server. DNS server translates the *hostname* into an *Internet Protocol* (IP) address. Crawler then attempts to connect to server host using specific port. After connection, crawler sends an *HTTP* request to the Web server to request a page (usually a GET request). To reduce the waiting for responses to requests, Web Crawlers use threads for fetching hundreds of pages at once.

A Web Crawler does the following for a Search Engine:

- Maintains a queue that stores URLs
- Repositories of Documents and URLs
- Filters and collects resource from the Internet for constructing database. Web databases are of two types:
  - Unstructured databases consisting of data objects, e.g., Texts, Images, Audio, and Video
  - Structured databases consisting of data objects such as a database for books

Search engines index billions of Web pages for generating multiterabyte database. Search engines rely on this database for efficient execution of user queries for searching contents in the Web. For instance, *Become.com* shopping search engine created using Java technology Web Crawler is massively scaled for obtaining information over 3 billion Web pages and writing them (more than 8 TB of growing data) on 30 fully distributed servers in 7 days [10]. *Become.com's* crawlers build searchable database based on what the crawler discovers on the Web Crawl [10]. *Become.com* replaced the Structured Query Language (SQL) database to enhance the performance by new database. The new database (most likely a NoSQL database) uses an internal format including APIs for both the Java and C++ language for better throughput than SQL [10]. Similarly, Google has very robust, state-of-the art C++ libraries that handle large-scale distributed systems.

Integration tools such as PHP, Servlets, Java Server Pages (JSP), ASP, etc. are used for integration of Web data. One example of Web data integration is QuickCode (erstwhile ScraperWiki), hosted environment for writing automated processes to scan public websites and extract structured information from the pages visited. QuickCode is a simple online editor for Ruby, Python, or PHP scripts, and automatically runs crawler as a background process [11].

Indexer produces data structures for fast searching of all words in the pages. Existing Web Crawlers use relational databases, which have completely different performance results for retrieving structured and unstructured data from a massive amount of data. However, for the storage of exponentially growing Web data NoSQL databases are the best fit. This chapter discusses the application of NoSQL databases in Web Crawler.

2. FEATURES OF WEB CRAWLERS

- Distributed—should have the ability to execute in a distributed environment across multiple machines.
- Scalable—capable of scaling up the crawl rate by adding extra machines and bandwidth.
- Performance and efficiency—capable of making efficient use of system resources, i.e., processor, storage, and network bandwidth. Crawlers downloading a site for the first time is likely to download all the available archives.
- Quality—the crawler should be capable of fetching "useful" best pages first.
- Freshness—should obtain fresh copies of previously fetched pages, basically news and RSS feed crawlers.
- Extensible—should be designed to be extensible to cope up with new data formats and new fetch protocols. To be extensible, the crawler architecture should be modular.

The primary challenge of a search engine is to return results that match a users' need as the annual global IP traffic is projected to reach 2.3 ZB by 2020 (nearly 95 times from 2005 [12]). Further, crawling the social Web data also known as the Big Data is another challenge.

Bad bots (crawlers) consume CDN (Content Delivery Network) bandwidth, take up server resources, and steal valuable content for misuse. Further, by downloading the complete content of a Web server, a bad Web Crawler might adversely affect the performance of the server. The spammer bots (Bad bots) hits around 31% of sites with Web forms such as *Contact Us*, *Discussion forums*, and *Review* [13].

3. TYPES OF WEB CRAWLER
3.1 Focused Web Crawler

A crawler cannot access pages or data using secure browsers like Tor (the Onion Router) and entire networks of VPN tunnels [14]. The ordinary crawlers can reach only a portion of the Web called the *publicly indexable Web* (**PIW**). Hidden Web/Deep Web are the content stored in databases/ Web pages behind security layers. The traditional crawlers are not capable of searching beneath the surface of the PIW; hence, the Deep Web remains unexplored. The size of Deep Web is estimated to be 500 times of the surface Web and contains 7500 TB of information compared to 19 TB of information in the surface Web [15]. According to Ref. [16], the share of Deep Web is Google.com (32%), Yahoo.com (32%), and MSN.com (11%). Hence, the sea of information hidden in the Deep/Hidden Web have a huge potential. For instance, **Facebook** data about people is used by **Facebook** for highly targeted advertisement. Facebook does not allow Google to crawl its content; hence, Google cannot reach the Facebook user data [17]. The following are the some more instances of huge Hidden/Deep Web:

(i) Large portions of the Web "***hidden***" in searchable databases

(ii) HTML pages dynamically generated in response to queries submitted via the search forms

(iii) In *virtual hosting* multiple websites are hosted on the same physical *IP address*; it becomes very difficult to identify the virtual hosts within a physical IP address.

Two approaches for **Deep Web** Crawling:

• Virtual Integration
• Surfacing

Deep Web is visited by using the Focused Crawling technique such as Form-Focused Crawler (**FFC**) and Adaptive Crawler for Hidden Web Entries (**ACHE**).

Juxtapose aka JXTA Search is intended for two complementary search types:

• Wide
• Deep

Deep search engines find information embedded in large product databases, e.g., Amazon or CNN news article databases [18].

3.2 Incremental Crawler

The incremental crawler revisits only the pages that have changed or having the high probability of change, instead of refreshing the entire collection. The existing collection is updated on a periodic basis, leading to save network bandwidth and overall effectiveness of the search engine. The primary issue in incremental crawling is defining metrics for performance. Most of the strategies for incremental crawling are heavily dependent upon polling (due to the uncoordinated nature of the Web). However, incremental crawler using polling is fairly inefficient [19].

3.3 Distributed Crawler

Many crawlers are employed to distribute the process of Web Crawling, in order to have the most coverage of the Web. A central server manages the communication and synchronization of the nodes, as it is geographically distributed. Distributed crawling needs increased computing nodes and storage capability to increase crawling efficiency. It basically uses PageRank algorithm for its increased efficiency and quality search. Types of distributed Web Crawling are:

- Dynamic Assignment: With Dynamic Assignment, a central server assigns new URLs to different crawlers dynamically. This allows the central server dynamically balance the load of each crawler.
- Static Assignment: Here a fixed rule is applied from the beginning of the crawl that defines how to assign new URLs to the crawlers. A hashing function can be used to transform URLs into a number that corresponds to the index of the corresponding crawling process. To reduce the overhead due to the exchange of URLs between crawling processes, the exchange is done in batch.

3.4 Parallel Crawler

It is difficult to retrieve the whole or a significant portion of the Web using a single process, hence becomes imperative to parallelize the crawling process, in order to finish downloading pages in a reasonable amount of time. This type of crawler is known as *parallel crawler*.

The main goals in designing a parallel crawler are:

- Maximize its performance (download rate)
- Minimize the overhead from parallelization

The parallel crawler can be intrasite or distributed. The agents of a parallel distributed crawler can communicate either through a LAN or through a WAN.

UbiCrawler is a distributed crawler capable of running on any kind of network [20]. The Hadoop has been proved to be inevitable tool for parallel crawling of data using MapReduce programming model.

3.5 Traditional Web Crawlers

Traditional (Generic) Web Crawler crawls documents and links belonging to a variety of topics, whereas focused crawlers use some specialized knowledge to limit the crawl to pages pertaining to specific topics. Components of a traditional Web Crawlers are:

- **Crawlers**: Collects documents by recursively fetching links from a set of starting pages. Each crawler has different policies since the pages are indexed by various search engines differently. Crawlers should obey Robot Exclusion Protocol. Specific document can be shielded from a crawler by adding the line:
  <META NAME="ROBOTS" CONTENT="NOINDEX">
  - Inclusion: URL regular expression ^https://.*\.example(?:corp){0,1}\.com crawls only HTTPS URLs in example.com and examplecorp.com domains. Anything ending with example.com is crawled, but http://www.example.com.tw is not to be crawled [21].
  - Exclusion: URL regular expression (?i:\.doc|\.ppt)$ excludes files ending with .doc and .ppt.
- **Indexer**: Processes pages, decides which of them to *index*, and builds various data structures representing the pages (inverted index, Web graph, etc.). Might also build additional structures (LSI).
- **Query Processor**: Processes user queries and returns matching answers in an order determined by a ranking algorithm.

For instance, the Google's **PageRank algorithm** returns pages from the Web's collection of 25 billion documents that match search criteria [22].

3.6 Open Source Web Crawlers

Apache Software Foundation (ASF), the incubators of nearly 150 Open Source projects and initiatives, launched Apache **Nutch** (a highly extensible and scalable open source Web Crawler project) for catching up the NoSQL trends and adopts a table-like representation (The Apache Software Foundation, Blogging in Action [23]). Nutch uses Lucene to build the index.

The open source ***StormCrawler*** is a SDK (Software Development Kit) for building distributed Web Crawlers with **Apache Storm** [24]. ***StormCrawler*** is a collection of resources and libraries that developer can use to build their own crawlers [25]. The Yahoo! crawler **Slurp** is capable of indexing microformat content [26].

Four main components of Nutch Architecture are:
1. Crawler
2. Web database (WebDB, LinkDB, segments)
3. Indexer and
4. Searcher

YaCy comprises a **Web Crawler**, Indexer, Index Library, User Interface, and a P2P (Peer to Peer) network. The crawler harvests the Web Page which is the Input for the Indexer. The indexer Indexes the pages and put them into the Index Library locally. The local Index Library is merged into a global search Index library which is an integrated NoSQL database [27].

In **Ebot** the URLs are saved to NoSQL database, supporting MapReduce queries that can query via ***RESTful HTTP*** requests or using any programming languages.

Snowden accessed 1.7 million files including documents on internal NSA networks and internal "wiki" materials [28]. The URLs that need to be analyzed are sent to Advanced Message Queuing Protocol (AMQP) queues, making it possible to run several crawlers in parallel and stop and start them without losing URLs (*http://www.freecode.com/tags/nosql* [29]).

MapReduce is a large-scale data processing platform initially used to analyze the text data and build the text index. Hence, Web Crawler application fits perfectly with MapReduce programming model [30]. Fig. 2 illustrates how MapReduce framework is used in Web Crawling.

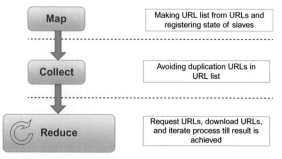

Fig. 2 MapReduce framework for Web Crawling.

3.7 IoT Web Crawlers

An interesting area of research in the new type of search engine under development is Resource Directory (RD) and Hypercat metadata specification defined by Internet Engineering Task Force (IETF) for IoT [31]. Each Hypercat catalog uses HTTPS, REST, and JSON to expose any number of URIs [32].

The IoT devices can be Public IoT and special IoT. Public IoT devices are connected to the Internet. In Public IoT device the IoT sensor/devices (node) are accessed by HTTP.

Shodan search engine looks for devices connected to the Internet using service banners (a block of text about the service being performed). Devices connected to the Internet such as printers are controlled remotely; Web cameras and other devices can be detected by **Shodan** [33].

Shodan collects data from the following ports [34]:

1. HTTP port 80
2. FTP port 21
3. SSH port 22
4. Telnet port 23
5. SNMP port 22
6. SNMP port 161
7. SIP port 5060
8. RTSP port 554

Shodan crawlers discover collect metadata about the devices, *while* **Google** crawls for websites. The **ZoomEye** collects information from Internet about the nodes all over the world using distributed Web Crawler. These nodes are websites and devices.

The Web 3.0 crawling and indexing technologies will be based on human–machine interaction. The gray area of IoT Web Crawler is that some crawlers also test for certain vulnerabilities such as argument injection vulnerability in PostgreSQL (CVE-2013-1899) [35].

4. APIs OF WEB CRAWLER

The website https://www.programmableweb.com/sites/default/files/pw-logo.png lists 81 new APIs. Table 1 shows some of the popular Web Crawlers related to this chapter [38].

Table 1 Popular Web Crawlers

Sl No	Name of Web Crawler	Remarks
1	Beevolve Web Crawler API	Social media monitoring
2	Swiftype Web Crawler API	Search engines for websites. Developers can implant it on their website to design their own site-specific search results. The API is a RESTful protocol and returns results in JSON Swiftype functionalities include indexing, searching, and more. The documentation includes Ruby and Python kits
3	Semantics3 Expanded UPC Search API	Expanded UPC Search API pulls information from a valid UPC scan for making available online Semantics3 API gives customers the ability to retrieve information about millions of products using UPCs (Universal Product Codes) and bar codes [36]
4	Spinn3r API	Web service for indexing the blogosphere
5	Extractiv API	Combines fast Web Crawler with powerful natural language processing technology. Extractive APIs use RESTful protocol and responses can be formatted in XML and JSON
6	Crawler4j	Open source Web Crawler for Java. Provides simple interface for crawling the Web. Using crawler4j multithreaded Web Crawler can be set up in few minutes [37]
7	80Legs	For cost-effective distributed Web Crawling **80Legs's**, Web-crawler-as-a-Service API can be used
8	CouchDB	RESTful JSON API to access data stored in this NoSQL using HTTP

5. CHALLENGES OF WEB CRAWLER DESIGN

The following characteristics of the Web make Web Crawl design difficult [39]:

- **Volume**: in August 1991, when World Wide Web Project was launched the number of website was 1 which have already crossed 1.1 billion [40,41]. In the first quarter of 2016, the total number of registered **Domain Names** was approximately **326.4** million worldwide [42]
- **Rate of change**: rate of changes in the Web is very rapid. Around 23% of Web pages change daily, while the rate of change of commercial Web pages is 40% [43]

- *Dynamic page generation*
- *Crawling Multimedia Content*
- *I/O-related waits* are major performance problem in designing a crawler. Network I/O is not only very slower than disk I/O but also unpredictable in terms of request response duration.

Single request to a website involves not only an HTTP request but also DNS resolution. Hence, improperly designed Crawlers can create lots of troubles [44]. In worst-case situations, crawling carried out by multiple **worker threads**, e.g., 500 threads for a big crawl by sending too many requests to a single server, can lead to Distributed Denial of Service (DDoS) attack.

- *Links* are not easy to extract and record. Keeping track of the URLs that have been visited is a major component of a good crawler design.
- *Database* used at the backend of a crawler is another key limitation in the design of the crawler. Web Crawlers using SQL databases are found comparatively slow.

The following section discusses about the application of NoSQL databases by Web Crawlers.

6. APPLICATION OF NoSQL DATABASES IN WEB CRAWLING

There is a continuous growth in the amount of publicly available OWL (Web Ontology Language) and Resource Description Framework (RDF) datasets on the Web. The graph-based nature of RDF(S) data necessitates special data structures for the storage of Web data [45]. Lots of Semantic Web data retrieval (SWR) techniques are using NoSQL databases to increase processing power and data storage of search engines.

Hadoop is originally designed to handle crawling of billions of Web pages. The result of the crawl was stored at Hadoop's HDFS. Hadoop's distributed processing engine *MapReduce* is used for processing the Web-Crawled text data [46]. Another very important component of Hadoop for Web Crawler is the resource manager called YARN (Yet Another Resource Negotiator).

TalentBin is a Big Data analytics tool for Web development jobs. TalentBin crawls websites for personal information and reduces it to distinction identities. TalentBin stores it to maintain a candidates' database of Web resumes. *TalentBin API* gives developers access to personal information of the job seeker such as their *Location*, *Skills*, and *Company* by using the RESTful API [47].

The following are the types of NoSQL database in use [48].

1. **Graph stores** are used to store information about networks of data, such as social connections. Graph stores include **Neo4j** and triple stores like Fuseki.
2. **Document databases** pair each key with a complex data structure known as a document.
3. **Key–value stores** are the simplest NoSQL databases. Every single item in the database is stored as an attribute name (or "key"), together with its value. Examples of key-value stores are **Riak** and **Berkeley DB**.
4. **Wide–column stores** such as **Cassandra** and **HBase** are optimized for queries over large datasets, and store columns of data together, instead of rows.

The key-value database is suitable to store the HTTP information, because the HTTP header has a key-value-type structure, and keys cannot be estimated before analyzing the HTTP stream [49]. SimpleDB can also be used to store the status of Internet crawl with a limit of 10 GB per domain [50].

BigTable is a distributed, high–performance, fault-tolerant, NoSQL storage system build on top of Google File System (GFS). BigTable is mainly used by Google projects such as Web indexing, Personalized Search, Google Earth, Google Analytics, and Google Finance. Google's document storage system BigTable is customized for Storing, Finding, and Updating Web pages [51].

BigTable is logically organized into rows. A row stores data for a single Web page. BigTable shares characteristics of both row-oriented and column-oriented databases. Each table has multiple dimensions. The values are kept in a compressed form and optimized to the underlying file system (GFS). Rows are ordered in lexicographic order by row_key in Web databases. Such a key is often a URL. If the reverse URL is used as the row_key, the column used for different attributes of the Web page and the timestamp indicates data. Related implementations of BigTable are:

1. HBase
2. Hypertable
3. Apache Cassandra

HBase preserves some of the BigTable characteristics such as the ability of storing the results of crawling. HBase runs on top of the Hadoop Distributed File System (HDFS). HDFS uses commodity hardware to create multiple replicas of the same document on different machines. Storing data in HBase enables faster access as well as data cleaning and processing by using MapReduce [52].

Redis can be used to store the status, i.e., *running*, *paused*, and *canceled*, for all Web Crawls thread for a distributed Web Crawler. Each thread maintains a connection to a Redis server for each domain being crawled by the thread. A Redis key-value pair is used to keep track of the current position of the **URL** for that domain. The Redis sorted sets can be used to store the priority associated with each URL and hashes to store the visited status of the discovered URLs [53].

Apache **CouchDB** uses RESTful JSON API that makes it easy to access data stored in this NoSQL using HTTP. Another interesting feature is the ability to use JavaScript or Erlang-based incremental MapReduce operations to create "views" like RBDMS.

MarkLogic, an XML search platform, provides strong NoSQL database capabilities for search.

RaptorDB is a JSON-based, NoSQL document store database used for the backend store of websites. Knowledge of C# programming language is essential for using RaptorDB [54].

MongoDB uses JavaScript as its primary interface and user shell. MongoDB stores data using a binary serialization of the JavaScript Object Notation (BSON). It stores larger objects using GridFS, a specialized file storage mechanism. MongoDB database would be able to scale up with a highly parallel Web Crawler [55]. However, while reading data from the middle of a crawl job, MongoDB needs to walk the index from the beginning to the offset specified. MongoDB gets slow in browsing deep into a job with lots of data. Crawl logs need to be returned in the order it was written, which is possible to maintain in order in MongoDB [56].

The Web Crawler-indexed information can be stored in a **Neo4j** database for further analysis [57].

Scrapy-Redis, a Redis-based component for Scrapy, is used for starting multiple spider instances sharing a single Redis queue [58]. However, this comes with a large memory footprint. The possible alternative to store the visited status of the discovered URLs would be the use of Bloom filter (to track which *url* has already discovered) at the cost of erroneous results in the form of false positives (due to probabilistic nature of Bloom filters) [59]. Semantics3 uses bloom filters for keeping track of visited URLs and storing the state of the crawl (URLs) in NoSQL database like **Riak** or **Cassandra** (*Ferrit* is an API-driven focused Web Crawler written using programming language *Scala* using Akka, Spray, and Cassandra) [59,60].

Shen139 [61] compared the databases *MySQL* and *MongoDB* for full-text index-supported search engine OpenWebSpider. In the test environment they used VM with 3 Gb of RAM in Windows 8.1 with an index of about 15,000 pages. It was found that MongoDB was slower in responding to search queries in comparison to MySQL. However, in case of applications dynamically storing different field names, a NoSQL database is the best option.

Sleepycat database allows crawling and ranks the crawled URLs efficiently by using the crawler written using Java. Similarly, Berkeley DB (acquired by Oracle) Java Edition is the repository for *Heritrix*, the Web Crawler for Internet Archive [62]. Heritrix follows the robots.txt exclusion directives and META robots tags for not hampering the normal website activities [63]. Heritrix is multithreaded to make progress on many URIs in parallel during network and local disk I/O lags. Internet Archive [62] and other institutions are using Heritrix to perform focused and increasingly broad crawls [64].

Azure Blob (Binary Large Object) provides a NoSQL storage facility for Binary files, Video, Images, and PDF files collected from the Internet by a Web Crawler [65].

Google handles lots of (semi-)structured data such as URLs (Contents, crawl metadata, links (*SRC*, i.e., src="image.gif" and *HREF* attribute specifies the link's destination), anchors, PageRank); Googlebot (a search bot) used by Google finds pages on the Web and hands them off to the Indexer for later use.

Amazon *CloudSearch*, a managed service in the AWS Cloud, supports 34 languages with popular search features such as highlighting, autocomplete, and geospatial search [66,67].

SchemaCrawler, a free database schema discovery and comprehension tool, is capable of creating *Entity–Relationship* diagrams in Microsoft DOcument Template (DOT) (plain text graph description language). SchemaCrawler has powerful scripting ability, using JavaScript, Groovy, Ruby, or Python [68].

ModBot Web Crawler developed by Modmacro is capable of gathering thousands of data points to enhance the Google ranking of Modmacro and monitors the uptime of Modmacro *clients&#39*; websites. If a website crashes, ModBot sends out an immediate notification so the problem can be quickly remedied for reducing a *website&#39* downtime [69].

Lily, an open source data management platform, unifies Apache HBase, Hadoop, and Solr into a comprehensively integrated, interactive platform for real-time indexing and impressive search power of Apache Solr (an open source enterprise search platform) [70]. Lots of open source projects work with Solr to crawl websites and other repositories like Binary Document formats (Microsoft Office and PDF documents) [71]. It is possible to use MongoDB and Solr in an application to get a polyglot-persistent solution [72]. Since Solr has the properties of NoSQL such as **Schema-less**, capable of **Storing semi structured data**, and does not use **relationship between records**, Solr could be considered as NoSQL Document Store [73].

Solr can be integrated with Tika for indexing doc and pdf files [74].

Isomorphic applications (aka universal applications) address the problem of crawling and indexing by taking the best of two worlds, i.e., static HTML and dynamic JavaScript for providing fully indexable, fast, and user-friendly websites. Isomorphic applications execute the codes of the dynamic pages on the server to convert to static HTML pages and then leaving all the JavaScript responsible for user interactions to be run by the browser [75].

Polyglot persistence in Web Crawler design is possible. Relational database can be used to store:

1. **crawl lists** and **their corresponding views** (both *one to many*)
2. **crawl lists** and **their entries** (*both one to many*), etc.

NoSQL database such as MongoDB will provide the document storage for the Web Crawler [76]. Nutch Web Crawler stores its data in HBase. Nutch 2.x uses NoSQL databases such as Cassandra and Accumulo as backends [77]. HBase facilitates *MapReduce* analysis on the latest version of data or at the snapshot of recent *timestamp* [78]. **Mail.ru** is a Web search engine with a content system built on top of HBase. However, in this case the direct communication between HBase and Python was not found reliable [79].

Adhoc Web Crawler written in Python using the search function of GitHub can be used to search the open source repositories written in Java, PHP, and Python that matched a list of "keywords" previously defined. The Server side scripting language PHP also works as intermediary for AJAX for accessing specific data source. PHP is also capable of evaluating queries and sending requests to MySQL database [80]. Being open source MySQL and PHP are a perfect combination for application development for lots of organization. MySQL uses port number 3306 for connections using TCP/IP. Another opensource database management system suitable for using PHP is PostgreSQL.

7. CONCLUSION

Basically Crawling is an optimization problem. Site-specific crawling of various social media platforms, e-Commerce websites, Blogs, News websites, and Forums is a requirement for various business organizations to answer a search query from webpages. Indexing of huge number of webpage requires a cluster with several petabytes of usable disk. Since the NoSQL databases are highly scalable, use of NoSQL database for storing the Crawler data is increasing along with the growing popularity of NoSQL databases.

KEY TERMS AND DEFINITIONS

Robots Exclusion Protocol: robots.txt protocol was developed *for protection for site owners against heavy use of their server* and **network resources** by crawlers. The second objective was to **stop Web search engines from indexing undesired content**, such as test or semi-private areas of a website.

Put the file *robots.txt* at the root directory of the server, i.e., *http://www.w3.org/robots.txt*, for implementing *Robots Exclusion Protocol*.

Internet Archive: amassing of information generated by crawling into a single digital catalog.

JSONP (JSON with Padding) is a common method used to bypass the cross-domain policies in Web browsers.

Hypercat: it is a Hypermedia catalog format designed for exposing information about IoT assets over the Web. A catalog is an array of URI (Uniform Resource Identifier)s. Each resource in the catalog is annotated with metadata (RDF-like triples).

Meta crawling involves directly harvesting URLs from search engines without crawling the Web.

HTML crawling *aka* conventional Web Crawling harvests Web documents by extracting and following hyperlinks, and is useful in harvesting clusters of SWD (Semantic Web Document)s on the Web.

Mobile crawlers are capable of reducing the network traffic by performing Data analysis and data compression at the data source. Mobile crawlers transmit only relevant information in compressed form over the network.

Bloom filter is a probabilistic data structure used to test whether an element is a member of a set. Bloom filters are used to get small, approximate lists from a huge set. For Distributed Systems Bloom filters send URLs. Application of Bloom filter finding objects (Object Location, Geographical Region Summary Service), Data summaries (IP Traceback).

Resource Description Framework is an infrastructure that enables the exchange of metadata structured in an ontology. Developed by the World Wide Web Consortium (W3C) to provide a standard for defining an architecture for supporting the vast amount of Web metadata. RDF is implemented using XML. Since it is an application of XML, it inherits the syntax from XML.

RDF vocabularies includes *FOAF* (friend of a friend, a project devoted to linking people and information using the Web, *http://xmlns.com/foaf/spec/*), *SIOC* (Storage I/O Control), *SKOS* (Simple Knowledge Organization System, *https://www.w3.org/2004/02/skos/*), *DOAP* (Description of a Project, https://github.com/ewilderj/doap/wiki), *vCard* (electronic business cards), *DublinCore* (a vocabulary of 15 properties for use in resource description, originated at a 1995 invitational workshop in Dublin, *http://dublincore.org/documents/dces/*), *OAI-ORE* (Open Archives Initiative Object Reuse and Exchange (*OAI-ORE*) defines standards for description and exchange of Web resources), or *GoodRelations* (a lightweight ontology for annotating offerings and other aspects of e-Commerce on the Web) in order to make it easier for client applications to process Linked Data.

Google uses crawled RDF data for its *Social Graph API*, while Yahoo! provides access to crawled data through its *BOSS API*.

Ontology describes the structure of information at different levels of abstraction. A tree-like structure which can be used to describe semantics.

Search engine optimization (SEO) according to Wikipedia definition "*Search engine optimization (SEO) is the process of improving the visibility of a website or a Web page in search engines via the "natural" or un-paid ("organic" or "algorithmic") search results.*"

Spider Trap-is a defense mechanism to keep spiders/Crawlers out of a site in order to protect email addresses from being found and used for unsolicited mailings.

REFERENCES

[1] V. Granville, Top 2,500 data science, Big Data and analytics websites, Retrieved March 13, 2017, from: http://www.datasciencecentral.com/profiles/blogs/top-2-500-data-science-big-data-and-analytics-websites, 2014.

[2] M. Thelwall, Methodologies for crawler based web surveys, Int. Res. 12 (2) (2002) 124–138.

[3] B.B. Cambazoglu, V. Plachouras, F. Junqueira, L. Telloli, in: On the feasibility of geographically distributed web crawling, Proceedings of the 3rd International Conference on Scalable Information Systems, Institute for Computer Sciences (ICST), Social-Informatics and Telecommunications Engineering, 2008. p. 31.

[4] A. Jain, How to create a web crawler and storing data using Java [web log post], Retrieved April 7, 2017, from: http://mrbool.com/how-to-create-a-web-crawler-and-storing-data-using-java/28925, 2016.

[5] B. Jackson, Web crawlers and user-agents—top 10 most popular [web log post], Retrieved April 14, 2017, from: https://www.keycdn.com/blog/web-crawlers/, 2016.

[6] Alessandro, How to make a web crawler using Java?, Retrieved April 15, 2017, from: http://www.programcreek.com/2012/12/how-to-make-a-web-crawler-using-java/, 2014.

[7] Scrapy Developers, Scrapy at a glance, Retrieved April 15, 2017, from: https://doc.scrapy.org/en/1.0/intro/overview.html, 2015.

[8] D.K. Gaurav, G. Venugopal, R. Deodhar, S. Kannan, S. Sharma, S. Ganesh, S. Mannarswamy, Interfacing Cassandra and MongoDB NoSQL with Python, Retrieved April 16, 2017, from: http://opensourceforu.com/2016/04/interfacing-cassandra-and-mongodb-nosql-with-python/, 2016.

[9] R. Python, Web scraping with Scrapy and MongoDB, Retrieved April 15, 2017, from: https://realpython.com/blog/python/web-scraping-with-scrapy-and-mongodb/, 2014.

[10] J.J. Heiss, Become.com's web crawler: a massively scaled Java technology application, Retrieved April 16, 2017, from: http://www.oracle.com/technetwork/java/javase/become-138235.html, 2005.

[11] P. Warden, Big Data glossary, O'Reilly, Sebastopol, CA, ISBN: 978-449-0, 2011.

[12] Cisco, White paper: Cisco VNI forecast and methodology, 2015-2020, Retrieved March 19, 2017, from: http://www.cisco.com/c/en/us/solutions/collateral/service-provider/visual-networking-index-vni/complete-white-paper-c11-481360.html, 2016.

[13] Distil Inc., 2017 Bad bot report, Retrieved April 17, 2017, from: https://resources.distilnetworks.com/white-paper-reports/2017-bad-bot-report, 2017.

[14] J. Tope, How to access the Deep Web and the Dark Net in 2017, Retrieved March 24, 2017, from: https://www.cloudwards.net/how-to-access-the-deep-web-and-the-dark-net/, 2017.

[15] M.K. Bergman, White paper: the Deep Web: surfacing hidden value, J. Electron. Publ. 7 (1) (2001). https://doi.org/10.3998/3336451.0007.104.

[16] B. He, M. Patel, Z. Zhang, K.C.C. Chang, Accessing the Deep Web, Commun. ACM 50 (5) (2007) 94–101.

[17] D. Howlett, Facebook, Google, Twitter, Apple, LinkedIn—the new media oligarchs, Retrieved March 29, 2017, from: http://diginomica.com/2015/09/20/facebook-google-twitter-apple-linkedin-the-new-media-oligarchs/, 2015.

[18] S. Waterhouse, JXTA search: distributed search for distributed networks, Sun Microsystems, Inc., 2001.

[19] K. McCurley, Encyclopedia of database systems—incremental crawling, Retrieved April 23, 2017, from: https://static.googleusercontent.com/media/research.google.com/en//pubs/archive/34403.pdf, 2016.

[20] P. Boldi, B. Codenotti, M. Santini, S. Vigna UbiCrawler: a scalable fully distributed web crawler. Retrieved April 13, 2017, from: http://vigna.di.unimi.it/ftp/papers/UbiCrawler.pdf, n.d.

[21] Oracle, Secure enterprise search administrator's guide, Retrieved April 16, 2017, from: https://docs.oracle.com/cd/B19306_01/search.102/b32259/crawler.htm#BABJBBCD, 2005.

[22] D. Austin, Feature column from the AMS, Retrieved March 10, 2017, from: http://www.ams.org/samplings/feature-column/fcarc-pagerank, 2017.

[23] The Apache Software Foundation, Blogging in action, Retrieved February 26, 2017, from: https://blogs.apache.org/foundation/entry/the_apache_software_foundation_announces29, 2012. The Apache Software Foundation, Blogging in Action.

[24] A. Ankerholz, StormCrawler: an open source SDK for building web crawlers with ApacheStorm, Retrieved April 12, 2017, from: https://www.linux.com/news/stormcrawler-open-source-sdk-building-web-crawlers-apachestorm, 2016.

[25] DigitalPebble Ltd., A collection of resources for building low-latency, scalable web crawlers on Apache Storm, Retrieved March 18, 2017, from: http://stormcrawler.net/, 2017.

[26] G. Tummarello, P. Mika, Web semantics in the clouds, IEEE Intell. Syst. 23 (5) (2008) 82–87, https://doi.org/10.1109/MIS.2008.94.

[27] YaCy, YaCy—the peer to peer search engine: technology. Retrieved February 25, 2017, from: http://yacy.net/en/Technology.html, YaCy decentralized web search. n.d.

[28] TV-Novosti, Snowden used common web crawler tool to collect NSA files, Retrieved March 27, 2017, from: https://www.rt.com/usa/snowden-crawler-nsa-files-227/, 2014.

[29] Freecode, 24 Projects tagged "NoSQL", Retrieved February 25, 2017, from: http://www.freecode.com/tags/nosql, 2015.

[30] W.C. Chung, H.P. Lin, S.C. Chen, M.F. Jiang, Y.C. Chung, JackHare: a framework for SQL to NoSQL translation using MapReduce, Autom. Softw. Eng. 21 (4) (2014) 489–508.

[31] A. Carlton, Web search engines for IoT: the new frontier, Retrieved March 18, 2017, from: http://www.networkworld.com/article/3111984/internet-of-things/web-search-engines-for-iot-the-new-frontier.html, 2016.

[32] Hypercat Alliance Ltd., Hypercat standard is a hypermedia catalogue format designed for exposing information about the Internet of Things assets over the web. Retrieved April 12, 2017, from: http://www.hypercat.io/standard.html, n.d.

[33] X. Li, Y. Wang, F. Shi, W. Jia, Crawler for nodes in the Internet of Things, ZTE Commun. 13 (5) (2015) 46–50. (3rd ser.). https://doi.org/10.3969/j.issn.16735188.2015.03.009.

[34] Cybrary.IT, Shodan: the hacker's search engine, Retrieved April 16, 2017, from: https://www.cybrary.it/0p3n/intro-shodan-search-engine-hackers/, 2016.

[35] The MITRE Corporation, Common vulnerabilities and exposures, Retrieved March 29, 2017, from: https://cve.mitre.org/cgi-bin/cvename.cgi?name=CVE-2013-1899, 2017.

[36] Semantics3, Announcing expanded UPC and barcode lookup [web log post], Retrieved March 27, 2017, from: https://blog.semantics3.com/announcing-expanded-barcode-lookups-via-an-api-fd70fb73957d#.iwm9s7our, 2015.

[37] GitHub, Inc., Yasserg/crawler4j, Retrieved April 13, 2017, from: https://github.com/yasserg/crawler4j, 2017.

[38] W. Santos, 81 New APIs: Associated Press, Nike, eBay, Rdio and SendGrid, Retrieved March 28, 2017, from: https://www.programmableweb.com/news/81-new-apis-associated-press-nike-ebay-rdio-and-sendgrid/2012/05/27, 2017.

[39] D. Shestakov, Current challenges in web crawling, in: Proceedings of the ICWE 2013, 2013, pp. 518–521.

[40] InternetLiveStats.com, Total number of websites. Retrieved March 7, 2017, from: http://www.internetlivestats.com/total-number-of-websites/#trend, n.d.

[41] J. Desjardins, The 100 websites that rule the Internet, Retrieved March 10, 2017, from: http://www.visualcapitalist.com/100-websites-rule-internet/, 2017.

[42] VeriSign, Inc., Internet grows to 326.4 million domain names in the first quarter of 2016, Retrieved March 07, 2017, from: https://investor.verisign.com/releasedetail.cfm?releaseid=980215, 2016.

[43] S. Pandey, C. Olston, User-centric web crawling, Retrieved March 10, 2017, from: https://www.scribd.com/presentation/62692001/web-UCC, 2017.

[44] M. Seeger, W. Kriha, D. Buytaert, Building blocks of a scalable web crawler, (Master's thesis), Computer Science and Media Stuttgart Media University, 2010, pp. 1–134. Retrieved April 4, 2017, from: http://blog.marc-seeger.de/assets/papers/thesis_seeger-building_blocks_of_a_scalable_webcrawler.pdf.

[45] A.S. Butt, A. Haller, L. Xie, A taxonomy of semantic web data retrieval techniques, Proceedings of the 8th International Conference on Knowledge Capture, ACM, 2015p. 9.

[46] K. Hess, Hadoop vs. Spark: the new age of Big Data, Retrieved March 29, 2017, from: http://www.datamation.com/data-center/hadoop-vs.-spark-the-new-age-of-big-data.html, 2016.

[47] Programmableweb, TalentBin API, Retrieved March 28, 2017, from: https://www.programmableweb.com/api/talentbin, 2017.

[48] KDnuggets, Top NoSQL database engines, Retrieved March 17, 2017, from: http://www.kdnuggets.com/2016/06/top-nosql-database-engines.html, 2017.

[49] Y. Takano, S. Ohta, T. Takahashi, R. Ando, T. Inoue, Mindyourprivacy: design and implementation of a visualization system for third-party web tracking, Privacy, Security and Trust (PST), 2014 Twelfth Annual International Conference on, IEEE, 2014, pp. 48–56.

[50] L. Pizette, T. Cabot, Database as a service: a marketplace assessment, 2012). Retrieved March 28, 2017.

[51] C. Nita-Rotaru, CS505: distributed systems—BigTable. HBase. Megastore. Spanner. Retrieved April 16, 2017, from: https://cnitarot.github.io/courses/ds_Fall_2016/505_bt_hbase_spanner.pdf, n.d.

[52] W. Dou, X. Wang, D. Skau, W. Ribarsky, M.X. Zhou, in: Leadline: interactive visual analysis of text data through event identification and exploration, Visual Analytics Science and Technology (VAST), 2012 IEEE Conference on, IEEE, 2012, pp. 93–102.

[53] M. Nielsen, How to crawl a quarter billion webpages in 40 hours, Retrieved April 15, 2017, from: http://www.michaelnielsen.org/ddi/how-to-crawl-a-quarter-billion-webpages-in-40-hours/, 2012.

[54] P. Sarathi, A deep dive into NoSQL: a complete list of NoSQL databases, Crayon Data, 2015. Retrieved from: http://bigdata-madesimple.com/a-deepdive-into-nosql-a-complete-list-of-nosql-databases/ (March 18, 2017).

[55] P. Desmarais, I. Macdonald, G. Viger, M. Rabbat, Project writeup a turbocharged web crawler built on V8 [scholarly project], in: A Turbocharged Web Crawler Built on V8, 2011. Retrieved March 24, 2017, from: http://www.networks.ece.mcgill.ca/sites/default/files/projectwriteup.pdf.

[56] S. Evans, Why MongoDB is a bad choice for storing our scraped data, Retrieved April 13, 2017, from: https://blog.scrapinghub.com/2013/05/13/mongo-bad-for-scraped-data/, 2013.

[57] GitHub, Inc., A utility to crawl websites and import their pages and links as nodes and relationships into a Neo4j graph database, Retrieved March 24, 2017, from: https://github.com/fgavilondo/neo4j-webgraph, 2013.

[58] E. Rolando, Scrapy-Redis, Retrieved April 15, 2017, from: https://scrapy-redis.readthedocs.io/en/stable/readme.html, 2016.

[59] Semantics3, S. Varun, How we built our 60-node (almost) distributed web crawler [web log post], Retrieved March 27, 2017, from: https://www.semantics3.com/blog/2012/09/03/how-we-built-our-almost-distributed-web-crawler/, 2012.

[60] Reggoodwin, Reggoodwin/ferrit, Retrieved April 15, 2017, from: https://github.com/reggoodwin/ferrit, 2014.

[61] Shen139, Full-text search: MySQL vs MongoDB vs Sphinx, Retrieved April 15, 2017, from: http://www.openwebspider.org/2015/04/03/full-text-search-mysql-vs-mongodb-vs-sphinx/, 2017.

[62] G. Burd, Oracle Berkeley DB in life sciences: embedded databases for better, faster results, Oracle Corporation, 2007. Retrieved from: http://www.olsug.org/wiki/images/7/72/Burd_-_May_07_-_LifeSci.pdf (March 22, 2017).

[63] Atlassian, Users of Heritrix, Retrieved March 22, 2017, from: https://webarchive.jira.com/wiki/display/Heritrix/Users+of+Heritrix, 2015.

[64] G. Mohr, M. Stack, I. Rnitovic, D. Avery, M. Kimpton, Introduction to Heritrix, in: 4th International Web Archiving Workshop, 2004.

[65] M. Bahrami, M. Singhal, Z. Zhuang, in: A cloud-based web crawler architecture, Intelligence in Next Generation Networks (ICIN), 2015 18th International Conference on, IEEE, 2015, pp. 216–223.

[66] Amazon Web Services Inc., Common crawl on AWS, Retrieved April 16, 2017, from: https://aws.amazon.com/public-datasets/common-crawl/, 2017.

[67] Amazon Web Services, Inc., AWS | Amazon CloudSearch—search service in the cloud, Retrieved March 24, 2017, from: https://aws.amazon.com/cloudsearch/, 2017.

[68] S. Fatehi, SchemaCrawler, Retrieved March 24, 2017, from: http://sualeh.github.io/SchemaCrawler/, 2017.

[69] Benzinga, Modmacro announces successful completion of BETA launch for proprietary web crawler, Retrieved March 24, 2017, from: https://www.benzinga.com/pressreleases/17/03/p9184467/modmacro-announces-successful-completion-of-beta-launch-for-proprietary, 2017.

[70] Lily, Lily, Retrieved April 7, 2017, from: http://www.lilyproject.org/lily/index.html, 2017.

[71] M. Kehoe, Which search platform is right for your business?, Simpler Media Group, Inc., 2016. Retrieved from: http://www.cmswire.com/informationmanagement/which-search-platform-is-right-for-your-business/ (April 17, 2017).

[72] StackOverflow, NoSQL (MongoDB) vs Lucene (or Solr) as your database [web log post], Retrieved April 17, 2017, from: http://stackoverflow.com/questions/3215029/nosql-mongodb-vs-lucene-or-solr-as-your-database, 2014.

[73] Findbestopensource.com, Lucene/Solr as NoSQL database, Retrieved April 17, 2017, from: http://www.findbestopensource.com/article-detail/lucene-solr-as-nosql-db, 2016.

[74] Solr, Indexing files Like doc, pdf—Solr and Tika integration [web log post], Retrieved April 17, 2017, from: http://solr.pl/en/2011/04/04/indexing-files-like-doc-pdf-solr-and-tika-integration/, 2011.

[75] B. Goralewicz, How to combine JavaScript SEO with isomorphic JS, Retrieved March 24, 2017, from: https://www.searchenginejournal.com/javascript-seo-like-peanut-butter-and-jelly-thanks-to-isomorphic-js/183337/, 2017.

[76] F. Rappl, Azure WebState [web log post], Retrieved April 15, 2017, from https://www.codeproject.com/Articles/584392/Azure-WebState, 2013.

[77] Awesome Inc., NUTCH FIGHT! 1.7 vs 2.2.1, Retrieved March 10, 2017, from: http://digitalpebble.blogspot.in/2013/09/nutch-fight-17-vs-221.html, 2013.

[78] MyNoSQL, Cloudera adds HBase to CDH [web log post], Retrieved April 16, 2017, from: http://nosql.mypopescu.com/page/3424, 2015.

[79] A. Sibiryakov, Distributed frontera: web crawling at scale [web log post], Retrieved April 19, 2017, from: https://blog.scrapinghub.com/2015/08/05/distributed-frontera-web-crawling-at-large-scale/, 2015.

[80] S. Munzert, C. Rubba, P. Meissner, D. Nyhuis, Automated data collection with R: a practical guide to web scraping and text mining, Wiley, Chichester, 2015. from: http://kek.ksu.ru/eos/WM/AutDataCollectR.pdf.

FURTHER READING

[81] Oracle, 7 Understanding the Oracle ultra search crawler and data sources, Retrieved April 16, 2017, from: https://docs.oracle.com/cd/B14099_19/portal.1012/b14041/crawler.htm#CCHIHGEJ, 2004.

[82] Scrapinghub and many other contributors, Curated resources, Retrieved April 16, 2017, from: https://scrapy.org/resources/, 2017.

[83] T. Lipcon, What's new in CDH3b2: Apache HBase [web log post], Retrieved April 16, 2017, from: http://blog.cloudera.com/blog/2010/07/whats-new-in-cdh3-b2-hbase/, 2010.

[84] N. Pant, How to do faster indexing for website and blog in Google search?, Retrieved April 16, 2017, from: http://www.iamwire.com/2017/04/faster-indexing-website-blog-google-search/150529, 2017.

[85] D. Anderson, Why crawl budget and URL scheduling might impact rankings in website migrations, Retrieved April 16, 2017, from: http://searchengineland.com/crawl-budget-url-scheduling-might-impact-rankings-website-migrations-255624, 2016.

[86] F. Nimphius, ADF faces web crawler support in Oracle JDeveloper 11.1.1.5 and 11.1.2, Retrieved April 16, 2017, from: https://blogs.oracle.com/jdevotnharvest/entry/adf_faces_web_crawler_support, 2011.

[87] S. Tozlu, How to run Scrapy Spiders on Cloud using Heroku and Redis, Retrieved April 16, 2017, from: http://www.seckintozlu.com/1148-how-to-run-scrapy-spiders-on-the-cloud-using-heroku-and-redis.html, 2015.

[88] OpenSearchServer, Inc., OpenSearchServer, Retrieved April 16, 2017, from: http://www.opensearchserver.com/, 2017.

[89] O. Hartig, A. Langegger, A database perspective on consuming linked data on the web, Datenbank-Spektrum 10 (2) (2010) 57–66.

[90] M. Thelwall, The top 100 linked-to pages on UK university web sites: high inlink counts are not usually associated with quality scholarly content, J. Inf. Sci. 28 (6) (2002) 483–491.

[91] GitHub, Inc., BruceDone/awesome-crawler, Retrieved April 15, 2017, from: https://github.com/BruceDone/awesome-crawler, 2016.

[92] GitHub Pages, J. Roman, The Hadoop ecosystem table. Retrieved April 15, 2017, from: https://hadoopecosystemtable.github.io/, n.d.

ABOUT THE AUTHOR

Ganesh Chandra Deka is currently Deputy Director (Training) at Directorate General of Training, Ministry of Skill Development and Entrepreneurship, Government of India.

His research interests include ICT in Rural Development, e-Governance, Cloud Computing, Data Mining, NoSQL Databases, and Vocational Education and Training. He has published more than 57 research papers in various conferences, workshops, and International Journals of repute including IEEE & Elsevier. So far he has organized 08 IEEE International Conference as Technical Chair in India. He is the member of editorial board and reviewer for various Journals and International conferences.

He is the co-author for four text books in fundamentals of computer science. He has edited seven books (three IGI Global, USA, three CRC Press, USA, one Springer) on Bigdata, NoSQL, and Cloud Computing in general as of now.

He is Member of IEEE, the Institution of Electronics and Telecommunication Engineers, India and Associate Member, the Institution of Engineers, India.

CHAPTER FOUR

NoSQL Security

Neha Gupta, Rashmi Agrawal
Faculty of Computer Applications, Manav Rachna International Institute of Research and Studies Faridabad,
Faridabad, India

Contents

1.	Fundamentals of NoSQL Security	102
	1.1 Security Issues of NoSQL Databases	103
	1.2 Issues in Implementing Traditional Security Solutions to NoSQL Data Stores	107
	1.3 Comparison of Relational Database Security and NoSQL Security	108
2.	Security Features of Various NoSQL Databases	111
	2.1 Strength and Weaknesses of NOSQL Databases	111
3.	Injection Attacks on NOSQL Database	116
	3.1 Java Script Injections	117
	3.2 Union Queries	118
	3.3 Piggybacked Query	119
4.	Challenges in Designing, Implementing, and Deploying NoSQL Databases	120
5.	NoSQL Security Reference Architecture	122
	5.1 NoSQL Cluster Security	122
	5.2 Data Centric Approach to Security	125
6.	Techniques to Mitigate the Attacks on NoSQL Databases	126
7.	Proposed Security and Privacy Solutions for NoSQL Data Stores	128
8.	Conclusion	129
	Glossary	129
	References	130
	Further Reading	130
	About the Authors	131

Abstract

The increased demand of Big Data and Cloud Computing technologies has forced the organizations to move from relational databases to nonrelational databases like NoSQL. A NoSQL database, also referred as non-SQL or nonrelational or not only SQL database, is scalable, distributed, and highly reliable and has been designed to support the requirements of modern web-based applications. However, security remains a very challenging issue even in NoSQL database as security challenges that were inherent in the previous databases have not spared even NoSQL. The key for ensuring adequate security of the NoSQL database system is a detailed consideration of security issues for the NoSQL database solution and implementation of corresponding security

Advances in Computers, Volume 109
ISSN 0065-2458
https://doi.org/10.1016/bs.adcom.2018.01.003

mechanisms. The aim of this chapter is to discuss security issues inherent in NoSQL data-bases and further explore it to find the best security mechanism for this environment. In this chapter, we will deliberate upon various security threats of NoSQL databases, secu-rity architecture of NoSQL databases, and the steps that can be taken to secure the NoSQL database. A comparative study of Relational & NoSQL database is also given to provide a better understanding of the databases, their relative advantages, and the security pitfalls. The technology/technical terms used in this book chapter are explained wherever they appear and at the "Glossary" section. At the end of the chapter, references are included for further reading for the benefit of advanced readers.

1. FUNDAMENTALS OF NoSQL SECURITY

NoSQL databases have been designed to deal with large volumes of unstructured data and to give real-time performance. NoSQL databases are nonrelational distributed databases. NoSQL databases support enormous data storage across various storage clusters [1]. Because of the growing data and the infrastructure needs, major web 2.0 companies are adopting NoSQL databases. However, security is a major concern for NoSQL databases. Most of the NoSQL databases do not provide inbuilt security features in the data-base itself and developers need to insert security in the middleware. NoSQL systems are, by and large, a new generation of databases that focus on scale-out issues first and use the application layer to implement security features. Since NoSQL databases have not been designed with security as priority, so protection of NoSQL data stores are major concerns to various organizations that are using NoSQL databases [2]. Moreover, no two NoSQL database products are the same as they are designed to meet the explicit requirements of particular cloud-based applications [2].

NoSQL databases lack various security features such as authentication, authorization, and integrity, indicating that sensitive data is safer in tradi-tional DBMS [1]. Relational databases [3] have adopted highly protected mechanisms to provide the security services, but they also faces many secu-rity threats like Cross-site scripting, SQL injections, Root kits, etc. NoSQL has inherited the security issues of traditional RDBMSs along with its own due to the enhanced features of the NoSQL databases. NoSQL databases are always prone to security attacks due to the unstructured nature of data and the distributed computing environment. In NoSQL database environment, the nodes are distributed to enable parallel computing that increases the attack surface which results in implementation of more complex procedures for security. Apart from [4] distributed environment, data from a variety of nodes move from one node to another which leads to sharing of data and increases the risk of theft.

Some of the major security concerns are [5]:

1. Insufficient encryption support for the data files
2. Weak authentication between the client and servers
3. Granular access control
4. Very simple authorization without the support for RBAC (Role-based Access Control)
5. Vulnerability to SQL injection
6. End-point input validation/filtering
7. Insecure computation
8. Insecure data storage and communication
9. Privacy preserving data mining and analytics
10. Attribute relationship methodology
11. Attribute encryption

Security experts have also realized the danger of relying on perimeter security as Java Script or JSON can be used to attack the firewalls to get unauthorized access of the data from database data. In addition, granular access controls required to separate roles and responsibilities of the user are not provided by many systems. NoSQL database may become more vulnerable to attack if the learning curve of the intruder is over and the intruder is able to identify hidden software weaknesses.

Data protection and access control are some of the key issues of security in NoSQL. To prevent unauthorized access to data stores and to maintain availability of secure data, data storage and transaction logs are also devised.

Attribute relationship [6] methodology is a popular method to impose security in NoSQL databases. Protecting the valuable information is the primary objective of this methodology. In this methodology, the attribute with higher relevance is considered as the key element of information extraction and is given more importance than other remaining attributes.

Another suitable data access control method for NoSQL databases is attribute-based encryption. Attribute encryption [6] is a method that allows data owners to encrypt data under access policy such that only those users who have permission to access data can decrypt it.

1.1 Security Issues of NoSQL Databases

NoSQL databases were designed to handle large datasets to meet the requirement of Data Analytics and less emphasis was given on security during the design phase. Most of the NoSQL databases do not provide embedded security features in the database itself and developers have to insert security-related solutions in the middleware. Also clustering feature of NoSQL databases add challenges in implementing security practices. NoSQL databases provide a very thin layer of security, and to make NoSQL databases secure the vendors configure bottom–up security solutions and solve security issues on the ad hoc basis. Also NoSQL databases use digest-based authentication or HTTP basic which is vulnerable to man-in-the-middle attack. NoSQL databases can be made secure by creating a virtual secure layer using a framework or by implementing a middleware. Middleware induces object-level security at the column level, ensuring no direct access to the data. However, data can only be accessed using the controls configured within the middleware.

1.1.1 Distributed Environment and Data Duplicity
In a distributed environment the database is distributed over a computer network that allows applications to access data from local and remote databases [7]. A centrally controlled distributed database periodically synchronizes all the data and ensures that updates and deletes performed on the data are automatically reflected in the data stored elsewhere.

In a distributed environment, there are several distributed nodes on which NoSQL databases run. These parallel nodes increase the attack surface which makes the system complex to secure. The possibility of unauthorized access to the database increases due to multiple entry points. These entry points may be from a remote location or from a client home location.

NoSQL databases segment data horizontally and distribute them across multiple servers. Data from multiple nodes travel to and from NoSQL database environment which is distributed across multiple servers.

Maintaining replicated wreck of data is computationally very expensive and more prone to error and also amplifies the risk of theft. Distributed Environment is generally more prone to security risks because of the lack of central security management system. Cassandra and Dynamo databases are vulnerable to security risk in distributed environment because of Gossip membership protocols.

1.1.2 Authentication
Authentication mechanism in a NoSQL database is enforced at the local node level but fails across all commodity servers [8]. Due to this NoSQL databases are exposed to brute force attacks, injection attacks, and relay

attacks, which lead to information leakage. The reasons for these attacks are weak authentication mechanism.

In NoSQL database environment, Kerberos can be used to authenticate the clients and the data nodes, but malicious clients may gain unauthorized access by duplicating the Kerberos ticket. Strict algorithms need to be designed to enforce strict authentication norms.

1.1.3 Safeguarding Integrity

Safeguarding integrity is a security requirement to ensure protection of data from unauthorized modification because of insertion, deletion, or update of data in the database. NoSQL databases lack confidentiality and integrity and to safeguard integrity in these databases is a major concern. The heterogeneous nature of NoSQL databases makes it difficult to protect the integrity of data. As NoSQL databases don't have a schema, so permissions on a table, column or row, cannot be segregated. This can lead to multiple copies of the same data, which makes it hard to keep data consistent, particularly as changes to multiple tables cannot be wrapped in a transaction where a logical unit of insert, update, or delete operations is executed as a whole [9]. So to maintain transactional integrity is also very difficult in NoSQL databases. Because of the intricacy in enforcing integrity constraints to NoSQL databases, these databases can never be used for financial transactions. Schema less nature of the database and absence of central control make it more difficult to enforce integrity constraints in these databases.

1.1.4 Fine-Grained Authorization and Access Control

Authenticating users is the first step in protecting the data. Once the identity of the user is verified, access is granted to the user for part of or the entire database. Authorization plays a very important role in any database. Relational databases store related data and allow authorization at the table level. NoSQL databases have no schema and store heterogeneous data together. Therefore, it is difficult to implement authorization on a table as a whole. Fine-grained authorization enables object-level security. Object-level security is further classified into row-level security and fields-level security. If the data is stored in tables, then row-level or cell-level security is applied. If the data or metadata is entered in forms, field-level security is applied and so on. Fine-grained access control is not allowed due to schema less nature of NoSQL databases. Most of the NoSQL databases allow Column Family-level authorization.

In NoSQL database, data is grouped according to their security level. The data may be classified as confidential, secret, or even may not be classified at all. So we need to implement access control row wise or column wise.

Role-based access control is difficult to implement as NoSQL database has a schema-less structure. Different types of data are stored in one big database in these types of databases. As heterogeneous data is stored together in one database in comparison to relational models, this becomes a challenge.

1.1.5 Protection of Data at Rest and in Motion

Data at rest means data that has been flushed out from the memory and written to the disk. Data in motion means data that is in communication or is being exchanged during a communication. Data in motion is categorized into two categories:

(a) Client-node communication

(b) Internode communication

Most of the NoSQL databases do not employ any technique to protect the data at rest. Only a few provide encryption mechanisms to protect data. To safeguard the data in storage, encryption techniques are used and are referred as de facto standards of encrypted data. Encryption makes the data unintelligible [8] and hence of no use to malicious intruder. Most of the industry solutions lack horizontal scaling while offering encryption services.

The popular NoSQL databases offer following encryption services for protection of data.

1. *Data at Rest*:

   (a) Cassandra uses TDE (Transparent Data Encryption) technique to protect data at rest. This feature helps to protect data at rest. This feature helps to protect sensitive data. In Cassandra databases, encryption certificates are stored locally, so a secured file system is required to implement TDE. Also the commit log of Cassandra Database is not encrypted, which also leads to breach of security.

   (b) MongoDB does not provide any method to encrypt the data file. Data files can be encrypted at the application layer before writing the data to the database which require strong system security.

2. *Data in Motion*:

   (a) Client-node communication: This is not encrypted in Cassandra. Encryption is done by generating valid server certificates at the SSL layer.

MongoDB does not support SSL client-mode communication. To encrypt the data using SSL client-node communication, MongoDB needs to recompile by configuring SSL communication.

(b) Internode communication: Cassandra doesn't support encrypted internode communication.

Using Cassandra.yaml file, Server encryption options can be edited to configure internode SSL communication. MongoDB doesn't supports internode communication at all.

1.1.6 Privacy of User Data

Privacy of user data is the main challenge for Web 2.0 and NoSQL databases. NoSQL stores large amount of sensitive information. Maintaining privacy of this information is the key concern for any database administrator.

Clients access NoSQL databases through various nodes and resource managers. Even if a single location is comprised, then the malicious data will propagate to the entire system as there is no central security management. Distributed nature of database also leads to compromised security. Major privacy issues are related to unauthorized access, vendor lock-in, data deletion, backup, vulnerabilities, isolation failure, inadequate monitoring, and audit.

1.1.7 Lack of Expertise and Buggy Applications

NoSQL is an emerging technology and there are very few experts who holistically understand the security aspects of NoSQL databases. Lack of standard security model for NoSQL databases leads to intricate implementation of security controls. Because of the cloud-based implementation of NoSQL databases, third-party security solutions are integrated into NoSQL databases which make applications even more complex.

Also cloud APIs are frequently updated, leading to introduction of new bugs that add another security hole in the application. Therefore, industry needs security experts to take care of NoSQL databases.

1.2 Issues in Implementing Traditional Security Solutions to NoSQL Data Stores

The increased demand of Big Data and cloud computing technologies has forced the organizations to move from relational databases to nonrelational databases like NoSQL. Although NoSQL databases have magnified the databases in terms of volume, velocity, and veracity, the security concerns have also magnified due to below-mentioned reasons:

(a) Relational databases have inbuilt security features, while NoSQL databases have a very thin security layer. NoSQL databases majorly rely on external or vendor-enabled security which is to be implemented on middleware.

(b) Clustering features of NoSQL databases add additional challenges to the robustness of security practices.

(c) Data in NoSQL databases are stored as plain text and no encryption mechanism is implemented to encrypt the text.

(d) Passwords in NoSQL databases are encrypted using MD5 or PBKDF2 algorithms which are not very secure algorithms.

(e) Most of the NoSQL databases address security issues on ad hoc basis leading to first attack and then resolve policy.

(f) Weak authentication and weak password storage methods expose NoSQL to replay and brute force attacks resulting in leakage of information.

(g) HTTP basic or Digest-based authentication mostly used by NoSQL databases is highly vulnerable to replay attack and man in the middle attack. NoSQL databases do not support authentication and authorization when running in shared mode.

(h) Inability of NoSQL databases to enforce authentication across the cluster nodes.

(i) Authorization mechanisms are inefficient in NoSQL databases. Authorization techniques vary from database to database. Authorization is primarily applied at higher layers and not on lower layers. Authorization is implemented for every database level rather than at the collection level.

(j) NoSQL databases are vulnerable to various injection attacks like JSON, schema injection, array injection, REST injection, view injection, GQL injection, etc. because of loosely coupled lightweight protocols and mechanisms.

(k) NoSQL databases are prone to Insider Attacks. Relaxed security mechanisms lead to insider attacks. These attacks even remain unnoticed due to poor logging and log analysis methods.

(l) A NoSQL database does not support inline auditing.

1.3 Comparison of Relational Database Security and NoSQL Security

Relational databases are collection of tables having relations with data categories and constraints. Relational databases uses SQL or MySQL as the tool to access the data and is based upon ACID properties. NoSQL databases are nonrelational databases that provide elastic scaling and are designed using low-cost hardware. NoSQL databases are schema free, distributive, and store huge amount of data. Relational databases [3] have adopted highly secure mechanisms to provide the security services, but they also face many security threats like Root kits, SQL injections, Cross-site scripting, etc. NoSQL has inherited the security issues of traditional RDBMSs along with its own due to the enhanced features of the NoSQL databases. NoSQL databases are always prone to security attacks due to the unstructured nature of data and the distributed computing environment. In NoSQL database environment, the nodes are distributed to enable parallel computing that increases the attack surface which results in implementation of more complex procedures for security. Apart from [4] distributed environment, data from a variety of nodes move from one node to another which leads to sharing of data and increases the risk of theft. Table 1 illustrates the differences between Relational databases and NoSQL databases [11] on the basis of various parameters listed.

Table 1 Comparison of Relational Database Security and NoSQL Security

S. No.	Point of Difference	Relational Databases	NoSQL Databases
1	Transaction Reliability	Guarantees very high transaction reliability as they fully support ACID properties	Do not guarantee very high reliability as they range from BASE to ACID properties
2	Performance	Caching has to be done with special infrastructure support	Performance is enhanced by caching data into system memory
3	Indexing	Index available on multiple column	Single index, key–value store
4	Data Model	Data model is very specific and well organized. Columns and rows are described by well-defined schema	Data model does not use the table as storage structure and is schema less. Data model is very efficient in handling unstructured data as well
5	Scalability	Scalability is greatest challenge in relational databases due to the dependency on vertical scalability	NoSQL databases depend on horizontal scalability

Continued

Table 1 Comparison of Relational Database Security and NoSQL Security—cont'd

S. No.	Point of Difference	Relational Databases	NoSQL Databases
6	Cloud	Not suitable for cloud environment as these databases do not support data search on full content. Relational databases are also very hard to extend beyond a limit	Well suited for cloud databases. All characteristics of NoSQL databases [10] are highly desirable for cloud databases
7	Handling Big Data	Big Data handling is a challenging issue for relational databases	NoSQL databases are designed to handle Big Data
8	Data Warehouse	When the size of stored data increases, problems related to performance degradation raises	NoSQL databases are not designed to serve data warehouse. NoSQL databases are faster than data warehouse
9	Complexity	Complexity arises due to nonfixture of data into tables	NoSQL databases have the capabilities to store unstructured data
10	Crash Recovery	They guarantee crash recovery through recovery manager	NoSQL databases use replication method as backup to recuperate from crash
11	Authentication	Relational databases come with authentication mechanism	NoSQL database does not have strong authentication mechanism and are dependent on external method for this
12	Data Integrity	Relational databases ensure data integrity	A NoSQL database does not support data integrity at every occasion
13	Confidentiality	Data confidentiality is a well-known feature of relational database	Generally data confidentiality is not achieved in NoSQL databases
14	Auditing	Relational databases provide mechanisms to audit database	Most of the NoSQL databases do not provide mechanism for auditing the database

2. SECURITY FEATURES OF VARIOUS NoSQL DATABASES

NoSQL databases are generally categorized on the basis of data storage model into the following four groups [12]:

Key-value databases: Uninterrupted arbitrary data values are stored in these database systems that can be evoked later using a key (hash). These databases are schema less, allow easy scaling of data, and are implemented using simple APIs.

Column databases: The storage structures of these databases are similar to key-value databases, except that the key used by these databases is a combination of column and row. Key can also be a time stamp, which can point to one or multiple columns (Column Family). The column family databases are more or less like a table commonly found in a relational database.

Document databases: These databases are designed to store documents having one or more self-contained named fields, like JSON or BSON format. These types of documents are dynamic in nature and provide freedom to modify the document with the ability to add or remove fields of existing documents. Indexing is used on named fields for faster data retrieval.

Graph databases: These types of databases follow a flexible graph model that can be scaled across multiple machines. This model is appropriate for data with relations that are best represented as a graph, such as network topologies, social relations, road maps, or public transport links.

Ebay, Facebook, Mozilla, Netflix, and Twitter mostly use column databases like Cassandra and HBase, while MongoHQ and Lotsofwords use document databases like MongoDB and CouchDB. Today, the NoSQL databases are rapidly growing and deployed in many internet companies and other enterprises. Although performance and scalability requirements of big users and Big Data are achieved through NoSQL databases, security is always a concern for the users. Section 2.1 will discuss security strengths and weaknesses of popular NoSQL databases such as MongoDB, Cassandra, CouchDB, HBase, Redis, and Neo4j.

2.1 Strength and Weaknesses of NOSQL Databases

In this section, we will discuss the strength and weaknesses of various NOSQL databases.

2.1.1 Cassandra

Cassandra combines two different technologies of Big Data, the distributed system technologies from Amazon DynamoDB and the data model from Google's Big Table [2]. Cassandra is eventually consistent and based on a Peer-to-Peer (P2P) model without a single point of failure. Like Big Table, Cassandra provides a Column Family-based data model richer than typical key/value systems (Abramov).

Cassandra is preferred as extremely robust and decentralized system. Following are the strengths and weaknesses of the Cassandra database:

Strengths

(1) Cassandra provides great performance under large data sets. Cassandra is designed for the system where immediate insert and updates are required and this is done using a different time stamp.

(2) Due to the distributed architecture of Cassandra database, it supports linear and elastic scalability.

(3) The architecture of Cassandra is more robust as there is no master/slave and hence no single-node failure point. It also stores duplicate copy of all written data.

(4) Cassandra offers an IAuthenticate interface with two implementation— Default implementation and Simple Authenticator. In default implementation, the requirement to authenticate the database is turned off, whereas Simple Authenticator requires setting up users and passwords.

Weaknesses

(1) Cassandra Query Language (CQL) is vulnerable to SQL injection attacks as it is a parsed language.

(2) All the nodes in Cassandra can communicate freely with each other as there is no master/slave, and for their communication, no encryption or authentication mechanism has been used.

(3) Communication between the client and the database is unencrypted and hence an attacker can monitor the database traffic.

(4) It uses the MD5 hashing algorithm which is not cryptographically secure and comparatively easy to find matching text for the attacker.

(5) No internal mechanism for intrusion detection.

(6) No internal mechanism for nonrepudiation

(7) Security auditing is not available.

2.1.2 MongoDB

So far, MongoDB is the most popular NOSQL database in many countries as this database supports very interesting features such as its unique data model, availability, and scalability which makes it suitable for content management. MongoDB is suitable for the application where the data can be fitted easily in the document data model, but MongoDB is actually not a replacement of relational database because this does not support the applications having relational data such as social media. Following are the strength and weaknesses of MongoDB database:

Strengths

(1) MongoDB supports horizontal scalability with the help of sharding. Sharding is a technique which distributes the data into physical partitions to automatically balance the data into cluster.

(2) A configuration called a replica set is supported by MongoDB which is used for data redundancy, disaster recovery, and fault tolerance. When the primary node of a replica set is not available due to any reason, MongoDB automatically supports the failover process. This makes MongoDB a highly available database.

(3) MongoDB supports the authentication mechanisms such as AD, LDAP, and certificates.

(4) MongoDB provides a cost-effective solution as it reduces cost on hardware and storage.

(5) In MongoDB, passwords are hashed using MD5 which makes use of locks for concurrency control.

Weaknesses

(1) MongoDB consumes too much disk space and the replicas set in MongoDB also have a limit of only 12 nodes.

(2) There is no immunity to injection attacks, which are possible via Java Script or string concatenation.

(3) When MongoDB is run in the shared mode, it does not support authentication and hence no support for authorization.

(4) Binary wire-level protocol for the client interface is neither encrypted nor compressed in MongoDB.

(5) The data files in MongoDB are unencrypted and there is no method for automatic encrypting of these files. This makes easy access for the attacker to the file system.

2.1.3 Redis

Redis (Remote Dictionary Server) is a key-value, networked, in-memory, single-threaded, open-source database which is also known as a data structure server. Redis is used as a way to boost the application at any point of time. It is most often used for catching or for session storage. Redis is considered as the world's fastest database which reduces the application complexity. Redis can be used as a search engine, as a time series database, as a graph database, or as a JSON store, and can serve many more purposes. In this section, we will briefly discuss the strength and weaknesses of the Redis database:

Strengths

(1) Redis databases are rich in a diverse set of data types as compared to various other key-value data stores.

(2) Redis is frequently used as a very fast cache where responsiveness of the data is of prime importance.

(3) Redis has a master–slave distributed system which is known as Redis Sentinel to guarantee high availability.

(4) Redis is a single-threaded application, which means that single command can be executed at one point of time, making fast use of cache.

(5) Redis has an inbuilt replication which is used for availability as well as for its scalability. Redis can reproduce data to any number of slaves.

Weaknesses

(1) The authentication is not provided by default in Redis and is sent and received by all IP addresses on port 6739.

(2) No support is provided by Redis for configuration security, data encryption, and auditing mechanism.

(3) As Redis uses memory dump to create snapshots, persistence can impact performance.

(4) Only basic security options are provided by Redis. It does not provide any access control and also it does not support any encryption mechanism and this must be implemented using a separate layer SSL proxy.

(5) Redis is not suitable for the applications which are intended for complex queries.

2.1.4 CouchDB

CouchDB is a document-oriented, open-source database developed to support web-oriented data. It uses JSON to store document data. It is a multimaster application which was initially released in 2005. The file layout of CouchDB and commitment system of the database marks all the ACID properties. In the following paragraph, we briefly present strength and weaknesses of CouchDB database:

Strengths

(1) To communicate easily with the database, CouchDB has an HTTP-based REST API, which helps.

(2) There is easy-to-use replication in CouchDB which makes easy to copy, share, and synchronize the data between databases and machines.

(3) By default CouchDB listens only a loopback network interface.

(4) In CouchDB, users can be segregated into admin and normal user, thus restricting access. It also defines a set of requests which only a super user/admin can do.

Weaknesses

(1) Authentication mechanism uses plain text passwords which is not very secure.

(2) Passwords in CouchDB are hashed using the PBKDF2 hashing algorithm and are sent over the network using the SSL layer.

(3) CouchDB does not provide automatic logging and only a basic-level auditing is provided for viewing the logs.

(4) CouchDB does not provide support for automatic backups of database logs and replicas in CouchDB.

2.1.5 HBase

Also known as Hadoop database, HBase is a NoSQL column-oriented, distributed database used extensively for random read and write access. HBase is used extensively for online analytical applications. One example of HBase implementation can be in banks where real-time data is updated in ATM machines. Operations such as data reading and processing will take less time as compared to traditional relational models. Here we briefly discuss strength and weaknesses of HBase database security:

Strengths

(1) HBase supports user authentication and token-based authentications for tasks related to map reduce.

(2) Authentication of the user is done by simple authentication (Security Layer) using Kerberos on per connection basis.

(3) Login support up to data node level is also provided by HBase.

(4) ACLs (Access Control Lists) are used to manage authorizations in HBase or by Coprocessors with column family-level granularity and on per user basis.

Weaknesses

(1) HBase storage architecture has "*Single Point of Failure*" feature with no exception handling mechanism associated with it.

(2) High-level monitoring and auditing features are not available in HBase.

(3) Security features are slowly improving in HBase for the different users to access the data.

(4) HBase uses master and slave. Even for 5000 nodes only one master is there and if it goes down, it may take too long to recover.

(5) No immunity, injection attacks are possible in HBase.

2.1.6 Neo4j

Neo4j is a *native* graph database which is deliberated to store and process graphs to manage interconnected data. Graph databases help to find relationships between data and extract their true value. Each piece of data has an explicit connection resulting in unparalleled speed and scale. Neo4j's native graph processing is also known as "index-free adjacency." Following are the strengths and weaknesses of Neo4j database:

Strengths

(1) ACID transactions in Neo4j ensure that data is fully consistent and reliable for global enterprise applications.

(2) It supports the concept of multiple users and a new role-based access control framework.

(3) Neo4j provides a new plug-in architecture for building and deploying users' own authentication and authorization controls.

Weaknesses

(1) We can't share Neo4j. That means that we have whole dataset in ONE server and only vertical scalability is possible.

(2) Neo4j has some upper bound limit for the graph size and can support tens of billions of nodes, properties, and relationships in a single graph.

(3) No security is provided at the data level and there is no data encryption.

(4) Security auditing is not available in Neo4j.

3. INJECTION ATTACKS ON NOSQL DATABASE

SQL injection is a widely known attack in SQL database where a malicious user tries to change the existing data using web forms. If the value of a variable is set through an input field in a base statement and the statement is concatenated with the queries, then such kind of applications are susceptible to SQL injection attack. For example, consider the following query:

```
SELECT * FROM accounts WHERE username = '$username' AND
password = '$password'
```

If this query is not properly handled, then the attacker may pass an "admin" value in the username field with bypassing the requirement of password in the admin account and the query will look like as follows:

```
SELECT * FROM accounts WHERE username='admin' - AND password="
```

Different query languages are used by NOSQL databases which makes traditional SQL injections irrelevant for NOSQL databases. However, the NOSQL databases are not free from injection attacks and malicious users apply fundamentally improved techniques for this. The equivalent query of above-discussed SQL database in NOSQL MongoDB database is:

```
db.accounts.find({username: username, password:
password});
```

here, the attacker can still inject this query using JSON object as follows:

```
{
    "username": "admin",
    "password": {$ne: NULL}
}
```

Here in MongoDB $ne will check and select the documents where the value of password is not equal to null and hence will access admin documents.

To store customer data, web applications typically use NOSQL databases. The following figure represents a typical architecture of a NOSQL database which is used to store the data via web application. Fig. 1 shows that malicious users devise an injection with web access request that will allow the illegal database operation when processed by the client/protocol wrapper.

The mechanism of NOSQL injection attacks can be more understood by the following three major techniques.

3.1 Java Script Injections

In Java Script-based applications, injection problem occurs when unsanitized data is concatenated to build a new structure. Attacker sends simple text-based

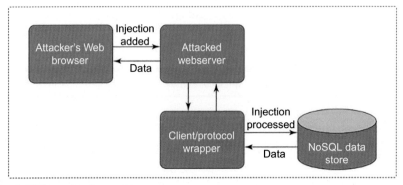

Fig. 1 Web application architecture.

attacks that exploit the syntax of the targeted interpreter. Almost any source of data can be an injection vector, including internal sources.

It is recommended to stick to APIs that do not involve string concatenation to avoid injection attack. JSON injection can happen if JSON is built by string concatenation before being sent to the server:

```
'{value1:"' + value + '"}'
```
In this case the value could inject a new key-value pair.
```
'{value1:"hello",value2:"user"}'
```
This can also happen while building unsafe URLs like as follows:
```
url: "http://mysite.com/search/" + searchTerm
```
and the URL after injection may become
```
url: "http://mysite.com/search/../something/wrong"
```

3.2 Union Queries

In Union query attack a UNION query is inserted into a susceptible parameter which results in the union of original query and the inserted query. UNION query is a well-known SQL injection attack in which the UNION operator combines the results of multiple queries to produce a result set with rows fetched from participating queries. The most common uses of union queries are to bypass authentication pages and extract data.

As an example, take a login form that receives username and password via http post at backend:

```
string query = "{ username: '" + post_ username + "',
password: '" + post_passport + ' " }"
```

If the user name is "happy" and password is "human" the query would look like

```
{ username: 'happy', password:  'human' }
```

A malicious user can turn this query to ignore the password such as

```
username='happy', $or: [ {}, {'a':  'a&password=' }],
$comment: 'successful MongoDB injection'
```

And it will construct the query like

```
{ username: 'happy', $or: [ {}, {  'a': 'a', password " } ],
$comment: 'successful MongoDB injection' }
```

If the username is correct the query would succeed. In SQL the similar query is

```
SELECT * FROM logins WHERE username =  'happy' AND (TRUE OR
('a'='a' AND password = ")) #successful MongoDB injection
```

The password of the query becomes redundant because an empty query is always true.

3.3 Piggybacked Query

This type of injection attack is different from other attacks because here the attacker injects additional queries along with the original query, and as a result, multiple database queries are received by the database. Here the attacker does not intend to modify the original query but wants to inject some additional query that piggybacks on the original query. The first query is executed in its normal way, while the subsequent queries are executed to satisfy the attack.

A set operation adds a key and its value using memcached. The set function uses two lines of input when it is called from the command line interface. First is key and its value and second is the data which should be stored. When it is called using PHP driver's set function, it looks like as

```
$memcached->set('key', 'value');
```

Research has been done that the driver fails to sanitize ASCII characters like carriage return (\r) and line feed(\n) which provides an opportunity to the malicious user to insert a new line in the key parameter and also append any unintended commands.

4. CHALLENGES IN DESIGNING, IMPLEMENTING, AND DEPLOYING NoSQL DATABASES

NoSQL databases are becoming an increasingly important part of the database landscape and can offer real benefits when used appropriately. However, organizations can use these NoSQL databases with caution after getting full awareness of the limitations and issues associated with these databases. For example, NoSQL databases can handle a massive amount of data for large web-based systems and can meet the demand of quick response time, but the efforts to install and manage such huge databases may increase the cost and can lead to compromise on security. Depending upon the type of NoSQL solution adoption used by various users and the challenges faced by them in terms of designing, implementing, and designing NoSQL databases, NoSQL adoption challenges are divided into two broad categories:

(a) Technical challenges

(b) Nontechnical challenges

Technical challenges are further divided into:

(i) Data Model differences

The emergence of various NoSQL databases like Cassandra, MongoDB, CouchDB, etc. has led to the development of various model-based query solutions. Each model has its own features and uses different versions of query language. For example, a query in CouchDB can be implemented using a pair of Java Script function, while Cassandra uses CQL that needs to be passed and understood by server. Every model of NoSQL database has its own tool to query the database and there is no common tool that can help in switching between data stores, and administrators are often left to write extensive codes to switch between data stores. Although codes can be written, the problem aggravates with the complexity of the database.

(ii) Lack of Security

A NoSQL database lacks embedded security features and the developer needs to entail security in the middleware. Most of the NoSQL systems focus on scale-out issues first and the use of application layer to implement security features later.

(iii) ACID Transaction Support

NoSQL databases are not the right solution for the transactions that requires complex, nested transactions that necessitate rollbacks and save points. Although Cassandra supports atomicity, durability, and isolation (AID) with consistency being tunable, still the support is dependent upon the need of the operation and the requirement of transaction support.

(a) Table Join Issues and Indexing Support

NoSQL databases are meant to process data with speed, so they don't support traditional join operation on a table. Due to the distributive nature of the data stores, implementation of joins is difficult. Coders have to write the join code as per the requirement of the database. Sometimes, it works but most of the time it leads to inconsistency in the database. Also NoSQL databases don't support indexing of data.

(b) Transactional Control

NoSQL databases are working on transactional control on the data written at one node and give user the flexibility of choosing the amount of inconsistency across multiple nodes. 100% consistency can be achieved by waiting for each write to be completed on each node. This process may introduce delay.

(iv) Open-Source Projects

Most of the NoSQL databases are open-source projects and are being supported by one or more firms (mostly startups not having global reach). These firms lack support resources and the credibility of big names like Oracle, IBM, Microsoft, etc.

(v) Connectivity with Common BI Tools

NoSQL databases have evolved to meet the scaling demand of web 2.0 applications. However, data in these applications require BI tools to improve the efficiency and competitiveness of these applications. But most of the NoSQL databases don't provide connectivity with BI tools. However, some features related to ad hoc query and analysis are available with the evolution of HIVE and PIG.

(vi) Schema Flexibility

NoSQL models don't require a fixed schema. Illustrating this fact means that NoSQL databases don't require a static column and row format. It means that a programmer can make decision on an entry of item whenever they need to store something. One entry can have 20 integers and other can have 12 strings attached to it. This scheme flexibility can speed development but leads to complexity also.

(vii) Data Administration Issues

NoSQL databases are designed to provide a zero admin solution, but the reality is NoSQL requires a lot of skill to install and efforts to maintain these databases. The complexity of the databases requires large administrative intervention to maintain the smooth working of the databases.

Nontechnical challenges are further divided into:

(i) Vendor Viability Concerns

Many small startups with very small customer base are coming up with the better versions of NoSQL databases. However, the IT executives are more interested to make a long-term commitment with the technology provider who will be around for a long time and have a large customer base.

(ii) Maturity

RDBMS system is stable and richly functional, while NoSQL alternatives are in their preproduction stage with many key features yet to be implemented. NoSQL databases will take time to mature and to be stable.

5. NoSQL SECURITY REFERENCE ARCHITECTURE

5.1 NoSQL Cluster Security

NoSQL database considers a different method for solving the problems related to Big Data. Virtually NoSQL does not provide any security within the NOSQL cluster as security of database and data is also dependent on the network and the applications that form a protective shell around the data. This security model is easy to implement as it will not result in performance or functional degradation. Disadvantage of using type of security model is misuse of data by credentialized user or exposure of system to the malicious user if application or firewall fails. Some of the built-in security tools available in NoSQL cluster are SSL/TLS for secure communication, Kerberos for node authentication, Data at rest security using transparent encryption, etc. However, these tools of cluster security are difficult to implement and are expensive. The intrinsic architecture and design considerations have been carefully taken into account in most of the NoSQL database. Figs. 2 and 3 represent the core architectural difference between RDBMS databases and NoSQL databases.

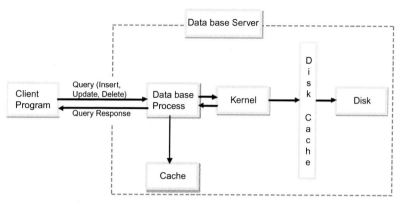

Fig. 2 RDBMS architecture—ACID complaint.

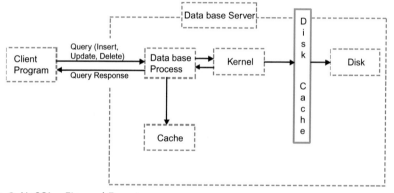

Fig. 3 NoSQL—Fire and Forget.

A typical NoSQL database makes compromises on some ACID properties. In general, a complete security architecture must cover the following points:

(1) User Access Management

(2) Logging Operations

(3) Data Protection

(4) Environmental and Process Control

5.1.1 User Access Management—Authentication

Authentication is the process of identifying an individual which is based on a username and password. In this context, various entities are

(a) Users needing access to database

(b) Administration

(c) Software Systems

(d) Physical and Logical nodes

Some of the best practices used for user access management are:

- **(i)** Create login credentials: To avoid creation of a single admin login which is shared by all users a separate security credential is given to each user.
- **(ii)** Centralized user access management: Providing ability to databases to manage authentication within the database itself. It may be done by integrating organizational identity management system.
- **(iii)** Enforcing password policies: Adhering at least minimum password complexity requirements stated by the organization.

5.1.2 Authorization

Authorization is defined as the process of giving individual access to the system objects which is based on the user identity. Authentication purely ensures that the individual is the same person who is claiming to be but has no clue about access right of the individual. Best practices of the authorization include:

(a) Granting minimum access to entities

(b) Grouping common access privileges into roles

(c) Controlling actions of individual entity

(d) Controlling access of sensitive data

5.1.3 Auditing

The process of auditing helps to detect the attempts of accessing unauthorized data. By creating audit trails, logs can be used for compliance. Best practices of auditing include:

(a) Tracking the changes in database configuration

(b) Tracking changes in data

5.1.4 Encryption

Encryption is defined as the translation of data into a secret code. It is said to be the most effective way to achieve data security. Some of the best practices used for encryption are:

(a) Enforcing connections to databases

(b) Encrypting data at rest

(c) Enforcing strong encryption

(d) Signing and rotating encryption keys

5.1.5 Environmental and Process Control

Protection of the underlying infrastructure is also very important. To ensure this, best practices used are:

(a) Installation of firewalls
(b) Network configurations
(c) Defining file permissions

5.2 Data Centric Approach to Security

As discussed earlier, NoSQL security is provided either by network or applications that surround the data or by the third-party tools available in a NoSQL cluster. Another approach to ensure security in NoSQL databases is data centric security where security controls are part of data not the database. It means protection of data takes place before data is moved into the database. Three basic tools that support data centric security are:

(a) Tokenization
(b) Masking
(c) Data element encryption

To ensure security tokenization substitute sensitive data with data tokens to ensure security. A token has no basic value; it only carries a reference to the original value to the database. Credit card processing systems use a tokenization technique to substitute credit card numbers (Fig. 4).

Masking technique replaces an original value of data with a random value to protect the data but preserve the original value of the dataset for analysis. For example, the name of a person can be changed with any other name, date of birth can be modified, etc. as shown in Fig. 5.

Data element encryption encrypts the data and only authorized users can decrypt the data using keys (Fig. 6).

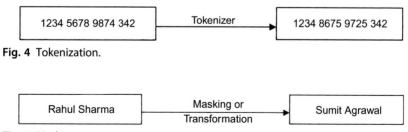

Fig. 4 Tokenization.

Fig. 5 Masking.

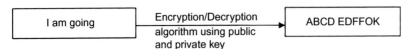

Fig. 6 Data element encryption.

6. TECHNIQUES TO MITIGATE THE ATTACKS ON NoSQL DATABASES

Mitigating security attacks on NoSQL databases is a challenging task as most of the NoSQL databases are open source and are founded upon the Hadoop framework. Hadoop framework is not a single technology but is a combination of various cooperating applications. Each of these applications has different security requirements and is modified as per the need of the data. Also the code analysis of application layer alone is not sufficient to ensure that all threats are mitigated. Apart from this the emerging cloud and Big Data systems are complex applications that execute multiple processes and use heterogeneous open-source tools and platforms. Another challenge is the development speed of modern code with DevOps methodologies that reduces the time between development and production. Therefore to mitigate the attacks on NoSQL databases and to achieve security, various parameters (listed below) need to be implemented in combination with each other. These parameters are:

1. Using Firewalls
2. Security Scanning
3. Protection of APIs
4. Auditing and Logging
5. Authentication
6. Input Validation
7. Access Control
8. Segregation of Duties
9. Encryption

It is necessary to implement the security mechanism within the NoSQL database environment rather than on a control point from where data and queries enter the database environment. Although we have discussed all the listed parameters earlier as well, a brief description is as follows:

1. *Firewalls*: Firewalls are used to protect the system from malicious and intentional attacks. Firewalls can be embedded close to the data store to strengthen the user authentication process. Once the authentication process is strengthening, other security factors will automatically be reinforced.

2. *Security Scanning*: SQL injection vulnerabilities can be detected by performing dynamic testing (DAST) or static testing (SAST) on the source code. Most of the NoSQL tools lack rules for detecting these injection attacks. DAST technology is more reliable as it uses backend inspection technology that improves detection reliability. Developer and coders recommend integrating these techniques into the system at all the phases of development to capture and fix the bugs during every running cycle of development. Security testing should not be done only during testing phase, hence improving reliability.

3. *Protection of APIs*: Mitigating the risk because of an exposed REST API or CSRF attacks can be done with the help of following methods:
   (a) Accepting only JSON in content type.
   (b) Limiting HTML forms to URL-encoded content type, so attackers can't use HTML forms for CSRF.
   (c) Using Ajax requests, which are blocked by the browser.
   (d) Ensuring JSONP (JSON with padding) and cross-origin resource sharing to be disabled in the server API, so no actions can be made directly from a browser.

4. *Audit and Logging*: Auditing of the databases needs to be done very frequently to identify mishaps as early as possible. Third-party tools like Scribe and Logstash can be integrated into the database to log the records of transactions. Auditing using Scribe helps in identifying theft of data and logging the transaction records will help in saving data misuse.

5. *Authentication*: Authentication can be implemented within the database or within the framework. Authentication keeps a strong check over the malicious user. Strong password mechanisms like user-defined passwords can also be a measure to take care of.

6. *Input Validation*: Input validation is required to filter Java Script which will help in the elimination of Java Script injection attacks and string concatenation.

7. *Access Control*: Access control helps in imposing restriction on data access. Access to data can be granted on the basis of security policy of the organization and the permission levels.

8. *Segregation of Duties*: Access to data can be granted on the basis of role and responsibility of the employee. Segregation of duties will help in restricting data theft and malpractices.

9. *Encryption*: It is required to transform the data into a format which can only be decrypted by users having an authorized key. It helps in protecting sensitive data.

7. PROPOSED SECURITY AND PRIVACY SOLUTIONS FOR NoSQL DATA STORES

A lot has been said about the security and privacy loopholes of NoSQL databases and the various techniques to mitigate the attacks have also been discussed. Below are few proposed security and privacy solutions for NoSQL data stores.

(a) *End-Point Input Validation/Filtering*: NoSQL data stores collect data from various sources, such as end-point devices. An important challenge is to validate the input while collecting the data. Validating the source of input to identify and filter malicious data is a challenging task as data is collected from untrusted input sources, specially with BYOD (bring your own device) model. To validate and filter large data sets effectively, algorithms need to be created.

(b) *Real-time Security and Compliance Monitoring*: Big Data technologies are monitored using real-time security mechanisms to provide faster processing and analysis of data that later on can be used to provide real-time anomaly detection based on scalable security analytics.

(c) *Scalable and Composable Privacy Preserving Data Mining and Analytics*: Invasion of privacy is a troubling manifestation of Big Data. Anonymizing of data is important for Big Data analytics as it will help in maintaining user privacy; therefore, it is important to design robust and scalable algorithms for preventing inadvertently privacy disclosures. Also these algorithms will help in collecting relevant information and increase user safety.

(d) *Implement Logic Filter in Application Space*: With granular access control mechanisms, applications have the ability to access data for many different purposes on behalf of a user. Any policies that prevent cross-purpose data combinations rely in some part on the application properly separating different uses of data.

(e) *Cryptographically enforced access control*: NoSQL databases store sensitive data unencrypted. Large datasets are difficult to encrypt because of their all-or-nothing retrieval policy. The problem of encryption can be solved by designing a cryptographically secure communication network where data can be communicated in a secured and agreed upon way.

8. CONCLUSION

In this chapter, we have discussed major security concerns in NoSQL databases. Data protection and access control are some of the key issues of security in NoSQL. Reasons for security threats in various NoSQL databases have also been discussed thoroughly in the chapter. Various security threats for NoSQL databases are—privacy of user data, distributed environment, authentication, fine-grained authorization and access control, safeguarding integrity, and protection of data at rest and in motion.

In NoSQL database, Kerberos is used to authenticate the clients and data node. In order to ensure fine-grained authorization, data is grouped according to their security level. Cassandra uses TDE technique to protect data at rest. In order to maintain a secure MongoDB deployment, it is required that administrators must implement controls for ensuring that application and users have only access to the data that they need. Various techniques to mitigate the attacks on NoSQL databases have been discussed along with proposed security and privacy solutions of NoSQL databases.

GLOSSARY

Distributed Databases A distributed database (DDB) is a database distributed over a computer network that allows applications to access data from local and remote databases. It is a collection of multiple, logically interrelated databases where the users can access the portion of the database that is relevant to their tasks without interfering with the work of others. A centrally controlled distributed database periodically synchronizes all the data and ensures that updates and deletes are performed on the data and will automatically be reflected in the data stored elsewhere.

Big Data Large quantity of unstructured, semistructured, and structured data that can be mined to be converted into useful information is known as Big Data. Big Data has the potential to improve the operational and the decision-making capability of any organization. The four Vs of Big Data are volume, velocity, variety, and value of data. Big Data Analytics reduces the volume and the variety of data and adds value to the data.

Cassandra Database It brings together the distributed system technologies from Amazon DynamoDB (based on the principles of Dynamo) and the data model from Google's Big Table [2]. Cassandra is eventually consistent and based on a Peer-to-Peer (P2P) model without a single point of failure. Like Big Table, Cassandra provides a Column Family-based data model richer than typical key/value systems (Abramov). Cassandra supports transaction logging and automatic replication.

MongoDB Database MongoDB, classified as a NoSQL database, is a cross-platform document-oriented database. It is an open-source database and has been written in C ++ Language. It provides high availability, high performance, and easy scalability. MongoDB database uses the concept of document and collection. A collection is said to be a group of MongoDB documents like an RDBMS table.

REFERENCES

[1] I.A.T. Hashem, I. Yaqoob, et al., The rise of big data on cloud computing: review and open research issues, Inf. Syst. 47 (2015) 98–115.

[2] G.C. Deka, Cloud database security issues and challenges, in: Handbook of Research on Securing Cloud-Based Databases With Biometric Applications, 2014.

[3] A. Mohamed, G. Obay, Relational vs. NoSQL databases: a survey, Int. J. Comp. Inform. Technol. 03 (03) (2014) 598–601.

[4] J. Ahmed, R. Gulmeher, NoSQL databases: new trend of databases, emerging reasons, classification and security issues, Int. J. Eng. Sci. Res. Technol. 4 (6) (2014) 176–184.

[5] P. Factor, NoSQL: Are You Ready to Compromise With Security, Available at: http://www.sqlservercentral.com/articles/Editorial/NoSQL+NoAuthorisation/98005/, 2013.

[6] S. Ebrahim, A.N. Mohammad, Survey on security issues in big data and NoSQL, ACSIJ 4 (4) (2015) 68–73. No. 16.

[7] H.S. Patil, Y.S. Mukhtar, Distributed database: an relevance to business organization, J. Inf. Oper. Manag. 2 (1) (2011) 21–24.

[8] L. Okman, In: Security issues in NoSQL databases, International Joint Conference of IEEE TrustCom-11/IEEE ICESS-11/FCST-11, 2011.

[9] C. Michael, NoSQL Security: Do NoSQL Database Security Features Stack Up to RDBMS?, http://searchsecurity.techtarget.com/tip/NoSQL-security-Do-NoSQL-database-security-features-stack-up-to-RDBMS, 2016. Accessed 4 October 2016.

[10] R.P. Padhy, M.R. Patra, RDBMS to NoSQL: reviewing some next-generation non-relational databases, Int. J. Adv. Eng. Sci. Technol. 11 (2) (2011) 45–52.

[11] C. Nance, T. Losser, In: NOSQL vs RDBMS—why there is room for both, Proceedings of the Southern Association for Information Systems Conference, Savannah, Georgia, USA, 2013.

[12] Fidelis Cybersecurity, Current Data Security Issues of NoSQL Databases, 2014).

FURTHER READING

[13] R. Cattell, Scalable SQL and NoSQL data stores, ACM SIGMOD Rec. 39 (4) (2010) 12–27.

[14] Chang F, Dean J, et al., "Bigtable: a distributed storage system for structured data", Proceedings of 7th USENIX Symposium on Operating Systems Design and Implementation (OSDI 06), available at: https://www.usenix.org/legacy/event/osdi06/tech/chang/chang.pdf, 2006.

[15] NoSQL Databases: A Step to Database Scalability in Web Environment, (PDF Download Available). Available from: https://www.Researchgate.net/publication/221237715_NoSQL_Databases_a_step_to_database_scalability_in_Web_environment. Accessed 10 April 2016.

[16] MongoDB Security (March 18, 2017), Retrieved from https://docs.mongodb.com/manual/tutorial/enable-authentication/#overview.

[17] Apache Cassandra (April 2, 2017), Retrieved from http://cassandra.apache.org.

[18] No SQL and No Security (March 23, 2017), Retrieved from, https://www.securosis.com/blog/nosql-and-no-security.

[19] MongoDB Customers (March 18, 2017), Retrieved from http://www.mongodb.com/industries/high-tech.

[20] MongoDB Sharding (March 18, 2017), Retrieved from http://docs.mongodb.org/manual/sharding/.

[21] MongoDB documentation on Security (March 19, 2017), Retrieved from http://docs.mongodb.org/manual/core/security-introduction/.

[22] K. Fan, Survey on NoSQL, Programmer 6 (6) (2010) 76–78.

[23] K. Yang, Secure and verifiable policy update outsourcing for big data access control in the cloud, parallel and distributed systems, IEEE Trans. 6 (4) (2016) 762–767.

[24] Okman L, et al., "Security issues in NoSQL databases, Trust, Security and Privacy in Computing and Communications (TrustCom)", IEEE 10th International Conference on, pp 541–547, 2011.

[25] K. Grolinger, H.A. Higashino, Data management in cloud environments: NoSQL and NewSQL data stores, J. Cloud Comput. Adv. Syst. Appl. 2 (1) (2013) 1.

[26] P. Noiumkar, T. Chomsiri, In: A comparison the level of security on top 5 open source NoSQL databases, The 9th International Conference on Information Technology and Applications (ICITA 2014), 2014.

[27] Boicea A, Radulescu A, "MongoDB vs Oracle—database comparison", Emerging Intelligent Data and Web Technologies (EIDWT), 2012 Third International Conference on, pp 330–335, 2012.

[28] A. Zahid, R. Masood, In: Security of sharded NoSQL databases: a comparative analysis, Conference on Information Assurance and Cyber Security (CIACS), 2014, pp. 1–8.

[29] N. Leavitt, Will NoSQL database live up to their promise? IEEE 10 (9162) (2010) 12–14.

[30] M. Stonebraker, M. Samuel, In: The end of an architectural era: (it's time for a complete rewrite), Proceedings of the 33rd International Conference on Very Large Data Bases, VLDB, 2007, pp. 1150–1160.

[31] MongoDB, MongoDB [Online], Available: http://www.mongodb.org/, 2016. Accessed 15 April 2016.

[32] F. Chang, J. Dean, et al., Bigtable: a distributed storage system for structured data, ACM Trans. Comput. Syst. 26 (2008) 4:1–4:26.

[33] M. TabrezQ, Security issues in distributed database system model, Int. J. Adv. Comp. Technol. 2 (12) (2013) 396–399.

[34] S. Chakraborty, In: Implementation of execution history in non-relational databases for feedback-guided job modeling, Proceedings of the CUBE International Information Technology Conference—CUBE 12, 2012.

ABOUT THE AUTHORS

Dr. Neha Gupta is currently working as an Associate Professor, Faculty of Computer Applications at Manav Rachna International University, Faridabad Campus. She has done her PhD from Manav Rachna International University, Faridabad. She has a total of 11 + years of experience in teaching and research. She is a Life Member of ACM CSTA and Tech Republic, and Professional Member of IEEE and CSI. She has authored and coauthored 23 research papers in SCI/SCOPUS/Peer-Reviewed Journals and IEEE/IET Conference Proceedings in areas of Web Content Mining, Mobile Computing, and Web Content Adaptation. Her research interests include ICT in Rural Development, Web Content Mining, Cloud

Computing, Data Mining, and NoSQL Databases. She is a technical program committee (TPC) member in various conferences across the globe. She is an active reviewer for International Journal of Computer and Information Technology and in various IEEE Conferences around the world. She is one of the Editorial and Review Board Members in International Journal of Research in Engineering and Technology.

Dr. Rashmi Agrawal is working as a Professor and Head of Department—Department of Computer Applications in Manav Rachna International University Faridabad. She is having a rich teaching experience of more than 14 years in the area of computer science and applications. She is UGC-NET(CS) qualified. She has completed MPhil, MTech, MSc, and MBA(IT). She has completed her PhD in the area of Machine Learning. Her area of expertise includes Artificial Intelligence, Machine Learning, Data Mining, and Operating System. She has published more than 20 research papers in various National and International conferences and Journals. She has organized various Faculty Development Programmes and participated in workshops and Faculty Development Programmes. She is actively involved in research activities. She is a lifetime member of CSI (Computer Society of India). She has been a member of Technical Programme Committee in various conferences of repute.

CHAPTER FIVE

Comparative Study of Different In-Memory (No/New) SQL Databases

Krishnarajanagar G. Srinivasa*, Srinidhi Hiriyannaiah†
*Ch Brahm Prakash Government Engineering College, Jaffarpur, New Delhi, India
†Ramaiah Institute of Technology, Bengaluru, India

Contents

1. Introduction 134
2. Advanced Database Processing 134
   2.1 In-Memory Database and Analytics 135
   2.2 Concepts of In-Memory Database 135
   2.3 Challenges With In-Memory Analytics 138
3. In-Database Analytics 138
   3.1 Types of In-Database Processing 139
4. NewSQL Databases 141
   4.1 Clustered, Parallel, Scalable SQL Databases 141
5. Case Study on Alteryx (Demonstration/Installation, Creation of Database/Record) 151
   5.1 Creating Table in Alteryx 151
6. Conclusions 153
References 153
About the Authors 155

Abstract

With the advancement and changes in the hardware, availability of data, collection of data at a faster rate, emerging applications; database systems have been evolving over the last few decades. The application areas where the databases are used have moved over from relational data to graph-based data and stream data. The expanding cloud computing services that use internet and require big data has come upfront due to data-intensive computing problems. Social media companies such as Amazon, Facebook, Google use the world wide web as a large distributed data repository. This large data repository on the web cannot be processed with traditional RDBMS systems. In-memory databases play a key role in the upcoming years for processing data that is huge in volume with wide variety of data formats. In this chapter, an overview around in-memory databases with advanced processing, techniques, case studies are presented.

Advances in Computers, Volume 109
ISSN 0065-2458
https://doi.org/10.1016/bs.adcom.2017.09.001

1. INTRODUCTION

With the advancement and changes in the hardware, availability of data, collection of data at a faster rate, emerging applications; database systems have been evolving over the last few decades. The application areas where the databases are used have moved over from relational data to graph-based data and stream data. The expanding cloud computing services that use internet and require big data has come upfront due to data-intensive computing problems. Social media companies such as Amazon, Facebook, Google use the world wide web as a large distributed data repository. This large data repository on the web cannot be processed with traditional RDBMS systems. In-memory databases play a key role in the upcoming years for processing data that is huge in volume with wide variety of data formats. In this chapter, an overview around in-memory databases with advanced processing, techniques, case studies are presented.

2. ADVANCED DATABASE PROCESSING

The old decade of relational database technology involves transaction processing that is characterized with atomicity (A), consistency (C), isolation (I), and durability (D) shortly known as ACID properties. The result of each transaction on the tables in the database should abide by these properties ensuring all or nothing, legal data for all transactions on the tables, independent transactions and failure tolerant for each transaction [1]. Cloud computing is a paradigm that is upcoming providing services such as infrastructure as a service (IaaS), platform as a service (PaaS), and software as a service (SaaS). Dynamic scalability is one of the advantages that cloud computing provides and is one of the essential requirement for databases [2].

Scalability is one of the main issues related to RDBMS systems. Most of the applications are distinguished by the properties such as massive scalability, latency, on-demand service, and easier programming model. RDBMS does not support these features in a cost-effective way. NoSQL databases, a new family to the SQL-based systems provides the advantage of scaling over the databases [3]. The key technique provided by NoSQL databases to achieve scalability is partitioning. In this section, we provide an overview of in-memory databases and its concepts for advanced database processing.

2.1 In-Memory Database and Analytics

A database is generally a collection of data such as, a list of employees, students. The storage for databases is disk-based. In the current world of e-commerce systems, data moves fast, and the database of products as well as customers become bigger and bigger. Sensors are becoming integral part of every product or item in the market. For example, it is possible to track every product online and status of the order placed. So, in order to gain the value of such information the processing of results has to be fast and available. If the time gets exceeded, the value of the information will no longer be valid.

Since, the databases storage are disk-based, each query will have to load the data from and to disk. The access time for small databases is relatively small and not noticeable. Disk-based systems maintain a buffer management to keep data in cache and reduce the access time for queries. An in-memory database consists of data that are primarily stored in main memory [4]. Compared to the disk-based systems, in-memory database systems maintain the data in the memory. It needs techniques to store and make it persistent and durable. Using the concept of storing the data in-memory has some advantages and disadvantages. Some of the concepts related to in-memory databases are discussed in the next section.

2.2 Concepts of In-Memory Database

The data in the in-memory database are stored in the main memory, the mechanisms that optimize the access time compared to disk-based systems are discussed with an example table "student_population" as shown in Table 1. The table consists of the attributes first name, last name, gender,

Table 1 Student Population Data

Column Name	Size of Each Record	Size for 500,000,000 Students
First name	200 bytes	$200 \times 500,000,000$
Last name		$= 10,000,000,000$ bytes
Gender		$= 10\,GB$
Country		
City		
Birthday		

country, city, and birthday. Assuming the scenario of a university where there at least lakhs of students and each record of the student taking 200 bytes. The various optimizations that in-memory databases provide are discussed in the next subsequent sections [5,6].

2.2.1 Optimized Indices

Optimized index technique can be used for reducing the access time. For the example of the "student_population" table a skip index can be applied on the column *country*. It can define borders for a specific attribute. In fact that the table will be sorted by this column, the records can be easily skipped depending on the query. If the search is based on a country column, the whole table is not looked but, it looks into the indexed column only. Less memory is needed for accessing the table and querying on a particular column that is indexed.

2.2.2 Data Compression

The basic problem of a developer or an engineer is to come up with a solution for storing the data that takes less space. Considering the example of "student_population" table, the columns country and city are limited which can be encoded as static values. Thus, these columns can be encoded into dictionary and reduce the size *f* the database [5]. Assuming the column takes 49 bytes per record, with the use of dictionary the memory can be saved upto 3 bytes per record. Similarly, for large tables the dictionary can be used to save the memory space for table and reducing the overall footprint of the database in-memory.

2.2.3 Data Layout

The layout in which the data is stored in the databases also does matter on the performance impact of the in-memory databases. Usually, the data stored in the databases are either row-format or column format [7,8]. Significant differences among the two different layouts are outlined as follows.

The data layout in row format is as shown in Fig. 1. The attributes P, Q, and R are next to each other and no jumps are necessary. If the column operations need to be performed on the same data layout then jumps are necessary. These jumps, visualized as an arrow, do occur, as reason of that column operation just operates with the same attribute. The columnar data layout is as shown in Fig. 2. In this case column operations does not need any jump but the row operations, jumps are needed to get all attributes.

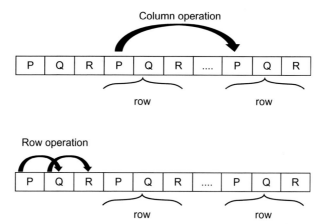

Fig. 1 Example of row store data layout.

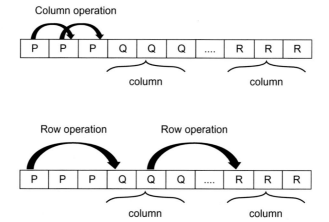

Fig. 2 Example of column store data layout.

The speed of the access time for in-memory databases depends on the data layout of the databases. SQL queries on the database determine the data layout that needs to be used. The row data layout allows to easily adding and modifies a record but unnecessary data might be read. Compared to that, a columnar data layout is suitable for read-mostly, read-intensive, and large data repositories. Most times programs think in records but can operate column wise, so there is a trade-off between them. A hybrid layout might be the best approach.

2.3 Challenges With In-Memory Analytics

In-memory databases provide mechanisms to reduce the time access for the databases. However, certain challenges do exist that are discussed as follows [9,10]:

- Main memory is a volatile memory. Since, the in-memory databases store the data in main memory power failure will lead to the loss of data. So, if the data need to stored for longer time then more cost needs to be spent on the power. However, it can be overcome using nonvolatile random access memory (NVRAM) [11]. It will retain the information even if the power is down. An alternative way is to keep a back up of data on slave systems.
- If the transfer of the data is over a network, then access time will be affected by additional latency of the transfer overhead. In this case, the in-memory needs to be used locally [12].
- There might be some modifications required at the code level [13] in a consistent manner where the database migration needs to be done from in-memory to disk-based and vice versa.

3. IN-DATABASE ANALYTICS

The process of building and developing a data model for programmer involves spending the effort on locating, extraction, integration, and preprocessing. Data sources are often scattered in the process consisting of multiple resources and variety of repositories. In this competitive technological era, where speed of the execution of the application is necessary, there is lack of agility in the data model and development process. In-database analytics is such a technology that fosters the need for integration of several repositories into one place for building data model and development [14]. In-database analytics involves data processing that can be conducted within the database itself by building analytic logic into the database itself. This process of carrying out analytics within the database reduces the time and effort required for transforming data and move back and forth to database and analytics applications. The analytics system for in-database processing consists of data warehouse providing the functions such as partitioning, scalability, optimization, and parallel processing [15]. This approach of in-database analytics helps in making better predictions and making business decisions such as trends, business risks, and opportunities to invest time and money to key areas in the market. Some of the applications where in-database analytics include fraud detection, risk management, pattern recognition, ad hoc analysis of reports, and statistical analysis of records.

3.1 Types of In-Database Processing

The general processing techniques for processing data involves extract, load, and transform (ETL) from the traditional RDBMS systems. In this process, the data present in the various sources of data warehouses are extracted, transformed into the required format, and loaded into the application for further analytics. In the case of in-database processing, additional capability of processing is required for facilitating and enriches the analytics. In this section, the approaches for in-database processing are discussed [4].

3.1.1 Distributed Data Marts Processing

The old approach of processing the data within a database involves the use of distributed data marts as shown in Fig. 3. In this process of building a data model and development, improvements, and scoring is normally performed outside the traditional RDBMS systems for proficient access to the analytics programs that resides outside of it. The shadow frameworks are normally used in SAS® applications as ordinary nonsocial information and overseen generally as a distribution center and server outside for information access. The main disadvantage of such database processing is the usage of shadow frameworks that acts as the centralized information server. The process of extracting the information from this social distribution center involves much time and cost. The second issue is the replication of the information to the central server and removing the information through a strung interface. The

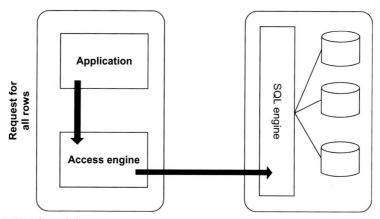

Fig. 3 Distributed data marts processing.

cost of replicating the copy of information to the procedures that require data can be critical.

In addition, the handling, extraction, and system cost for copying information from an information distribution center condition to the investigative condition is additionally critical regarding equipment usage. There is additionally a hazard factor, from a security viewpoint, related with information replication. The information expected to construct advanced expectation models must be definite. Frequently, this point by point information is extricated into working sets that are not full secured. In conclusion, and above all, the human cost for dealing with the copy informational indices is very high. This exertion is typically constrained upon the very assets that are profoundly talented explanatory modelers. The business offer for changing information administration time into scientific demonstrating time is noteworthy.

3.1.2 In-Database Repository Processing

In-database repository processing, uses the data available in the database itself to set up the information for analytics as shown in Fig. 4. By preparing inside the information distribution center it is conceivable to reduce the expenses of information duplication. Instead of making separate informational indices that are controlled using PROC SORT, PROC RANK procedures, this style of in-database processing helps analytics applications utilizing the SQL primitives efficiently. An additional preferred standpoint of the in-database handling approach is that it will have the capacity to use the parallel preparing ability of the information center. This design enables to abuse

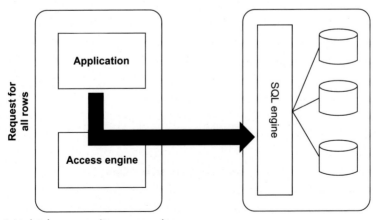

Fig. 4 In-database repository processing.

the greatly parallel preparing (MPP) execution capacity crosswise over tens, hundreds, or even a great many high volume (HV) servers. The in-database preparing structure additionally disposes of a lot of the information administration overhead connected with the old style approach. In any case, take note of that model improvement and model scoring are most viably executed with "smoothed" informational collections as opposed to the standardized (or star pattern) shapes regularly found in information distribution centers. In order to make out the changes from social frame to a systematic informational index structure, instruments are required to mechanize the rotating of information from a social shape into perception informational collections to sustain into display building and scoring forms. The sandboxes can be used as storehouses showing the improvement and scoring.

The main advantage of using in-database processing is scoring algorithms for applications can be used in parallel. The functions used in the query processing are translated into optimized C code and are fully parallelized across all servers in the massively parallel processing (MPP) architecture.

4. NewSQL DATABASES

With the rapid development of databases in the computer science, NewSQL is a new database that is touted superior to SQL. It was designed to preserve the atomicity, consistency, isolation, and durability (ACID) properties of the traditional RDBMS systems with addressing the issues of scalability and performance compared to NoSQL systems. NewSQL phrase was coined by Matthew Aslett, shorthand for scalable/high performance SQL systems. The main focus of NewSQL systems is to address the issue of scalability of online transaction processing (OLTP) workloads with support of programming model customizability and reduce management tasks in SQL databases like manual sharding. Enterprises who need to support in migrating to Big data platforms and applications from SQL systems can use NewSQL for scaling up the performance of OLTP workloads.

4.1 Clustered, Parallel, Scalable SQL Databases
4.1.1 Clustered Databases
A cluster consists of collection of components providing scalability and availability with low cost. Many high performance computing applications can be developed using clustering applications as the processing takes place on commodity nodes with low cost. Since, many commodity nodes are involved in a cluster, the responsibilities are shared across different nodes

and the architecture is distributed. There are two main basic types of architecture in clustered databases namely shared–nothing and shared–everything [16,17].

In a shared–nothing architecture, each node in the cluster performs computation on its own subset of data. The partitioning of the data can be in a form such that compute node places its data portion on its local disk, and all nodes are connected via a high-speed network. It resembles the storage area network wherein if one of the node fails, it makes the database completely inaccessible. Thus, the physical partitioning of the data should be such that if a node fails then, one of the other nodes can take over if there is any fault in the network. This type of architecture is followed by wide variety of enterprise products but has difficulty over installation and maintenance.

In the case of shared–everything architecture, there is no distinction among the compute nodes in the cluster. Any compute node can access and operate on the information of the database. The physical architecture generally involves network attached storage where all the nodes can communicate via a high-speed interconnect. Scalability of the architecture can be achieved based on the logical portioning. The downside of shared–everything architecture is there can be erroneous information communication when numerous nodes communicate with disk simultaneously.

4.1.2 Parallel Database

During the database operations, we encounter various operations such as loading, indexing, and queries. A parallel database system seeks to enhance the performance of such database operations using multiple CPUs and disks in parallel. Centralized architectures for databases are not powerful enough to handle such as parallelization of operations. In parallel processing of data, many operations can be performed in parallel simultaneously as compared to sequential processing where sequences of operations are performed one after the other. The different types of parallelism and architectures for parallel databases are discussed as follows.

4.1.2.1 Types of Parallelism in Parallel Databases

The different types of parallelism [18] that can be employed in parallel databases and query execution process are as follows:

- *Interquery parallelism*: Multiple queries are executed in parallel.
- *Interoperation parallelism*: A single query on a database may involve two or more operations to be performed. In this case, the dependency of

operations can be checked to see whether the operations can be executed in parallel or not based on the dependency.

- *Independent parallelism*: It is a form of interoperation parallelism where operations are independent to each other and can be executed in parallel. For example, consider a scenario wherein there are four tables X1, Y1, X2, and Y2. Here X1 and Y1, X2 and Y2 have joining dependencies, respectively. The two join operations within a single query can be performed in parallel using two processors and the final join can be done later.
- *Pipelined parallelism*: It is a form of interoperation parallelism where operations are in a pipelined fashion. For example consider a scenario where in there are three tables X1, X2, and X3 that need to be joined. The joining dependency is such that result of X1 join X2 has to be used for another join operation with X3. Here, the join operations are pipelined one after the other but one processor can be used for joining X1 and X2 and another processor can be used for joining the result with X3.
- *Intraoperation parallelism*: It is a form of parallelism where complex and large query operations can be executed in parallel with multiple processors. For example, the ORDER BY clause can be used to execute over a set of records in parallel on multiple processors.

4.1.2.2 Types of Architectures in Parallel Databases

The different types of architectures that can be used in parallel databases and query execution process are as follows:

- *Shared memory*: In this type of architecture in parallel databases, multiple processors share the main memory but having there own disk for storage. Since, the memory is shared among multiple processors, speed is greatly reduced if all of them are executing large complex queries and using the same memory.
- *Shared disk*: In this form of architecture, multiple processors share the disk for storage but each of them having their own memory for processing. Usually, such form architectures has multiple processors in each compute node for processing.
- *Shared nothing*: In this form of architecture, neither the main memory nor disk is shared among the multiple processors in the system. It is more prone to failure as fault in one of the compute nodes might replicate to other nodes as well in the system.

4.1.3 Scalable SQL Databases

SQL has been one of the widely used languages for query execution on databases. The main disadvantage is the scalability of SQL with RDBMS systems. Most of the social media companies such as Facebook, MySpace, and Twitter use NoSQL databases as the back-end for query processing and execution. Scalability can be met by following some of the requirements [2]. These requirements are discussed as follows:

- *User load*: The applications should support large number of users with a potential of million users.
- *Data load*: The systems should support large amount of data to be processes in number of petabytes that are produced by less or large number of users.
- *Computational load*: The systems should support large number of operations with large number of data.
- *Agility and scalability*: Since, the applications in social media are ad hoc and agile, there should be ease in changing data models, operations, and programming.

4.1.4 Scalable Architectures

Along with SQL, NoSQL and NewSQL support to address the scalability issues in the databases, the architectures behind the languages need to be stable to enable scalability and ensure adding/deleting computational nodes in the system [15,19]. Some of the architectures that support scalable systems are discussed as follows.

4.1.4.1 Functional Partitioning

In functional partitioning approach, a service-oriented approach is followed where in different services for each task is used as shown in Fig. 5. For

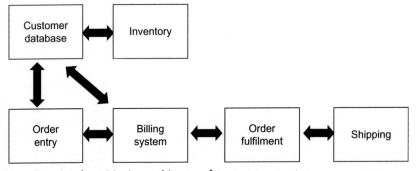

Fig. 5 Functional partitioning architecture for commerce system.

example, consider an e-commerce system that consists of different function-alities such as billing, inventory, customer database, and order fulfillment. Each of the services is independent in nature with certain dependencies. These dependencies can be resolved with less effort and time. A service-oriented paradigm for this approach helps to scale out by assigning separate resources to the services as they are needed. The drawback of this approach is the consideration of functions alone rather than the data usage and its depen-dencies. So, though it addresses the scalability in terms of functions per-formed in the system but not with respect to data.

4.1.4.2 Data Partitioning

In the case of data partitioning [20], application and its processing steps are distributed over a set of data and its partitions as shown in Fig. 6. For example in the scenario of e-commerce platform, data associated with inventory, cus-tomer database, billing can be partitioned into different datasets. The par-titioning of data can be based on the topology of the users in the system who are distributed over a geographical area. Social media sites often use geographical data partitioning in their database systems as the user audience is widely distributed both locally and globally.

4.1.4.3 Scale Out Architecture

In order to build distributed systems with scalability that performs more than the centralized applications, following requirements have to be satisfied popularly known as CAP (consistency, availability, and partitioning) theorem [3].

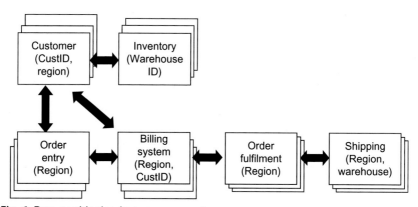

Fig. 6 Data partitioning in commerce systems.

- Consistency

    A distributed architecture for the database systems should provide a consistent view of data. Even, when the transactions are executed among the databases, it should be consistent among all of them.

- Availability

    The database system should be available for any transaction processing and modifications. Any changes that have to be rolled out to the database system, all the computing nodes in the distributed database system should be available. In addition, changes to the application have to be rolled out to all the partitions in a way that will not interfere with the consistency guarantees and requirements of the application.

- Partitioning

    The partitioning scheme used for the database systems needs to provide a functional, scalable, available, and fault-tolerant architecture. The computing nodes need to be redundant and resilient to network partitioning.

All of the three requirements in CAP theorem cannot be satisfied at a time. For example, there are times where the application requirement is strongly on the availability of all the nodes so the consistency has to be compromised. It means that stronger level of consistency cannot be guaranteed across all the partitions in the database system. In the other way round, where the application demands scalability, the computing nodes must operate synchronously with each other waiting for reply from the nodes even in the case of node failure. Since, the close coupling of data sources is a key requirement; availability needs to be compromised at this point. Some of the ways to build scalable applications with the database systems are discussed as follows [21]:

- Service-oriented, functional-oriented, and data-oriented partitioning schemes should be followed to ensure proper partitioning of data sources.
- The computing nodes can be mirrored to achieve high availability. Depending on the service-level guarantee around failover and read vs update frequency, each mirror will be managed either synchronously or asynchronously.
- Consistency can be achieved relaxing the ACID properties of the systems across the computing nodes. The techniques such as open nested transaction systems, optimistic concurrency control, specific partitioning can be used to reduce the risk of inconsistencies. For example, open nested multilevel transactions relax transactional isolation by allowing

certain local changes to become globally visible before the global transaction commits. In practice, such advanced transaction models have not yet been widely used, even though some transaction managers provide them.

4.1.5 NewSQL Environments

NewSQL refers to new scalable, high performance relational database systems that provide the properties of relational model and its databases in distributed manner [22]. Some of the market vendors in NewSQL include ScaleDB, Clustrix, ScaleBase, VoltDB, etc., providing features such as sharding, scalability in a distributed manner for SQL systems. The NewSQL system is often hosted in the organization environments rather than as a cloud service. If the database has to be provided as a service, it is supposed to provide the user with the peace of mind knowing that high availability, elastic scalability, and distribution (across clouds/zones) are all taken care of in an on–demand plug-and-play way. NuoDB is one of the NewSQL platforms is presented as a case study in the next section.

4.1.6 Case Study on NuoDB

NuoDB is one of the NewSQL platforms for data analysis is presented in this section describing the installation and some of the query examples related to it.

4.1.6.1 Installing NuoDB on Windows

NuoDB is platform independent and can be installed on Windows as well as Linux systems. Most of the production systems use NuoDB on Linux systems. It offers free community edition as well as professional or enterprise edition. The only requirement for NuoDB installation is Java run–time environment (JRE). JRE has to be installed based on the 32-bit or 64-bit machines usage to work with NuoDB. The links for installation of NuoDB are in Ref. [23].

4.1.6.2 Creating a Database in NuoDB

An automation console is provided by NuoDB for facilitating in creating a database and manage operations on it. A GUI system tray is provided from NuoDB or the address http://localhost:8888 can be accessed for creating a database. A log in window is prompted for sign in as shown in Fig. 7. Once, the user is signed in automation console appears as shown in Fig. 8.

Fig. 7 Log in window in NuoDB.

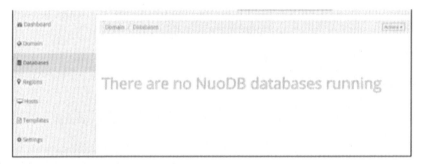

Fig. 8 Automation console in NuoDB.

The following steps are followed for creating a database in NuoDB:

- "Databases" option is selected in the left panel of the automation console and "Actions|Add Database" option can be selected for creating a database.
- A dialog box for creating managed database appears as shown in Fig. 9.
- The details such as database name, user, and password can be filled leaving the defaults. Once all the details are filled, database can be created using the submit option and database name with active status is created as shown in Fig. 10.

NuoDB includes one broker, one storage manager (SM), one transaction engine (TE), and one storage group. The automation console provides a brief description of the database created showing the distributed regions as shown in Figs. 11 and 12. It also helps in viewing the memory usage of each of the databases created in NuoDB as shown in Fig. 13.

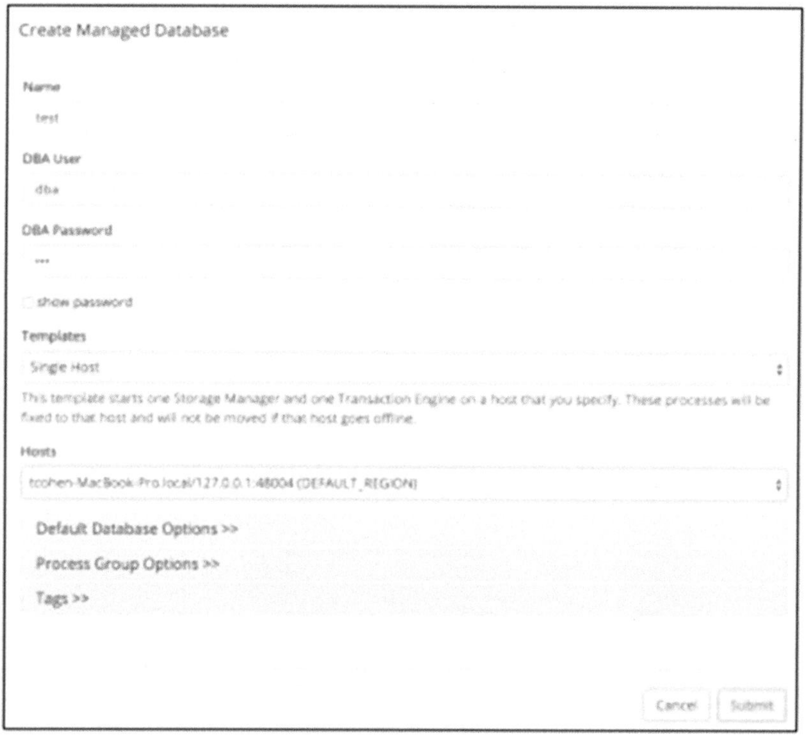

Create Managed Database

Name

test

DBA User

dba

DBA Password

***

show password

Templates

Single Host   ⇕

This template starts one Storage Manager and one Transaction Engine on a host that you specify. These processes will be fixed to that host and will not be moved if that host goes offline.

Hosts

tcohen-MacBook-Pro.local/127.0.0.1:48004 (DEFAULT_REGION)   ⇕

Default Database Options >>

Process Group Options >>

Tags >>

Cancel   Submit

Fig. 9 Create database window in NuoDB.

Fig. 10 Active status of database in NuoDB.

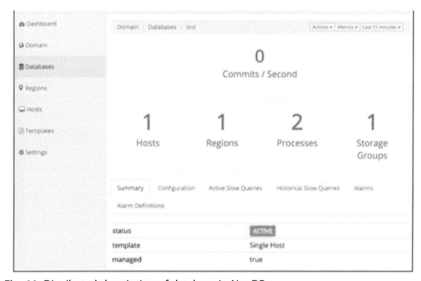

Fig. 11 Distributed description of database in NuoDB.

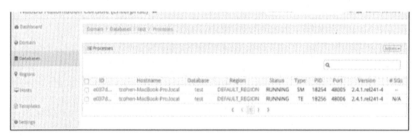

Fig. 12 Region processes of a database in NuoDB.

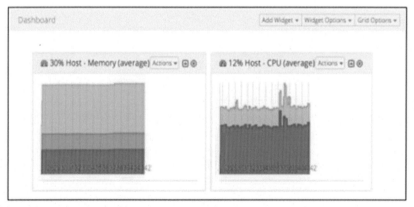

Fig. 13 Memory usage of databases in NuoDB.

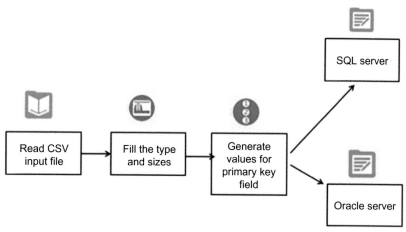

Fig. 14 Creating table in Alteryx.

5. CASE STUDY ON ALTERYX (DEMONSTRATION/ INSTALLATION, CREATION OF DATABASE/RECORD)

Alteryx[®] [24] software provides solutions for designing solutions to analytics problems and provides analytics as a service. In the back-end it uses database systems to provide analytics as a service. In this section, an overview of Alteryx designer is presented with creation of database and records.

Alteryx[®] designer provides a workflow approach for creating a database and table. Precreate SQL and Postcreate SQL options can be used in the input data tools and output data tools to create a table. The basic workflow for creating a table with primary key is as shown in Fig. 14. If the table needs to be created from a csv source file, the file is imported with import functionality, optimized, and primary keys are created with SQL.

5.1 Creating Table in Alteryx

To create a table in Alteryx, the following steps are followed with the help of alteryx designer workflow. In this example a student table is created with USN as the primary key.

- Primary Key for a new table.
- When creating a new table in Alteryx and then saving it on the database, the easiest way is to save the table first and then "alter" the table using Postcreate SQL to set the Primary Key as shown in Figs. 15 and 16.

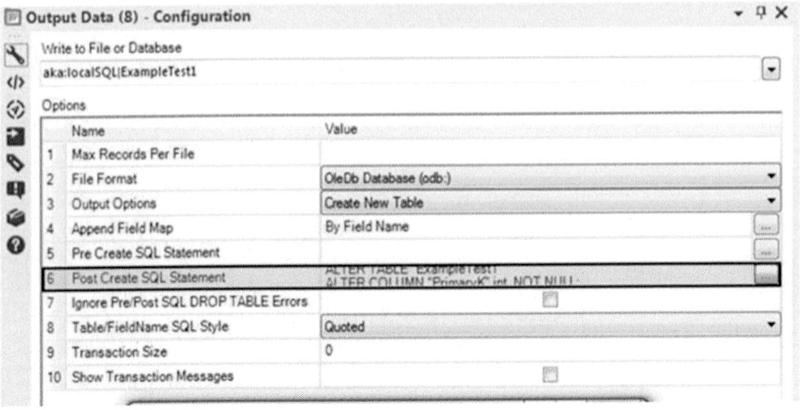

Fig. 15 New table in Alteryx.

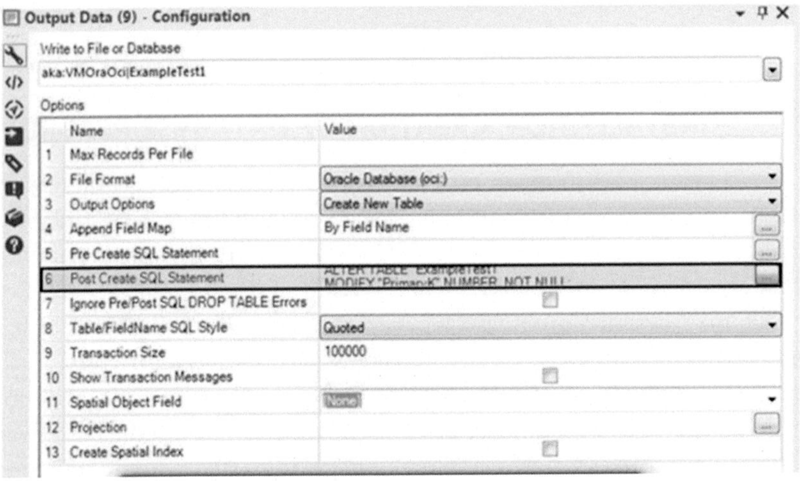

Fig. 16 New table with postcreate SQL in Alteryx.

- "Student_test" is the name of the table; replace it with the name of your table.
- "USN" is the name of field that you want to make as Primary Key.

The expressions for SQL server and Oracle version are as shown below.

For SQL Server—Expression 1a:

```
ALTER TABLE Student_test
ALTER COLUMN USN int NOT NULL;

ALTER TABLE Student_test
ADD PRIMARY KEY (USN);
```

For Oracle (10+)—Expression 1b:

```
ALTER TABLE "Student_test"
MODIFY "USN" NUMBER NOT NULL;

ALTER TABLE "Student_test"
ADD CONSTRAINT Student_pk PRIMARY KEY ("USN");
```

6. CONCLUSIONS

In-memory data management and analytics become an interesting research area both in the field of academics and industry as the memory will become the new disk for storage and analysis. This shift of the memory from disk to main memory leads to wide area of research opportunities and lot of improvement in response time and throughput. However, rethinking of the design in the databases for data layouts, indices, parallelism, concurrency, query execution, processing, etc., needs to address carefully. In this chapter, the focus was on the key design principles for in-memory database and processing. As discussed in Sections 2 and 3, in-memory database use main memory directly to store the data and for further analysis. Buffer management is not necessary in the case of in-memory databases as the data becomes more durable with persistent mechanisms. The data layout used by most of the in-memory databases use a columnar format so as to minimize the access time for getting the data.

REFERENCES

[1] G. Feuerlicht, J. Pokorny, Can relational DBMS scale-up to the cloud? in: R.J. Pooley, J. Coady, H. Linger, C. Barry, M. Lang (Eds.), Information Systems Development—Reflections, Challenges and New Directions, Springer, Berlin, 2012.

[2] D. Intersimone, The end of SQL and relational database? (Part 2 of 3), Computerworld (2010) 10 February, available at http://blogs.computerworld.com/15556/the_end_sql_and_ relational_database_part_2__3. Accessed 30 July 2012.

[3] E.A. Brewer, CAP twelve years later: how the 'rules' have changed, Computer 45 (2) (2012) 22–29.

[4] H. Zhang, G. Chen, B.C. Ooi, K.L. Tan, M. Zhang, In-memory big data management and processing: a survey, IEEE Trans. Knowl. Data Eng. 27 (7) (2015) 1920–1948.

[5] C. Binnig, S. Hildenbrand, F. Färber, Dictionary-based order-preserving string compression for main memory column stores, in: Proceedings of the 2009 ACM SIGMOD International Conference on Management of Data, ACM, 2009, June, pp. 283–296.

[6] J.J. Levandoski, P.Å. Larson, R. Stoica, Identifying hot and cold data in main-memory databases, in: 2013 IEEE 29th International Conference on Data Engineering (ICDE), IEEE, 2013, April, pp. 26–37.

[7] V. Sikka, F. Färber, W. Lehner, S.K. Cha, T. Peh, C. Bornhövd, Efficient transaction processing in SAP HANA database: the end of a column store myth, in: Proceedings of

the 2012 ACM SIGMOD International Conference on Management of Data, ACM, 2012, May, pp. 731–742.

[8] M. Kaufmann, D. Kossmann, Storing and processing temporal data in a main memory column store, Proc. VLDB Endowment 6 (12) (2013) 1444–1449.

[9] J. Pokorny´, Databases in the 3rd millennium: trends and research directions, J. Syst. Integr. 1 (1/2) (2010) 3–15.

[10] G.W. Burr, B.N. Kurdi, J.C. Scott, C.H. Lam, K. Gopalakrishnan, R.S. Shenoy, Overview of candidate device technologies for storage-class memory, IBM J. Res. Dev. 52 (4.5) (2008) 449–464.

[11] G.W. Burr, M.J. Breitwisch, M. Franceschini, D. Garetto, K. Gopalakrishnan, B. Jackson, B. Rajendran, Phase change memory technology, J. Vac. Sci. Technol., B: Nanotechnol. Microelectron.: Mater., Process., Meas., Phenom. 28 (2) (2010) 223–262.

[12] J.J. Yang, R.S. Williams, Memristive devices in computing system: promises and challenges, ACM J. Emerg. Technol. Comput. 9 (2) (2013) 11.

[13] S. Chen, P.B. Gibbons, S. Nath, in: Rethinking database algorithms for phase change emory, Proc. CIDR, 2011, pp. 21–31.

[14] S.M. Rumble, D. Ongaro, R. Stutsman, M. Rosenblum, J.K. Ousterhout, It's time for low latency, in: HotOS, vol. 13, 2011, p. 11.

[15] A. Kipf, V. Pandey, J. Böttcher, L. Braun, T. Neumann, A. Kemper, Analytics on fast data: main-memory database systems versus modern streaming systems, in: EDBT, 2017, pp. 49–60.

[16] J. Pokorny, in: NoSQL databases: a step to database scalability in web environment, Proc. of the 13th Int. Conf. on Information Integration and Web-Based Applications & Services (iiWAS) 2011, Ho Chi Minh City, Vietnam, ACM, New York, NY, 2011, pp. 278–283.

[17] R. Hecht, S. Jablonski, NoSQL evaluation: a use case oriented survey, in: 2011 International Conference on Cloud and Service Computing (CSC), IEEE, 2011, December, pp. 336–341.

[18] J. Ousterhout, P. Agrawal, D. Erickson, C. Kozyrakis, J. Leverich, D. Mazières, S.M. Rumble, The case for RAMClouds: scalable high-performance storage entirely in DRAM, SIGOPS Oper. Syst. Rev. 43 (4) (2010) 92–105.

[19] T. Mühlbauer, W. Rödiger, A. Reiser, A. Kemper, T. Neumann, ScyPer: elastic OLAP throughput on transactional data, in: Proceedings of the Second Workshop on Data Analytics in the Cloud, ACM, 2013, June, pp. 11–15.

[20] T. Mühlbauer, W. Rödiger, R. Seilbeck, A. Reiser, A. Kemper, T. Neumann, Instant loading for main memory databases, Proc. VLDB Endowment 6 (14) (2013) 1702–1713.

[21] A. Ailamaki, D.J. DeWitt, M.D. Hill, M. Skounakis, Weaving relations for cache performance, in: VLDB, vol. 1, 2001, pp. 169–180.

[22] R. Cattell, Scalable SQL and NoSQL data stores, ACM SIGMOD Rec. 39 (4) (2010) 12–27.

[23] NuoDB, https://www.nuodb.com.

[24] Alteryx, https://www.alteryx.com.

ABOUT THE AUTHORS

Krishnarajanagar G. Srinivasa is currently working as Associate Professor, Department of Information Technology at CBP Government Engineering College, Delhi, India. He received his PhD in Computer Science and Engineering from Bangalore University in 2007. He is the recipient of the All India Council for Technical Education—Career Award for Young Teachers, Indian Society of Technical Education—ISGITS National Award for Best Research Work Done by Young Teachers, Institution of Engineers (India)—IEI Young Engineer Award in Computer Engineering, Rajarambapu Patil National Award for Promising Engineering Teacher Award from ISTE—2012, IMS Singapore—Visiting Scientist Fellowship Award. He has published more than one hundred research papers in International Conferences and Journals. He has visited many Universities abroad as a visiting researcher—University of Oklahoma, USA, Iowa State University, USA, Hong Kong University, Korean University, and National University of Singapore are a few prominent visits. He has authored two books, namely File Structures using C++ by TMH and Soft Computer for Data Mining Applications LNAI Series—Springer. He has been awarded BOYSCAST Fellowship by DST, for conducting collaborative Research with Clouds Laboratory in University of Melbourne in the area of Cloud Computing. He is the principal Investigator for many funded projects from UGC, DRDO, and DST. His research areas include Data Mining, Machine Learning, and Cloud Computing.

Srinidhi Hiriyannaiah is a Research Scholar at VTU. He received his Master of Technology in Software Engineering from M.S. Ramaiah Institute of Technology, Bengaluru (VTU). He is currently working as an Assistant Professor in Department of Computer Science and Engineering at Ramaiah Institute of Technology, Bengaluru. He previously worked at IBM India Software Labs, Bengaluru. His main area of interest includes studies related to parallel computing, big data and its applications, information management and software engineering for education.

CHAPTER SIX

NoSQL Hands On

Rebika Rai*, Prashant Chettri[†]
*Sikkim University, Samdur, Gangtok, East Sikkim, India
[†]HRDD South, Government of Sikkim, Mangan, South Sikkim, India

Contents

1. Introduction   158
2. Document-Oriented Databases   159
   2.1 Introduction to MongoDB   159
   2.2 Introduction to CouchDB   176
   2.3 Introduction to Cloudant   184
3. Graph-Based Databases   197
   3.1 Introduction to Neo4j   198
   3.2 Introduction to OrientDB   212
4. Key–Value Databases   226
   4.1 Introduction to Redis   228
   4.2 Introduction to InfluxDB   236
   4.3 Introduction to Aerospike   238
5. Hybrid SQL–NoSQL Databases   258
   5.1 Introduction to MariaDB   258
Key Terminology and Definitions   274
References   275
Further Reading   276
About the Authors   277

Abstract

With the expansion of huge and complex real-time data that is wandering across the internet today, the dimensions of data transmitted are escalating exponentially with each passing years. This makes working with standard database systems or on personal computers difficult because of its inability to handle outsized, unstructured, and complicated data. Various institutes store and use massive amounts of data which are further utilized for generating reports to guarantee stability regarding the services they propose. However, the challenge is how to analyze, capture, share, store, transfer, visualize, query, update, and finally manipulate an impressive volume of data that have to be delivered through the internet to reach its destination intact maintaining its information privacy. Almost all the applications developed using any programming languages require some external component to store and access data. The components for the same could be a local network, a cloud file, or even a database. While sources like the network and cloud file systems store the unstructured data, the structured data

Advances in Computers, Volume 109
ISSN 0065-2458
https://doi.org/10.1016/bs.adcom.2017.08.004

are usually stored in a typical relational database management system (RDBMS). Usually it is a time-consuming process to define, structure, distribute, and access data from RDBMS through SQL and hence, an alternative was developed for this called the NoSQL ("Non SQL," "Nonrelational," or "Not only SQL") database. NoSQL encompasses a wide variety of different database technologies that were developed in response to the demands presented in building modern applications wherein developers are working with applications that create massive volumes of new, rapidly changing data types structured, semistructured, unstructured, and polymorphic data. Applications that once served a finite audience are now delivered as services that must be always-on, accessible from many different devices, and scaled globally to millions of users. Organizations are now turning to scale-out architectures using open source software, commodity servers, and cloud computing instead of large monolithic servers and storage infrastructure. There are several categories of NoSQL databases. Each of these categories has its own specific attributes and limitations. There is not a single solution which is better than all the others; however, there are some databases that are better to solve specific problems. This chapter provides NoSQL hands on and attention has been paid to various types of NoSQL databases such as MongoDB, CloudDB, OrientDB, Cloudant, Neo4j, Redis, Aerospike, InfluxDB, and MariaDB, focusing on the details such as installation steps, setting environment, and various operations performed on various types of NoSQL databases.

1. INTRODUCTION

In the computing system, there are massive data that come out every day from the web. A large segment of these data are handled by relational database management systems (RDBMS). The relational model is well matched to client–server programming and today it is predominant technology for storing structured data in web and business applications. However, there are various types of data that may not required fixed schema, avoid some form of join operations, and typically scale horizontally likely known as unstructured data, RDBMS fails to handle so. As we are aware, it is painless to access and capture data through third parties such as Facebook, Google+, and others. Personal user information, social graphs, geo location data, user-generated content, and machine logging data are just a few examples where the data have been increasing exponentially [1]. To avail the above service properly, it is required to process huge amount of data and SQL databases were never designed for it. The evolution of NoSQL databases is to handle these huge data properly and efficiently. The most common categories are: document-oriented databases, graph-based databases, column-based databases, key–value databases, and hybrid SQL–NoSQL databases [2].

The technology/technical terms used in this chapter are explained at the "Key Terminology & Definitions" section.

2. DOCUMENT-ORIENTED DATABASES

In the year 1970s when relational database came into picture, data schema to be worked upon were reasonably elemental and simple wherein the data items were to be arranged as a set of formally described tables with rows and columns. But with the need to store volumes and variety of data (unstructured) in recent years, nonrelational database technologies (document-oriented, graph-based, column-based, key–value, and hybrid) [3] have emerged to address the requirement that allow data to be grouped together more naturally and logically. One of the most popular ways of storing data is a document-oriented database, basically employed for storing, managing, and retrieval of semistructured data where each record and its associated data are considered of as a "document." A document-oriented database is also termed as a document store or simple document, is one of the kind of NoSQL database [4]. The main advantage of storing data as documents is document being the independent unit that makes the processing faster and easier to distribute data across multiple servers while preserving its locality. Various document-oriented NoSQL databases are: BaseX, MongoDB, Cloudant, Couchbase server, CouchDB, DocumentDB, Informix, OrientDB, PostgreSQL, RethinkDB, SimpleDB, Terrastore, RavenDB, ThruDB, SisoDB, RaptorDB, and many more. Documents are the main concept in document databases. It stores the data in JSON like documents having key–value pairs. MongoDB is a document database that stores data as a hierarchy of key–value pairs which allows branching at different levels (a maximum of three levels) and provide a rich query language [5] and constructs such as database, indices, etc., allowing for easier transition from relational databases.

In this section, three document-oriented NoSQL databases have been taken into account, i.e., MongoDB, CouchDB, and Cloudant.

2.1 Introduction to MongoDB

MongoDB is one of the leading open-source, cross-platform document-oriented NoSQL database written in C++, C, and JavaScript developed by a software company 10gen (MongoDB Inc.) and available in 32-bit and 64-bit versions for Windows and Unix-like environments. The word "Mongo" in its name MongoDB comes from the word humongous meaning enormous, referring to its ease of use, performance, and vastly scalable prospective. It works on the concept of collection and document where a document is a set of key–value pairs and collection is a group of MongoDB documents [6]. The various platform supported by MongoDB are Windows Vista and later; Linux, OS X 10.7; and later, Solaris and FreeBSD. The software company began developing MongoDB in 2007 as a component of a planned platform as a service product. In 2009, the company shifted to an open source development model, with the company offering commercial support and other services. In 2013, 10gen changed its name to MongoDB Inc.

2.1.1 Why and Where to Use MongoDB?

MongoDB is not a substitute for any traditional RDMS databases, however, considering the fact that the MongoDB stores the data in the form of BSON (binary representation of JSON) which provides the scaling feature to the application to be developed, developers might choose MongoDB over traditional RDBMS [7] based on factors for their application development such as:

(a) *Data insert consistency*: If there is a requirement of writing huge amount of data without losing some data, then MongoDB is the best suited because of its high data insertion rate.

(b) *Data corruption recovery*: In MongoDB, data repair can be done on a database level and there is an automatic command to do so. However, the command reads all the data and rewrites it to a new set of files which may be time consuming if the database is huge but it is preferred over losing the entire dataset.

(c) *Load balancing*: MongoDB supports faster replication and automatic load balancing configuration because of data placed in shards.

(d) *Avoid joins*: If the developer wants to avoid normalization and joins, MongoDB is the best suited.

(e) *Changing schema*: MongoDB is schema-less, hence adding new fields will not result in any issues.

(f) *Not relational data*: If the data to be stored need not to be a relational one, then MongoDB should be selected.

(g) *Mapping*: If the mapping of application data objects is to be done directly into the document-based storage, MongoDB is the best suited.

(h) *Creating database cluster*: If the database is geographically distributed and the user wishes to create cluster and speed up data queries among remotely located databases, MongoDB is suitable.

MongoDB can be used to manage the following:

(a) *Big Data/Content management*: Big Data do not just deal with numbers, dates, and strings but also comprises of audio, video, unstructured text, log files, 3D data, etc. Traditional database systems were designed with an objective to cater to lesser amount of data (structured) and function on a single server, making increase in the data capacity costly and restricted. As applications have emerged to address large volumes of users and varieties of data, the traditional use of the relational database has become burden rather than an enabling factor for organizations. MongoDB solves these problems and provides companies with the means to form remarkable business value.

(b) *User data management/Social infrastructure*: Users are the core component of any business and storing data related to users are fundamental theme of most of the applications developed by organizations. MongoDB provides the platform for accessing and updating user data.

(c) *Data hub*: Organizations generate massive amount of data and have a variety of tools to analyze, process, summarize, and monetize information in different locations. MongoDB is preferred by organizations to serve as the central data hub for all of their data because of its ease of use, scalability, low cost of ownership, and a universal repository in which data can easily be stored, processed, and served to other applications.

2.1.2 Features of MongoDB

There are many key features of MongoDB that make it a preferred database when approaching modern web application developments. Some of the features of MongoDB are listed below.

(a) MongoDB is easily installable, free, and open-source, published under the GNU Affero General Public License.

(b) MongoDB is a distributed database at its core, so high availability, horizontal scaling, and geographic distribution are built in and easy to use.

(c) MongoDB supports rich query for all the major operations further providing good text search features.

(d) It stores data in flexible, JSON-like documents, meaning fields can vary from document to document and data structure can be changed over time. The document query language supported by MongoDB plays a vital role in supporting dynamic queries.

(e) MongoDB enables substituting the complete document (database) or some specific fields in the document with the help of command called update( ).

(f) An index on any attribute of a MongoDB record can be easily set with respect to which, a record can be instantly sorted or ordered.

(g) The data in the document can be easily accessed and analyzed using the concept of Ad hoc queries, indexing, and real-time aggregation.

(h) The document model maps to the objects in the application code, making data easy to work with thereby making no need of mapping the application objects to the data objects explicitly.

(i) Performance tuning is absolutely easy compared with any of the relational databases.

(j) It enables faster access of the data due to its nature of using the internal memory for the storage.

(k) GridFS specification of MongoDB supports storage of very large files. However, if load increases that require more of the processing power or storage space; it can be distributed to other nodes across computer networks using the mechanism termed as Sharding. It has an automatic load balancing configuration because of data placed in shards.

(l) The various programming languages are supported by MongoDB such as C, C++, Java, JavaScript, PHP, Python, Haskell, C#, Perl, Ruby, Scala, etc.

(m) MongoDB permits the developer to make use of the single programming language both at Client as well as Server side.

(n) MongoDB supports Master–Slave replication.

(o) MongoDB supports fixed-size collections using command db. createCollection called capped collections to maintain insertion order and once the specified size is reached, it starts behaving like a circular queue.

(p) MongoDB management service (MMS) is a powerful web tool that allows us tracking databases, machines, and also backing up the data. MMS also tracks hardware metrics for managing a MongoDB deployment. It shows performance in a rich web console to help user optimize their deployment. It also provides features of custom alerts which help to discover issues before the MongoDB instance will be affected.

(q) MongoDB supports multiple storage engines, such as Wired Tiger storage engine and MMAPv1 storage engine. Storage engines manages how data are saved in memory and on disk.

(r) MongoDB supports for geospatial indices. It allows users to store x- and y-coordinates within documents. To find the documents which are \$within a radius or \$near a set of coordinates.

2.1.3 Relationship of RDBMS Terminology With MongoDB

Relational database was proposed with an objective to endow the user with a layer of abstraction. Data model of RDBMS is Tables/Relations, with primary key to uniquely identify a record and foreign key to inter-link the tables thereby performing join on tables. RDBMS has several advantages such as abstraction, multiuser access, automatic optimization for searching, ACID properties enabling transaction support in extremely easy querying language, etc. However, as the data grow exponentially, scalability became a major issue for RDBMS leading to the introduction of nonrelational database. The major difference between MongoDB (NoSQL) and relational database is the way they handle data [8]. In relational databases, data are stored in form of traditional two-dimensional rows and column formation with a typical schema design comprising of several number of tables and relationship between them while MongoDB allows storage of any type of data with no concept of relationship. The differences between MongoDB and other SQL databases have been tabulated in Table 1. The terminology used in RDBMS and MongoDB have been tabulated in Table 2.

2.1.4 Advantages of MongoDB Over RDBMS

MongoDB has gained a significant importance and as of now it is one of the prominently utilized databases [9]. The various advantages of MongoDB over RDBMS are as follows:

(a) *Schema less*: It is a document database in which one collection holds different documents. Number of fields, content, and size of the document can differ from one document to another.

(b) No complex joins used in MongoDB.

(c) MongoDB supports dynamic queries on documents using a document-based query language that's nearly as powerful as SQL.

(d) Conversion/mapping of application objects to database objects not needed.

(e) Uses internal memory for storing the (windowed) working set, enabling faster access of data.

(f) Performance tuning is absolutely easy compared to any of the relational databases.

(g) MongoDB supports storage of very large files.

Table 1 Difference Between SQL Database and MongoDB

Parameters	SQL Database	MongoDB
Database	Relational database	Nonrelational database
Language	Structured query language (SQL)	JSON query language
Document structure	Table based	Collection based (Key–Value pair)
	Row based	Document based
	Column based	Field based
Key support	Foreign key	No support for foreign key
Triggers	Support for triggers	No support for triggers
Schema	Predefined schema	Dynamic schema
Hierarchical data storage	Not fit for hierarchical data storage	Best fit for hierarchical data storage
Scalability	Vertically scalable (increase RAM)	Horizontally scalable (add servers)
Properties/ Theorem supported	ACID (atomicity, consistency, isolation, and durability)	CAP theorem (consistency, availability, and partition tolerance)

Table 2 Terminology Used in RDBMS and MongoDB

RDBMS	MongoDB
Database	Database
Tuple/Row	Document
Table/Relation	Collection
Column/Attribute	Field
Table join	Embedded documents
Primary key	Primary key (Key_id by default provided)

(h) MongoDB also provides pluggable storage engine API that allows third parties to develop storage engines for MongoDB.

(i) MongoDB can run over multiple servers as the data are duplicated to keep the system up and also keep its running condition in case of hardware failure.

(j) MongoDB supports horizontal scalability by adding more servers.

(k) MongoDB is the best fit for hierarchical data storage.

(l) MongoDB is easy to scale by adding commodity hardware.

2.1.5 Installation Steps of MongoDB

The installers for MongoDB are available in both the 32-bit and 64-bit formats. The 32-bit installers are the best suited for development and test environments; 64-bit installers are recommended for production environments in regard to the amount of data that can be stored within MongoDB. The various platform supported by MongoDB are Amazon Linux 2013.03 and later; Debian 7 and 8; Ubuntu 12.04, 14.04, and 16.04; Windows Vista and later; OS X 10.7 and later; Windows Server 2008R2 and later; SLES 11 and 12; RHEL/CentOS 7.0 and later; RHEL/CentOS 6.2 and later; Solaris 11 64-bit, etc., and some have been tabulated in Table 3. Furthermore, this section highlights the installation steps of MongoDB. MongoDB builds that is available for Windows are:

(a) MongoDB for Windows 64-bit runs only on Windows Server 2008 R2, Windows 7 64-bit, and newer versions of Windows. This build takes advantage of recent enhancements to the Windows Platform and cannot operate on older versions of Windows.

(b) MongoDB for Windows 64-bit Legacy runs on Windows Vista, and Windows Server 2008, and does not include recent performance enhancements.

The various steps involved to install MongoDB on Windows have been highlighted below:

(a) Download MongoDB from official MongoDB website. Choose Windows 32 bits or 64 bits as per the requirements. Unzip the downloaded folder, extracts to the preferred location. Click on the run button to start installation as depicted in Fig. 1.

(b) Click on the "Next" button as depicted in Fig. 2.

(c) Click on the "Next" button to agree to the End User License Agreement as depicted in Fig. 3.

Table 3 Some Platforms Supported by MongoDB

Platform	3.4 Community and Enterprise	3.2 Community and Enterprise	3.0 Community and Enterprise
Debian 7 and 8	Yes	Yes	Yes
Ubuntu 12.04, 14.04, 16.04	Yes	Yes	Yes
Windows Vista and Later	Yes	Yes	Yes
OS X 10.7 and later	Yes	Yes	Yes

Fig. 1 Installation step 1 for MongoDB.

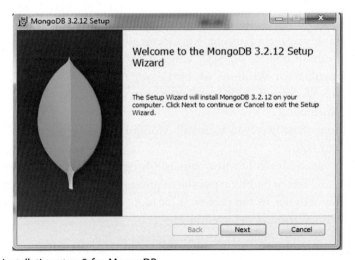

Fig. 2 Installation step 2 for MongoDB.

(d) Click on the "Complete" button to install all of the components as depicted in Fig. 4.

(e) Furthermore, click on "Install" button to start the installation as depicted in Fig. 5.

(f) Installation starts and once done click on the "Finish" button as depicted in Figs. 6 and 7, respectively.

(g) Check the folder "MongoDB" in C:\ProgramFiles as depicted in Fig. 8.

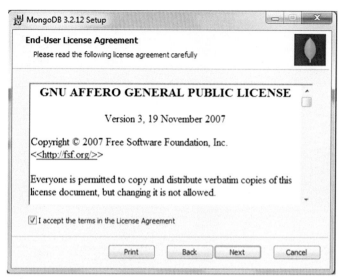

Fig. 3 Installation step 3 for MongoDB.

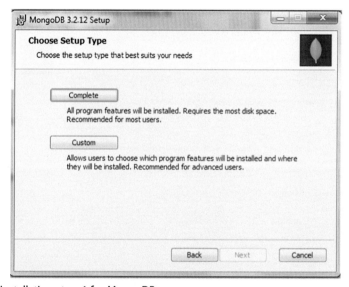

Fig. 4 Installation step 4 for MongoDB.

2.1.6 Setting Environment for MongoDB

(a) After successfully installing MongoDB as specified above, the user need to create directories for database and log files as depicted in Figs. 9 and 10.

(b) Add $MongoDB/bin to Windows environment variable, so that user can access the MongoDB's commands in command prompt easily.

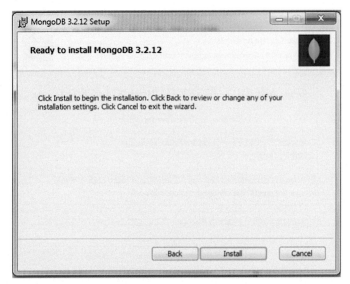

Fig. 5 Installation step 5 for MongoDB.

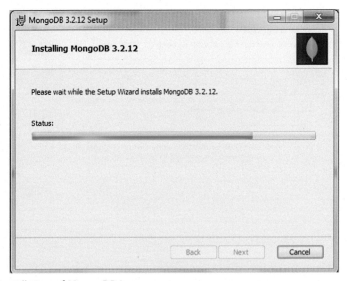

Fig. 6 Installation of MongoDB in progress.

In MongoDB, it contains only 10+ executable files (exe) in the bin folder, check files under $MongoDB\bin folder as depicted in Fig. 11.

(c) To start MongoDB, run **mongod.exe**. For example from the Command Prompt: C:\ProgramFiles\MongoDB\bin\mongod.exe. This

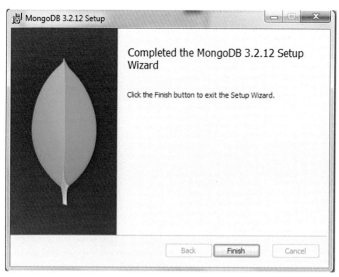

Fig. 7 Final step for installation of MongoDB.

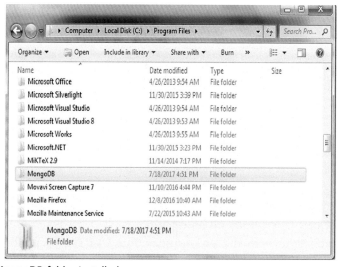

Fig. 8 MongoDB folder installed.

starts the main MongoDB database process. The waiting for connections message in the console output indicates that the mongod.exe process is running successfully.

(d) To connect to MongoDb, run mongo.exe. For example from the Command Prompt: C:\ProgramFiles\MongoDB\bin\mongo.exe.

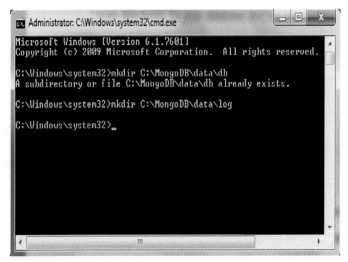

Fig. 9 Creating directories for database and log files.

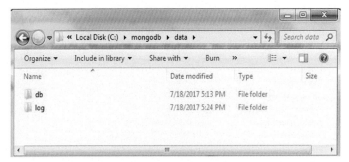

Fig. 10 Created directories for database and log files.

2.1.7 Various Operations on MongoDB

This section deals with the various operations that can be performed on MongoDB.

(a) The **use** command to define database.

This command will return the existing database if the database already exists else will create rather define a new database [5]. By issuing use command it enables switching from default database to defined database.

Syntax:

```
use Database_Name
```

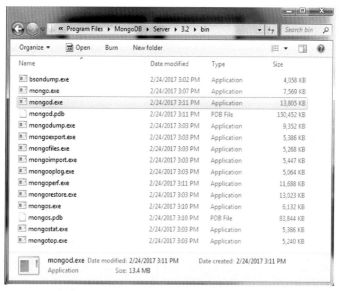

Fig. 11 Files under $MongoDB\bin folder.

Example:

`>use Rebikadb`

[*Note*: This command will switch from default database "db" to "Rebikadb" that is defined. However, MangoDB does not create Rebikadb on issuing the command, until user saves something inside the defined database.]

(b) The **show** command to show all existing database.

This command will display all the existing and available databases.

Syntax:

`show dbs`

[*Note*: If the newly defined database Rebikadb is not displayed as the outcome of show command or is not present in the displayed list, the user needs to insert at least one document into the database created.]

For checking the currently selected database, the command used is **db** that will display the current database that has been selected, for example Rebikadb.

Example:

`>db`

(c) The **dropDatabase** command to drop existing database.

This command will drop the existing database. It is used to delete the selected database and if the user has not selected any database, then this command will delete the default database called as "test."

Syntax:

```
db.dropDatabase( )
```

Example:

```
>use Rebikadb
Switched to db Rebikadb
>db.dropDatabase( )
```

Further when the user use the show dbs command, Rebikadb will not appear in the list displayed.

(d) The **createCollection( )** command to create collection.

This command is used to create collection. In MongoDB, databases hold collections of documents.

Syntax:

```
db.createCollection(name, options)
```

where **name** is name of collection to be created of type string and **options** specify configuration of collection such as memory size, indexing, and type of document. However, **options** is an optional parameter.

Example:

```
>db.createCollection("Student") or
>db.createCollection("Student", { capped : true, autoIndexId :
true, size : 6142800, max : 10000 } )
```

[*Note*: **options**, i.e., the second parameter as mentioned is optional. However, if the user wishes to specify, the various options that can be used are: capped, autoIndexId, size, and max. Here, capped, autoIndexedId are of type Boolean and size, max are of type number.]

In MongoDB, user need not create collection. MongoDB creates collection automatically, when user insert some document and specifying the collection name at that instance works.

(e) The **show collections** command to show collections.

For checking the created collection the command used is **show collections** that will display all the collections that exist.

Example:

```
>show collections
```

(f) The **insert( )** command to insert data.

This command enables the user to insert data into MongoDB collection. This command can be used to insert single document or multiple documents in a single query.

Syntax: For inserting single document
```
db.Collection_Name.insert(document)
```
Example:
```
> db.Student.insert({
_id: ObjectId(7df78ad8902c),
Name: "Prashant Chettri",
Age: 22,
RegnNo: "301123"
Address: { City: "Namchi",
           State: "Sikkim",
           Country" "India"
         },
 Course: { CourseName: 'Computer Engineering',
           CourseDuration: "3 Years"
         }
         })
```

Here **Student** is the collection name, as created above using command **createCollections**. If the collection does not exist in the database, then MongoDB will create this collection, and then insert a document into it. In the inserted document, if user does not specify the _id parameter, then MongoDB assigns a unique ObjectId for this document. _id is 12 bytes hexadecimal number unique for every document in a collection. 12 bytes are divided as follows:

_id: ObjectId (4 bytes timestamp, 3 bytes machine id, 2 bytes process id, 3 bytes incrementer).

To insert multiple documents in a single query, user can pass an array of documents in insert( ) command.

Example:
```
>db.Student.insert([.
{
Name: "Prashant Chettri",
Age: 22,
RegnNo: "301,123".
Address: {.
     City: "Namchi",
     State: "Sikkim",
     Country" "India"
     },
```

```
Course: {
      CourseName: "B.Tech",
      CourseDuration: "4 Years"
      }
},
{
Name: "Rebika Rai",
Age: 21,
RegnNo: "301124"
Address: {
      City: "Gangtok",
      State: "Sikkim",
      Country" "India"
      },
Course: {
      CourseName: 'MCA',
      CourseDuration: "2 Years"
      }
}
])
```

(g) The **find( )** command to query data from MongoDB collection.
This command permits the user to display all the documents in a unstructured way by retrieving data from MongoDB collection as specified in the query. Queries can return all documents in a collection or only the documents that match a specified criterion. User can specify the criteria in a document and pass as a parameter to the **find( )** method.

Syntax: To return all documents in a collection
```
db.Collection_Name.find( )
```
Example:
```
>db.Student.find( )
```
This query will return all the documents in the collection Student.

Syntax: To return specified document in a collection based on the criteria
```
db.Collection_Name.find({ <field1>: <value1>, <field2>: <value2>,
    ... })
```
Example:
```
>db.Student.find({Name: 'Prashant Chettri'})
```
The result set includes only the matching documents.

(h) The **update( )** command to update document in a collection.

This command is used in order to update specified values of the collection in MongoDB.

Syntax:

```
db.Collection_Name.update( )
```

The **update( )** command specified above will take the fieldname and the new value as an argument to update the document.

Example:

```
>db.Student.update(
                {
RegNo: "301124"
                },
                $set:
                {
                Name: "Deeyan Chettri"
                }
                )
```

This query will update the attribute **Name** of the collection **Student** for the document with **RegNo 301124**.

(i) The **remove ( )** command to delete documents of a collection.

The command **remove ( )** deletes a single document or all documents that match a specified condition.

Syntax:

```
db.Collection_Name.remove( )
```

Example:

```
>db.Student.remove({Name: "Prashant Chettri"})
```

The query will remove all the collections with Name: Prashant Chettri. The user has the flexibility to delete only the first collection if there are multiple collections in the database, the user need to set parameter in **remove( )** method. For example:

```
>db.Student.remove({Name: "Prashant Chettri"},1)
```

However, if the user does not specify deletion criteria, the MongoDb will delete all the collections from the database. For example:

```
> db.Student.remove( )
```

(j) The **drop( )** command to drop collection.

This command permits the user to drop the specified collection from the database. The user may first switch to the particular database using **use** command and further check the available connections in the particular database using **show** command and drop the connection.

Syntax:

```
db.Collection_Name.drop( )
```

Example:

```
db.Student.drop( )
```

The query will drop the collection Student from the database and will return true, if the selected collection is dropped successfully, otherwise it will return false.

2.2 Introduction to CouchDB

CouchDB is an acronym for Cluster Of Unreliable Commodity Hardware DataBase. CouchDB is NoSQL open source document stores database software developed by Apache software foundation, implemented in the concurrency-oriented language Erlang, and that focuses on ease of use providing an architecture that "completely embraces the web." It uses JSON to store data, JavaScript as its query language using MapReduce, and HTTP for an API. CouchDB was first released in 2005 started by Damien Katzand later became an Apache Software Foundation project in 2008. The distributed nature and flat address space of the database will enable node partitioning for storage scalability (with a map/reduce style query facility) and clustering for reliability and fault tolerance. The platform supported by CouchDB includes most POSIX systems, such as GNU/Linux and OS X. Aslo, Windows is officially supported.

2.2.1 Why and Where to Use CouchDB?

The user need not worry about the structure of data as the data are stored in the flexible document-based structure in CouchDB [10]. Users are provided with powerful data mapping, which allows querying, combining, and filtering the information. CouchDB provides easy-to-use replication, using which you can copy, share, and synchronize the data between databases and machines. It has an HTTP-based REST API, which helps to communicate with the database easily. And the simple structure of HTTP resources and methods (GET, PUT, and DELETE) are easy to understand and use. It is designed to handle changeable traffic graciously. CouchDB will absorb a lot of concurrent requests without falling over wherein all the requests all get answered. When the spike is over, CouchDB will work with regular speed again. Furthermore, it imposes a set of bounds on the programmer thereby allowing the programmer to create applications that could not deal with scaling up or down of the hardware of an application. Its internal architecture is fault-tolerant, and failures occur in a controlled environment and are dealt with gracefully. Single problems do not cascade through an entire server system but stay isolated in single requests.

CouchDB is suitable for numerous areas of an application due to its characteristics such as incremental MapReduce and replication [11]. However, it is the best suited for applications such as online interactive document and data management tasks. These are the sort of workloads experienced by the majority of web applications and further when combined with CouchDB's HTTP interface makes it considerably more appropriate for the web.

2.2.2 Features of CouchDB

CloudDB comprises of several features that makes it one of the most suited and acceptable NoSQL databases. Some features are listed here under.

(a) *Document storage*: CouchDB is a document storage NoSQL database wherein documents are the primary unit of data, each field is uniquely named and contain values of various data types such as text, number, Boolean, lists, etc. In these documents, there is no set limit to text size or element count. CouchDB provides an API called RESTful HTTP API for reading and updating (add, edit, and delete) database documents.

(b) *ACID semantics*: CouchDB provides ACID semantics and it does this by implementing a form of multiversion concurrency control (MVCC), meaning that CouchDB can handle a high volume of concurrent readers and writers without conflict.

(c) *Compaction*: Compaction is an operation to avail extra disk space for the database by removing unused data. While performing compaction operation on a particular file, a file with the extension .compaction is created and all the active/actual data are copied (cloned) to that file, when the copying process is finished then the old file is discarded. The database remains online during the compaction and all updates and reads are allowed to complete successfully. CouchDB allows compaction.

(d) *Views*: To solve the problem of filtering, organizing, and reporting on data that have not been decomposed into tables, CouchDB provides a view model. Views are the method of aggregating and reporting on the documents in a database and are built on-demand to aggregate, join, and report on database documents. Because views are built dynamically and do not affect the underlying document, user can have as many different view representations of the same data as desired.

(e) *Built for offline*: CouchDB can replicate to devices (like smartphones) that can go offline and handle data sync for you when the device is back online.

(f) CouchDB guarantees eventual consistency to be able to provide both availability and partition tolerance.

(g) CouchDB was designed with bidirection replication (or synchronization) and offline operation in mind. That means multiple replicas can have their own copies of the same data, modify it, and then sync those changes at a later time.

(h) CouchDB stores data as "documents," as one or more field/value pairs expressed as JSON. Field values can be simple things like strings, numbers, or dates; but ordered lists and associative arrays can also be used. Every document in a CouchDB database has a unique id and there is no required document schema.

(i) *Map/Reduce views and indices*: The stored data are structured using views. In CouchDB, each view is constructed by a JavaScript function that acts as the Map half of a map/reduce operation. The function takes a document and transforms it into a single value that it returns. CouchDB can index views and keep those indices updated as documents are added, removed, or updated.

(j) *HTTP API*: All items have a unique URI that gets exposed via HTTP. It uses the HTTP methods POST, GET, PUT, and DELETE for the four basic CRUD (Create, Read, Update, and Delete—CRUD) operations on all resources.

(k) CouchDB also offers a built-in administration interface accessible via Web called Futon.

2.2.3 Relationship of RDBMS Terminology With CouchDB

Unlike a relational database, a CouchDB database does not store data and relationships in tables. Instead, each database is a collection of independent documents. Each document maintains its own data and self-contained schema. An application may access multiple databases, such as one stored on a user's mobile phone and another on a server. Document metadata contains revision information, making it possible to merge any differences that may have occurred while the databases were disconnected. Terminology used in RDBMS and CouchDB has been listed in Table 4.

2.2.4 Advantages of CouchDB Over RDBMS

The various advantages of CouchDB over the RDBMS are as follows:

(a) CouchDB's design borrows heavily from web architecture and the concepts of resources, methods, and representations. It augments this

Table 4 Terminology Used in RDBMS and CouchDB

RDBMS	MongoDB
Database	Database
Table/Relation	Document
Column/Attribute	Field
Table join	Embedded documents
Primary key	Primary key (Key_id by default provided)

with powerful ways to query, map, combine, and filter your data. Add fault tolerance, extreme scalability, and incremental replication. It provides a different way to model data.

(b) CouchDB is a great fit for common applications like this because it embraces the natural idea of evolving, self-contained documents as the very core of its data model.

(c) CouchDB is a storage system useful on its own. You can build many applications with the tools CouchDB gives you. But CouchDB is designed with a bigger picture in mind. Its components can be used as building blocks that solve storage problems in slightly different ways for larger and more complex systems.

(d) CouchDB replication is one of these building blocks. Its fundamental function is to synchronize two or more CouchDB databases. This may sound simple, but the simplicity is key to allowing replication to solve a number of problems: reliably synchronize databases between multiple machines for redundant data storage; distribute data to a cluster of CouchDB instances that share a subset of the total number of requests that hit the cluster (load balancing); and distribute data between physically distant locations.

(e) The CouchDB API is designed to provide a convenient but thin wrapper around the database core.

(f) A table in a relational database is a single data structure. If user wants to modify a table example, update a row then the database system must ensure that nobody else is trying to update that row and that nobody can read from that row while it is being updated. The common way to handle this uses what's known as a lock. If multiple clients want to access a table, the first client gets the lock, making everybody else wait. Instead of locks, CouchDB uses MVCC to manage concurrent access to the database.

(g) CouchDB can validate documents using JavaScript functions similar to those used for MapReduce. Each time you try to modify a document, CouchDB will pass the validation function a copy of the existing document, a copy of the new document, and a collection of additional information, such as user authentication details. The validation function now has the opportunity to approve or deny the update.

2.2.5 Installation Steps of CouchDB
The installers for CouchDB are available in both the 32-bit and 64-bit formats. The 32-bit installers are the best suited for development and test environments; 64-bit installers are recommended for production environments in regard to the amount of data that can be stored within MongoDB. The various platform supported by MongoDB are Amazon Linux 2013.03 and later; Debian 7 and 8; Ubuntu 12.04, 14.04, and 16.04; Windows Vista and later; OS X 10.7 and later; Windows Server 2008R2 and later; SLES 11 and 12; RHEL/CentOS 7.0 and later; RHEL/CentOS 6.2 and later; Solaris 11 64-bit, etc.

(a) To download CouchDB, click on the link http://couchdb.apache.org from the official website of CouchDB. After, the link has been clicked, the homepage of the CouchDB is linked with as depicted in Fig. 12.

(b) Click on Windows (x64) for 64-bit or Windows (x86) for 32 bit depending on the requirements of the user as depicted in Fig. 13.

(c) Click on NEXT button to continue with the installation process as depicted in Fig. 14.

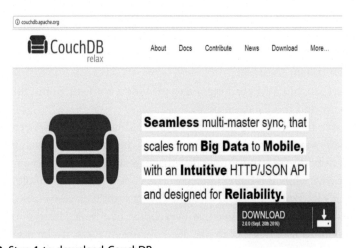

Fig. 12 Step 1 to download CouchDB.

Fig. 13 Step 2 to download CouchDB.

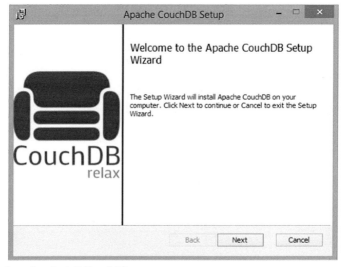

Fig. 14 Step 1 to install CouchDB.

(d) Accept the end user license agreement and click on NEXT button to continue with the download process as depicted in Fig. 15.

(e) Click next to install to the default folder or change to choose another folder as required as depicted in Fig. 16.

(f) Click on Install to begin the installation as depicted in Fig. 17.

(g) The installation now is in progress as shown in Fig. 18.

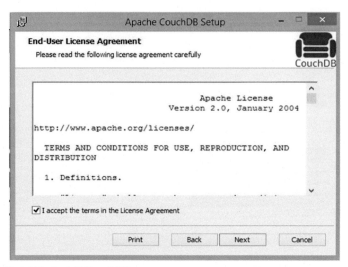

Fig. 15 Step 2 to install CouchDB.

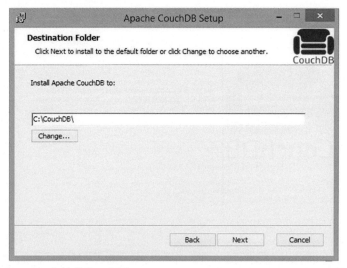

Fig. 16 Step 3 to install CouchDB.

(h) Click on Finish to complete the installation of CouchDB as depicted in Fig. 19.

2.2.6 Setting Environment for CouchDB
Open the browser and type in the url: 127.0.0.1:5984/_utils/ to get the page as depicted in Fig. 20.

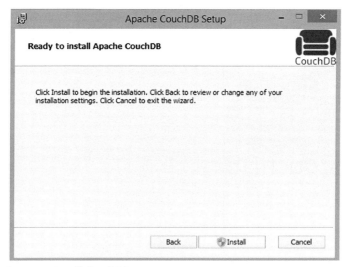

Fig. 17 Step 4 to install CouchDB.

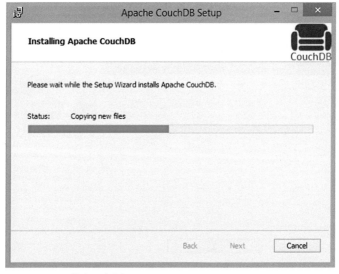

Fig. 18 Step 5 to install CouchDB.

2.2.7 Various Operations on CouchDB

(a) To create new database as depicted in Figs. 21 and 22.

(b) To delete the created database as depicted in Figs. 23 and 24.

(c) To add the documents to the database created as depicted in Figs. 25–31.

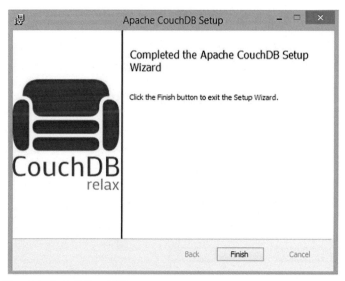

Fig. 19 Step 6 to install CouchDB.

Fig. 20 To start up with CouchDB.

2.3 Introduction to Cloudant

Cloudant is a document-oriented NoSQL database as a service (DBaaS), an IBM software product that scales globally, runs nonstop, and handles a wide variety of data types such as JSON, full-text, and geospatial, founded by Alan

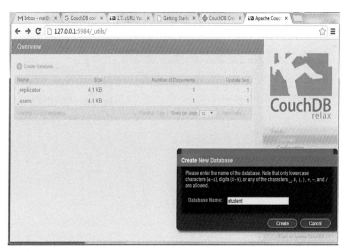

Fig. 21 Creating new database.

Fig. 22 New database created.

Hoffman, Adam Kocoloski, and Michael Miller. It scales databases on the CouchDB framework and provides hosting, administrative tools, analytics, and commercial support for CouchDB and BigCouch. Cloudant's distributed CouchDB service is used the same way as standalone CouchDB, with

Fig. 23 To delete database created.

Fig. 24 Delete database created.

the added advantage of data being redundantly distributed over multiple machines. Cloudant NoSQL DB provides access to a fully managed NoSQL JSON data layer that's always on. This service is compatible with CouchDB, and accessible through a simple to use HTTP interface for mobile and web application models.

Fig. 25 Creating document in the database.

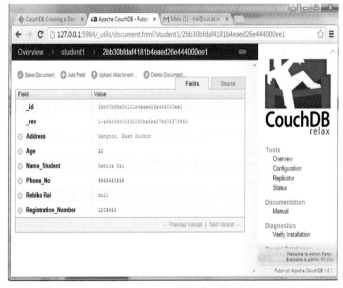

Fig. 26 Created fields with values associated in the database.

2.3.1 Why and Where to Use Cloudant?

Considering a scenario wherein the user builds in an application and is being used by multiple users. However, if the popularity of the application crashes the database resulting in loss of customers, downtime, and even loss of

Fig. 27 New fields and values added in the database.

Fig. 28 Updating value of the field in the database.

revenues, the developer will have to fix the database rather than concentrating on the application. Taking into account such scenario, the data persistence layer of the developed application can be resilient and highly accessible with IBM Cloudant that provides DBaaS. The data in Cloudant is steadily hosted and globally administered by Big Data experts all the way through.

Fig. 29 Edited value highlighted in the database.

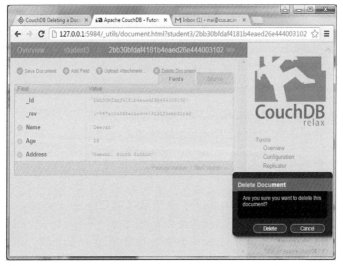

Fig. 30 Deleting the document from the database.

Above all, when the user sign up in order to use Cloudant, developers gets service level agreement (SLA) and application programming interface (API), and Cloudant engineers on the other hand handles and monitors all the ongoing databases and infrastructure so that the developer can just

Fig. 31 Uploading the attachment in the database.

concentrate on building an application rather than solving issues related to database. The Cloudant RESTful API makes every document in your Cloudant database accessible as JSON through a URL, which makes Cloudant so powerful for web applications.

2.3.2 Features of Cloudant

Cloudant is widely used by the developer for developing web applications wherein the requirements of the developer in terms of architecture might change or expand. The various features of Cloudant have been listed below.

(a) Cloudant provides a RESTful API to CRUD documents.

(b) Cloudant is designed to ensure that the flow of data between an application and its database remains uninterrupted and highly performant.

(c) It is designed to run across many servers in database clusters resulting in high-availability and fault tolerance.

(d) Unlike traditional database systems, all replicas of the data are available for both reads and writes. Also, it replicates and synchronizes data automatically and in few milliseconds.

(e) The core storage engine is based on Apache CouchDB, an open source JSon document store. Cloudant and CouchDB shares a common replication protocol with which with the click of a button developer can synchronize the copy of their Cloudant data to remote CouchDB.

(f) Additional data management features with managed service options are available with Cloudant such as Full-text search, Advanced geospatial capabilities, native mobile software libraries, etc.

(g) Getting started with Cloudant service is easy by just signing up with the free account in a public cloud.

(h) Cloudant has all the components to manage application data even in the situation where the developer architecture changes or grows as per the requirement.

(i) Cloudant allows the creation of indices by using MapReduce.

(j) Cloudant query provides a declarative way to define and query indices.

(k) Cloudant NoSQL DB saves to disk three copies of every document to three different nodes in a cluster. Saving the copies ensures that a working failover copy of your data are always available, regardless of failures.

(l) Cloudant NoSQL DB is accessed by using an HTTP API. Where the API endpoint requires it, the user is authenticated for every HTTPS or HTTP request Cloudant NoSQL DB receives.

(m) Cloudant NoSQL DB customers on dedicated environments can white list IP addresses to restrict access to only specified servers and users.

(n) Cloudant NoSQL DB gives user the flexibility to choose or switch among the different providers as your SLA and cost requirements change.

2.3.3 Setting Environment for Cloudant

As already mentioned above, starting Cloudant service is easy by just signing up with the free account in a public cloud. To do so, following steps need to be taken care of.

(a) First, the user needs to sign-up for Cloudant database by using the following link: https://cloudant.com/sign-up as depicted in Fig. 32.

(b) The user needs to further log in into the account after the account is created as shown in Fig. 33.

(c) The dashboard will appear comprising of several tabs as depicted in Fig. 34. Now, the user can perform operations with the click of the tab as desired.

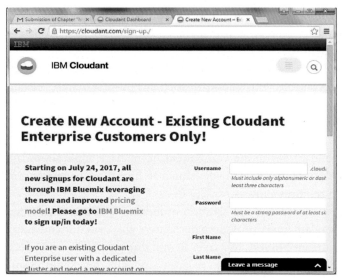

Fig. 32 To sign up for Cloudant database.

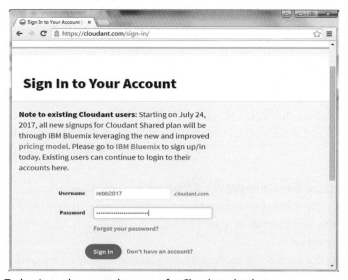

Fig. 33 To log in to the created account for Cloudant database.

2.3.4 Various Operations on Cloudant

There are several tabs available in the dashboard of Cloudant database, the user can click on the tab and perform operations such as database creation, database deletion, creating new documents, replicating database, view

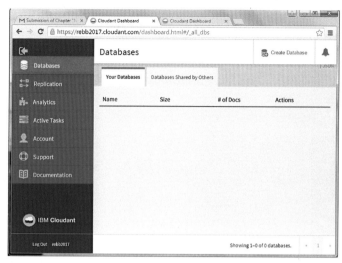

Fig. 34 Cloudant dashboard.

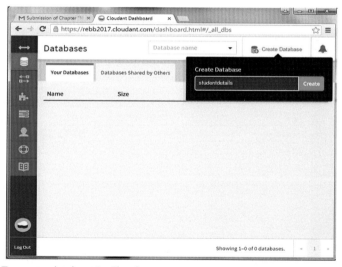

Fig. 35 To create database in Cloudant.

database, create new view, query indices, new search indices, new geo-spatial indices, view replicated database, etc. Some of the operations have been depicted in this section.

(a) Click on the tab "**Create Database**," enter the database name as desired and click on the create button. The database will be created successfully as depicted in Figs. 35–37.

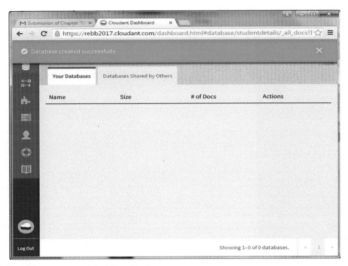

Fig. 36 Database created in Cloudant.

Fig. 37 The created database studentdetails in Cloudant.

(b) To "**create new document**" in the created database, click on the tab "New Doc." This will create a new document under the database created wherein the user needs to type in the query as shown in Figs. 38 and 39.

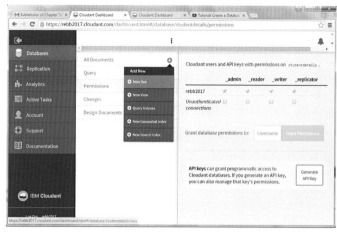

Fig. 38 Creating new document in Cloudant.

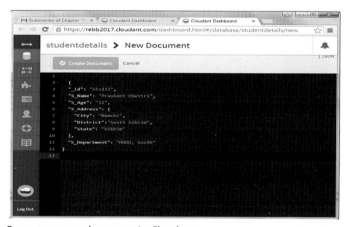

Fig. 39 Query to create document in Cloudant.

(c) To "**replicate database**" in Cloudant, to create the replica of the existing database; click on the tab "Replication." The user needs to fill in certain details such as Replication source, Source Name, Replication Target, New Database name, etc., and click on "Start Replication." The user will be asked to enter local account password and finally the replicated database created will be displayed on the console. Figs. 40–42 depict the operation performed.

(d) To "delete database" in Cloudant, click on the option delete Database_Name as depicted in Figs. 43 and 44.

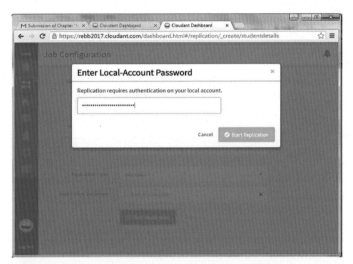

Fig. 40 To replicate the existing database in Cloudant.

Fig. 41 Local account password to replicate database in Cloudant.

Fig. 42 List of replicated database in Cloudant.

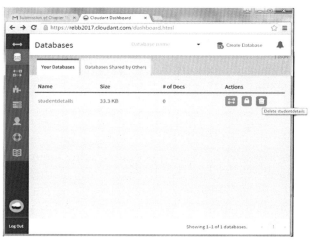

Fig. 43 To delete the database in Cloudant.

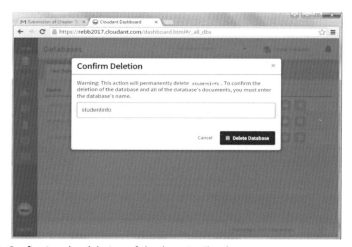

Fig. 44 Confirming the deletion of database in Cloudant.

3. GRAPH-BASED DATABASES

A graph database, also called a graph-oriented database or graph-based database, is a variety of NoSQL database that employs graph theory to store, map, and query relationships. A graph database is basically an assortment of nodes and edges wherein each node signifies an entity (such as a person or business) and each edge represents a connection or relationship between two nodes. Every node in a graph database is defined by a unique identifier, a set of outgoing edges and/or incoming edges, and a set of properties expressed as key/value pairs. Each edge is defined by a unique identifier, a starting-place and/or ending-place node, and a set of properties. It supports a very flexible and fine-grained data model that permits the user to model and manage rich domains in a trouble free and insightful way further supporting CRUD operations working on a graph data model. The fine-grained model also means that there is no fixed boundary around aggregates, so the scope of update operations is provided by the application during the read or writes operation. The well known and tested concept of transactions groups a set of updates of nodes and relationships into an atomic, consistent, isolated, and durable (ACID) operation. Unlike other databases, relationships take first priority in graph databases. This means that the application does not have to infer data connections using things like foreign keys or out-of-band processing, such as MapReduce, etc. There are two important properties of graph database technologies: Graph Storage and Graph Processing Engine. There are many graph databases such as Neo4J, Infinite Graph, GraphDB, AllegroGraph, InfoGrid, HyperGraphDB, OrientDB, FlockDB, VelocityGraph, DGraph, Grapholytic, ArangoDB, ThingSpan, etc. Graph databases like Neo4j [3] fully support the transactional concepts including write-ahead logs and recovery after abnormal termination [12]. So the user never loses their data that have been committed to the database.

In this section, two graph-based NoSQL databases have been taken into account, i.e., Neo4j and OrientDB.

3.1 Introduction to Neo4j

Like Oracle database has query language SQL, Neo4j has CQL as query language wherein CQL stands for Cypher query language (CQL). Sponsored by Neo Technology, Neo4j is an open source NoSQL graph database implemented in Java and Scala. With development starting in 2003, it has been publicly available since 2007. The source code and issue tracking are available on GitHub, with support readily available on Stack Overflow and the Neo4j Google group. Neo4j is used today by hundreds of thousands of companies and organizations in almost all industries. Use cases include matchmaking, network management, software analytics, scientific research, routing, organizational and project management, recommendations, social networks, and more. Neo4j implements the Property Graph Model efficiently down to the storage level [13]. As opposed to graph processing or in-memory libraries, Neo4j provides full database characteristics including ACID transaction compliance, cluster support, and runtime failover, making it suitable to use graph data in production scenarios.

3.1.1 Why and Where to Use Neo4j?

Whenever the application is to be developed, the user needs to ensure that the right database is chosen. Unlike other databases, relationships take first priority in graph databases. This means user's application does not have to infer data connections using things like foreign keys or out-of-band processing, such as MapReduce. The various reasons for choosing the Neo4j databases are:

(a) Neo4j is the world's best and first graph database.

(b) Neo4j has the largest and most vibrant community of graph database enthusiasts that contributes to the Neo4j ecosystem.

(c) Neo4j delivers the lightning-fast read and write performance that user requires, while still protecting your data integrity.

(d) Index-free adjacency shortens read time and gets even better as data complexity grows. Get reliably fast transactions with ultra-high parallelized throughput even as your data grow.

(e) Mature UI with intuitive interaction and built-in learning, time-tested training ecosystem to meet your needs, a wealth of training materials bringing years of deployment experience to your desktop, an expert-authored books for in-depth learning.

(f) Neo4j is the only graph database recognized by key analysts [14] (Forrester, Gartner, and others) to have enough production applications to warrant inclusion in reports. Clustering and data replication demanded by transactional and operational applications.

3.1.2 Features of Neo4j

Some notable features that make Neo4j very popular among users, developers, and Database Administrators (DBAs) have been listed below:

(a) Materializing of relationships at creation time, resulting in no penalties for complex runtime queries.

(b) Constant time traversals for relationships in the graph both in depth and in breadth due to efficient representation of nodes and relationships.

(c) All relationships in Neo4j are equally important and fast, making it possible to materialize, and use new relationships later on to "shortcut" and speed up the domain data when new needs arise.

(d) Compact storage and memory caching for graphs, resulting in efficient scale-up, and billions of nodes in one database on moderate hardware.

(e) Written on top of the JVM.

(f) *CQL*: Neo4j provides a powerful declarative query language known as Cypher. It uses ASCII art for depicting graphs. Cypher is easy to learn and can be used to create and retrieve relations between data without using the complex queries like joins.

(g) SQL-like easy query language Neo4j CQL.

(h) *Data model (flexible schema)*: Neo4j follows a data model named native property graph model. Here, the graph contains nodes (entities) and these nodes are connected with each other (depicted by relationships). Nodes and relationships store data in key–value pairs known as properties.

(i) *Scalability and reliability*: You can scale the database by increasing the number of reads/writes, and the volume without affecting the query processing speed and data integrity. Neo4j also provides support for replication for data safety and reliability.

(j) Built-in web application: Neo4j provides a built-in Neo4j Browser web application. Using this, user creates and queries the graph data.

(k) *Indexing*: Neo4j supports indices by using Apache Lucence.

(l) It supports UNIQUE constraints.

(m) It contains a UI to execute CQL commands: Neo4j Data Browser.

(n) *ACID properties*: Neo4j supports full ACID (atomicity, consistency, isolation, and durability) rules.

(o) It supports full ACID (atomicity, consistency, isolation, and durability) rules.

(p) It uses native graph storage with native GPE (Graph Processing Engine).

(q) It supports exporting of query data to JSON and XLS format.

(r) It provides REST API to be accessed by any Programming Language like Java, Spring, Scala, etc.

(s) It provides JavaScript to be accessed by any UI MVC Framework like Node JS.

(t) It supports two kinds of Java API: cypher API and native Java API to develop Java applications.

(u) It supports programming languages such as. Net, Clojure, Elixir, Go, Groovy, Haskell, Java, JavaScript, Perl, PHP, Python, Ruby, and Scala.

(v) *Drivers*: Neo4j can work with: REST API to work with programming languages such as Java, Spring, Scala, etc.; Java Script to work with UI MVC frameworks such as Node JS; it supports two kinds of Java API: cypher API and native Java API to develop Java applications. In addition to these, user can also work with other databases such as MongoDB, Cassandra, etc.

3.1.3 Advantages of Neo4j Over RDBMS

The various advantages of Neo4j over RDBMS are as follows:

(a) *Flexible data model*: Neo4j provides a flexible simple and yet powerful data model, which can be easily changed according to the applications and industries.

(b) *Real-time insights*: Neo4j provides results based on real-time data.

(c) *High availability*: Neo4j is highly available for large enterprise real-time applications with transactional guarantees.

(d) *Connected and semistructured data*: Using Neo4j connected and semistructured data can be easily represented.

(e) *Easy retrieval*: Neo4j enables not only representation but also easily retrieves (traverse/navigate) connected data faster when compared to other databases.

(f) *CQL*: Neo4j provides a declarative query language to represent the graph visually, using Ascii art syntax. The commands of this language are in human readable format and very easy to learn.

(g) *No joins*: Neo4j does NOT require complex joins to retrieve connected/related data as it is very easy to retrieve its adjacent node or relationship details without joins or indices.

The various terminology used in RDBMS and MongoDB have been tabulated in Table 5.

Table 5 Terminology Used in RDBMS and Neo4j

RDBMS	MongoDB
Table/Relation	Graph
Tuple/Row	Nodes
Column/Attribute	Properties and its values
Constraints	Relationship
Joins	Traversal

3.1.4 Installation Steps of Neo4j

The installers for Neo4j are available in both the 32-bit and 64-bit formats. Neo4j runs on Linux, Windows, and Mac OS X. There are desktop installers for Community Edition available for Mac OS X and Windows. There are also platform-specific packages and zip/tar archives of both Community Edition and Enterprise editions. Neo4j requires a Java Virtual Machine (JVM), to operate. Community Edition installers for Windows and Mac include a JVM for convenience. All other distributions, including all distributions of Neo4j Enterprise Edition, require a preinstalled JVM. The various platform supported by Neo4j are Ubuntu 14.04, 16.04; Debian 8, 9; CentOS 6, 7; Fedora, Red Hat, Amazon Linux, Windows Server 2012.

(a) Download Neo4j from official site using https://neo4j.com/. On clicking, this link will take you to the homepage of neo4j website as depicted in Fig. 45.

(b) Click on Download button on the top right-hand side. The downloaded page is generated, click on the community edition of Neo4j depicted in Fig. 46.

(c) This will redirect to download community version of Neo4j software compatible with different operating systems. Choose Windows 32 bits or 64 bits as per the requirements as depicted in Fig. 47. A file named neo4j-community_windows-x32_3_2_2.exe will be downloaded in the system.

(d) Double click on the .exe file that is downloaded to install Neo4j server and click on run button as shown in Fig. 48.

Fig. 45 Step 1 to download Neo4j.

Fig. 46 Step 2 to download Neo4j.

Fig. 47 Step 3 to download Neo4j.

Fig. 48 Step 1 to install Neo4j.

(e) After clicking on run button, the installation process starts and the status of installation are shown in Fig. 49.

(f) Once the installation process completes, choose the destination directory where the user prefers to install Neo4j and click next as depicted in Fig. 50.

(g) Click on next button to continue the setup of Neo4j as depicted in Fig. 51.

(h) Accept the license agreement to continue the setup of Neo4j as depicted in Fig. 52.

Fig. 49 Step 2 to install Neo4j.

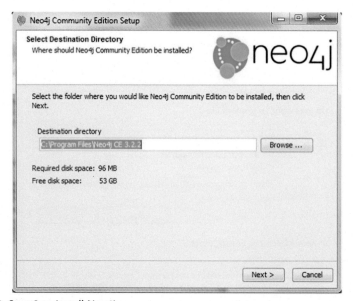

Fig. 50 Step 3 to install Neo4j.

Fig. 51 Step 4 to install Neo4j.

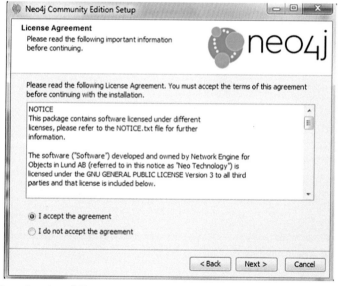

Fig. 52 Step 5 to install Neo4j.

(i) Create shortcuts of Neo4j as per the desired location and further click on next to continue the installation as depicted in Figs. 53 and 54, respectively. Finally click on finish button to complete the entire process of installation as shown in Fig. 55.

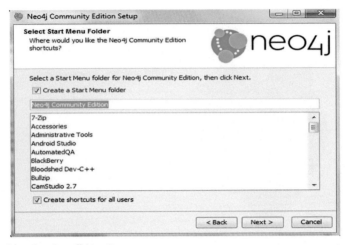

Fig. 53 Step 6 to install Neo4j.

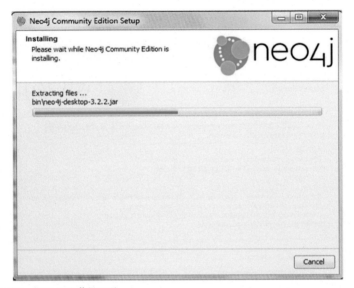

Fig. 54 Step 7 to install Neo4j.

3.1.5 Setting Environment for Neo4j

The various parameters to be considered for setting the environment for Neo4j are depicted in this section.

(a) Click the Windows start menu and start the Neo4j server by clicking the start menu shortcut for Neo4j. First, choose the location where the database will be stored and click on start button as depicted in Fig. 56.

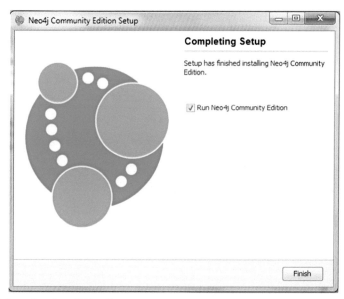

Fig. 55 Step 8 to install Neo4j.

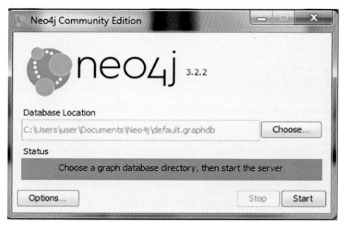

Fig. 56 Starting the Neo4j server.

(b) The server starts on the click of the start button as depicted in Fig. 57.

(c) Open the browser and type in URL http://localhost:7474/ as Neo4j provides an in-built browse application to work with Neo4j as highlighted in Fig. 58. Type in the password to connect and start working.

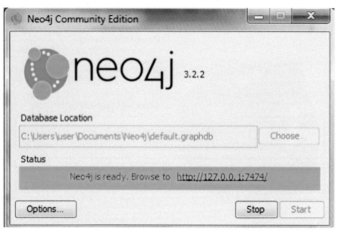

Fig. 57 Start of Neo4j server.

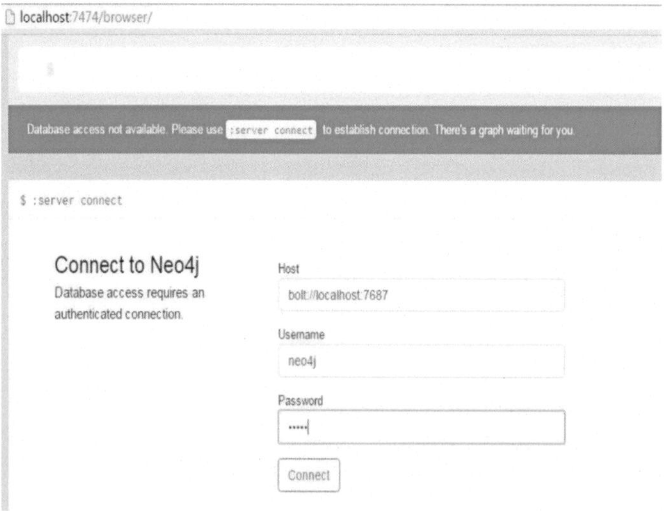

Fig. 58 In-built browse application to work with Neo4j.

(d) After clicking connect button, a built-in browser app of Neo4j with a dollar prompt as shown in Fig. 59 appears where after the $ the query can be triggered to get the desired output.

3.1.6 Various Operations on Neo4j
This section deals with the various operations that can be performed on Neo4j using CQL.

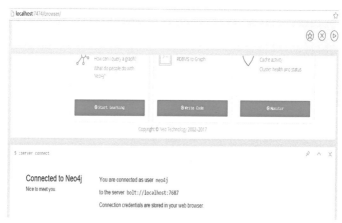

Fig. 59 A built-in browser app of Neo4j.

(a) The **create** command to create node.

This command can be used to serve the following: create a single node, create multiple nodes, create a node with a label, create a node with multiple labels, create a node with properties, and returning the created node. The query to be type and executed in the dollar prompt.

[*Note*: Semicolon (;) at the end of each query is optional.]

Syntax: Creating a single node.

```
CREATE (Node_Name_1);
```

Example:

```
CREATE (Rebika);
```

Syntax: Creating multiple nodes.

```
CREATE (Node_Name_1), (Node_Name_2), ...... (Node_Name_n);
```

Example:

```
CREATE (Rebika), (Prashant), (Deeyan);
```

Syntax: Creating a node with a label.

```
CREATE (Node_Name_1: Label_1);
```

Example:

```
CREATE (Rebika: Student_1);
```

Syntax: Creating a node with multiple labels.

```
CREATE (Node_Name_1: Label_1, Label_2, ...... Label_n);
```

Example:

```
CREATE (Rebika: Student_1, Person, Human);
```

Syntax: Creating a node with properties.

```
CREATE (Node_Name_1: Label_1{Key_1:Value_1, ...... Key_n: Value_n} );
```

Example:

```
CREATE (Rebika: Student_1 {Name: "Rebika Rai", RegNo: 302,234, Address:
    "Sikkim"});
```

Syntax: Verifying and returning the created node.

```
MATCH (Node_Name) RETURN (Node_Name);
```

Example:

```
MATCH (Rebika) RETURN (Rebika);
```

This query returns all the nodes in the database.

(b) The **CREATE** command to create relationship.

In Neo4j, a relationship is an element using which two nodes of a graph can be connected. These relationships have direction, type, and the form patterns of data. This command can be used to serve the following: create a relationship, create a relationship between the existing nodes, and create a relationship with label and properties. Relationship using the **CREATE** command is specified in the square braces "[]" depending on the direction of the relationship it is placed between hyphen "–" and arrow "→" as shown in the following syntax.

Syntax: Creating a relationship.

```
CREATE (Node_Name_1) - [: RelationshipType] → (Node_Name_2);
```

Example:

```
CREATE (Rebika) - [r: Friend_Of] → (Prashant);
```

Syntax: Creating a relationship between the existing nodes using CREATE and MATCH commands.

```
MATCH (a: LabelofNode_Name_1), (b: LabelofNode_Name_2)
      WHERE   a.Node_Name_1   =   "NameofNode1"   AND   a.Node_Name_2
   = "NameofNode2"
CREATE (a) - [ : RelationshipType] → (b);
RETURN a,b;
```

Example:

```
MATCH (a: Student1), (b: Student2)
      WHERE a.Name = "Rebika" AND b. Name= "Prashant"
CREATE (a) - [ r: Friend_Of] → (b);
RETURN a,b;
```

Syntax: Creating a relationship with label and properties.

```
CREATE (Node_Name_1) - [LabelofNode_Name_1: RelationshipType
      {Key_1:Value_1, Key_2:Value_2,...Keyn:Value_n}] → (Node_Name_2);
```

Example:

```
MATCH (a: Student1), (b: Student2)
      WHERE a.Name = "Rebika" AND b. Name= "Prashant"
CREATE (a) - [ r: Friend_Of {Position: 2, Marks: 90.25}] → (b);
RETURN a,b;
```

(c) The **Set** command to add properties to node or relationship.

This command is used to add new properties to an existing Node or Relationship, and also enables the user to add or update existing Properties values. This command can be used to serve the following: set a property, remove a property, setting multiple properties, set a label on a node, and set multiple labels on a node.

Syntax: Set a property in a node.

```
MATCH (Node_Name_1: Label_1{Properties .......})
SET Node_Name_1.propery = value_1
RETURN (Node_Name_1)
```

Example:

```
MATCH (Rebika: Student_1 {Name: "Rebika Rai", RegNo: 302234,
    Address: "Sikkim"})
SET Rebika.highestscore=98
RETURN Rebika;
```

Syntax: Removing a property in a node.

```
MATCH (Node_Name_1: Label_1{Properties .......})
SET Node_Name_1.propery = NULL
RETURN (Node_Name_1)
```

Example:

```
MATCH (Deeyan: Student_2 {Name: "Deeyan Chettri", RegNo: 302236,
    Address: "Sikkim"})
SET Deeyan.Address=NULL.
RETURN Deeyan;
```

[*Note*: An existing property can be removed by passing NULL as value to it.]

Syntax: Setting multiple properties.

```
MATCH (Node_Name_1: Label_1{Properties .......})
SET Node_Name_1.propery_1 = value_1, Node_Name_2.propery_2 = value_2
RETURN (Node_Name_1)
```

Example:

```
MATCH (Deeyan: Student_2 {Name: "Deeyan Chettri", RegNo: 302236})
SET Deeyan.Address= "Sikkim", Deeyan.highestscore="95"
RETURN Deeyan;
```

Syntax: Setting a label on an existing node.

```
MATCH (n{Properties .......})
SET n = Label
RETURN (n)
```

Example:

```
MATCH (Deezeena: {Name: "Deezeena Chettri", RegNo: 302,237, Address:
    "Gangtok"})
```

```
SET Deezeena: Student_4
RETURN Deezeena;
```
Syntax: Setting multiple labels on a node.
```
MATCH (n{Properties .......})
SET n = Label_1:Label_2
RETURN (n)
```
Example:
```
MATCH (Hrikesh: {Name: "Hrikesh Chettri", RegNo: 302238,
    Address: "Namchi"})
SET Hrikesh: Student_5: Person: Human.
RETURN Hrikesh;
```

(d) The **delete** command to delete all nose and relationships.
In Neo4j, this command can be used to serve the following: delete all nodes and relationships, delete a particular node.
Syntax: Deleting all the nodes and relationships.
```
MATCH (n) DETACH DELETE n
```
This query enables the user to delete all the existing nodes and relationship from Neo4j database and makes it empty.
Syntax: Deleting a specific node.
```
MATCH (Node_Name_1: Label_1{Properties .......})
DETACH DELETE Node_Name_1;
```
Example:
```
MATCH (Hrikesh: {Name: "Hrikesh Chettri", RegNo: 302238,
    Address: "Namchi"})
DETACH DELETE Hrikesh;
```

3.2 Introduction to OrientDB

OrientDB is an open source NoSQL database management system written in Java. It is a multimodel database, supporting graph, document, key/value, and object models, but the relationships are managed as in graph databases with direct connections between records. It supports schema-less, schema-full, and schema-mixed modes. OrientDB's has a Native Graph Database engine compliant with the Apache TinkerPop standard. It supports schema-less, schema-full, and schema-mixed modes and includes SQL among its query languages which reduces the learning curve for those new to OrientDB. In OrientDB all vertices and edges are documents. OrientDB also supports relationships. With improved auditing and authentication, password SALT, and data-at-rest encryption, OrientDB is the most secure open source NoSQL database on the market. It supports various languages such as Scala, .Net, Java, PHP, Ruby, Python, and C [15]. Both community and Enterprise editions of OrientDB can run on any operating system that implements the JVM. OrientDB requires Java with 1.7 or later version.

3.2.1 Features of OrientDB

The main feature of OrientDB is to support multimodel objects, i.e., it supports different models like document, graph, key/value, and real object with separated API for all four models. Some of the features of OrientDB have been listed here under.

(a) *Quick installation*: OrientDB can be installed easily and quickly in terms of seconds.

(b) *Graph structured data model*: Native management of graphs is possible in OrientDB and is fully compatible with Apache TinkerPop, Gremlin open source graph computing framework.

(c) *Fully transactional*: It supports ACID transactions guaranteeing that all database transactions are processed reliably and in the event of a crash all pending documents are recovered and committed.

(d) *SQL support*: It supports SQL queries with extensions to handle relationships without SQL join, manage trees, and graphs of connected documents.

(e) *Web technologies*: It supports HTTP, RESTful protocol, and JSON additional libraries or components.

(f) *Distributed*: It provides full support for multimaster replication including geographically distributed clusters.

(g) *Run anywhere*: It is implemented using pure Java allowing it to be run on Linux, OS X, Windows, or any system with a compliant JVM.

(h) *Embeddable*: Local mode to use the database bypassing the Server. Perfect for scenarios where the database are embedded.

(i) *Apache 2 license*: It is always free for any usage. No fees or royalties required to use it.

(j) Full server has a footprint of about 512 MB and commercial support is available from OrientDB.

(k) *Pattern matching*: Introduced in version 2.2, the Match statement queries the database in a declarative manner, using pattern matching.

(l) *Security* features introduced in OrientDB 2.2 provide an extensible framework for adding external authenticators, password validation, LDAP import of database roles and users, advanced auditing capabilities, and syslog support.

(m) *Teleporter.* The relational databases to be quickly imported into OrientDB in few simple steps.

(n) *Cloud ready*: OrientDB can be deployed in the cloud and supports the following providers: Amazon Web Services, Microsoft Azure, CenturyLink Cloud, Jelastic, and DigitalOcean.

3.2.2 Relationship of RDBMS Terminology With OrientDB

The difference in the terminology used in RDBMS and OrientDB is tabulated in Table 6.

The various advantages of OrientDB is highlighted below:

(a) The first Multimodel open source NoSQL DBMS that combines the power of graphs with documents, key/value, reactive, object-oriented, and geospatial models into one scalable, high-performance operational database.

(b) OrientDB was engineered from the ground up with performance as a key specification. It is fast on both read and writes operations. It stores up to 120,000 records per second, no more joins: relationships are physical links to the records, better RAM use, traverses parts of or entire trees and graphs of records in milliseconds, and traversing speed is not affected by the database size.

(c) With master–slave architecture, the master often becomes the bottleneck. With OrientDB, throughput is not limited by a single server. Global throughput is the sum of the throughput of all the servers.

Table 6 Terminology Used in RDBMS and OrientDB

RDBMS	OrientDB Document Model	OrientDB Graph Model	OrientDB Key/ Value Model	OrientDB Object Model
Table	Class or cluster	Class that extends "V" (vertex) and "E" (edges)	Class or cluster	Class or cluster
Row	Document	Vertex	Document	Document or vertex
Columns	Document field	Vertex and edge property	Document field or vertex/ edge property	Document field or vertex/ edge property
Relationship	Link	Edge	Link	Link

(d) No more multiple systems are required as OrientDB supports four different features, i.e., graph, document, object, and reactive model.

(e) OrientDB is written entirely in Java and can run on any platform without configuration and installation. It is a drop-in replacement for the most common existing graph databases in deployment today.

(f) There is absolutely no cost associated with using OrientDB Community Edition as OrientDB Community is free for commercial use, comes with an Apache 2 Open Source License. And eliminates the need for multiple products and multiple licenses.

3.2.3 Installation Steps and Setting Environment for OrientDB

OrientDB installation file is available in two editions: (a) Community Edition released by Apache under 0.2 license as an open source and (b) Enterprise released as a proprietary software, which is built on community edition. It serves as an extension of the community edition. The installers for OrientDB are available for various operating systems that implements the JVM. OrientDB requires Java with 1.7 or later version. OrientDB comes with built-in setup file to install the database on your system. It provides different precompiled binary packages (tarred or zipped packages) for different operating systems. The various steps to download and install OrientDB are provided as follows:

(a) Download OrientDB files from the link http://orientdb.com/download as depicted in Fig. 60.

Fig. 60 Download step 1 for OrientDB.

(b) The orientdb-community-2.1.9.zip file will appear in Download folder. Furthermore, extract the zip file using the zip extractor and the folder appears in the location as depicted in Fig. 61.

(c) Click in the folder bin and to verify the OrientDB database server installation, the user needs to run the server, run the console, and run the studio as depicted in Fig. 62.

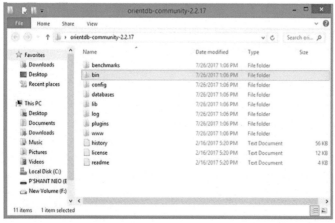

Fig. 61 Step 1 for starting OrientDB.

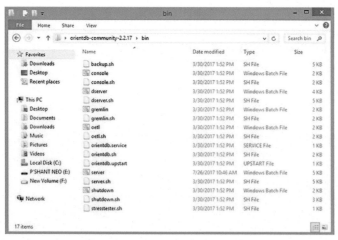

Fig. 62 Step 2 for starting OrientDB.

Fig. 63 Step 3 for starting OrientDB.

Fig. 64 Step 4 for starting OrientDB.

(d) Click on the server option given in Fig. 62 that will open the command prompt as depicted in Fig. 63 then click on the console option that will generate the window as shown in Fig. 64. Lastly click on the console option to get as shown in Fig. 65.

3.2.4 Various Operations on OrientDB

This section deals with the various operations that can be performed on OrientDB.

(a) The **create** command to create database (Fig. 66).

(b) The **connect** command to connect to the database (Fig. 67).

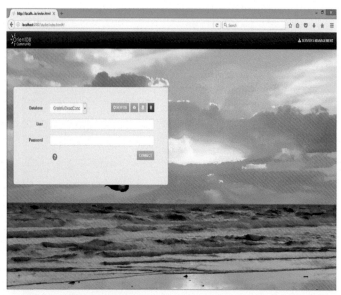

Fig. 65 Step 5 for starting OrientDB.

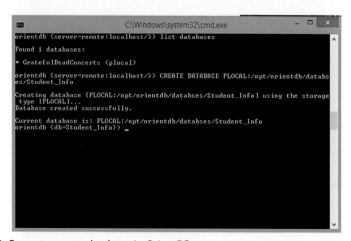

Fig. 66 Query to create database in OrientDB.

(c) The **disconnect** command to disconnect from database (Fig. 68).
(d) The **connect** command to connect to the server (Fig. 69).
(e) The **list** command to list the available databases (Fig. 70).
(f) The **info** command to get the information about the particular database (Fig. 71).
(g) The **freeze** command to freeze the database specified (Fig. 72).

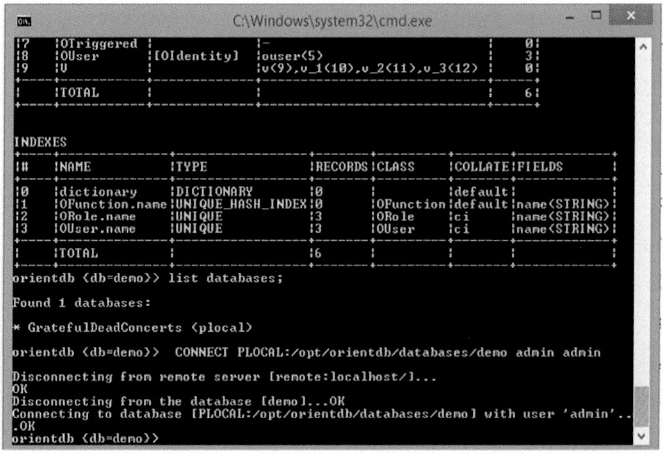

Fig. 67 Query to connect to the database in OrientDB.

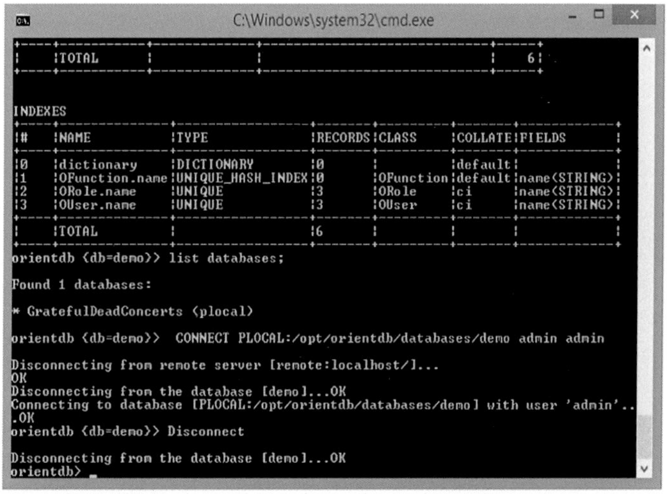

Fig. 68 Query to disconnect from the database in OrientDB.

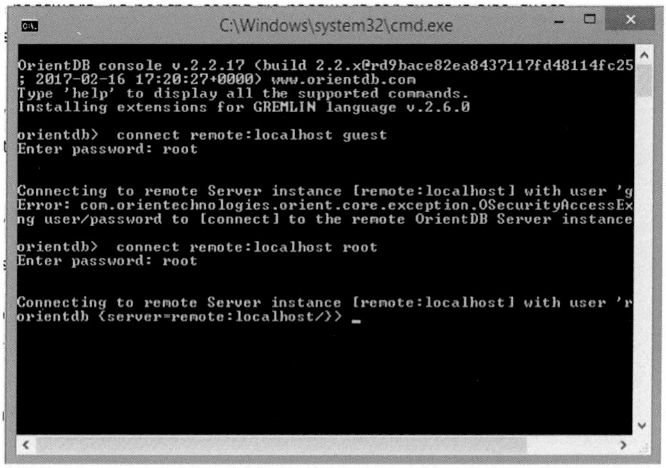

Fig. 69 Query to connect to the server in OrientDB.

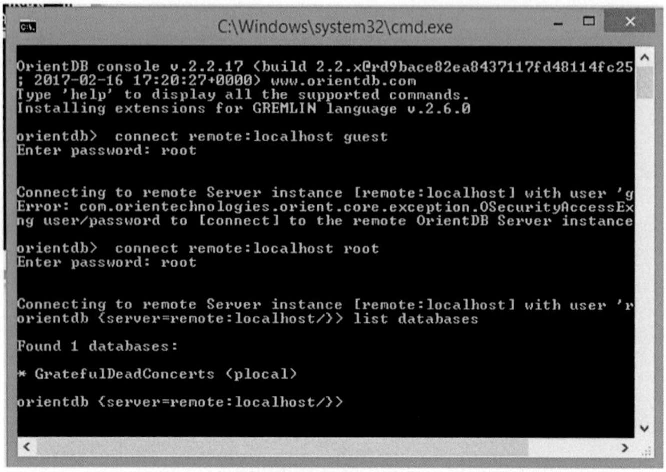

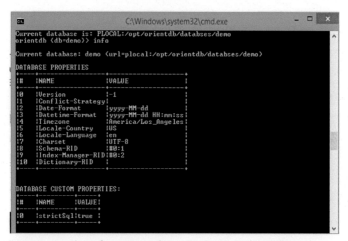

Fig. 70 Query to list the databases in OrientDB.

Fig. 71 Query to get the information of the databases in OrientDB.

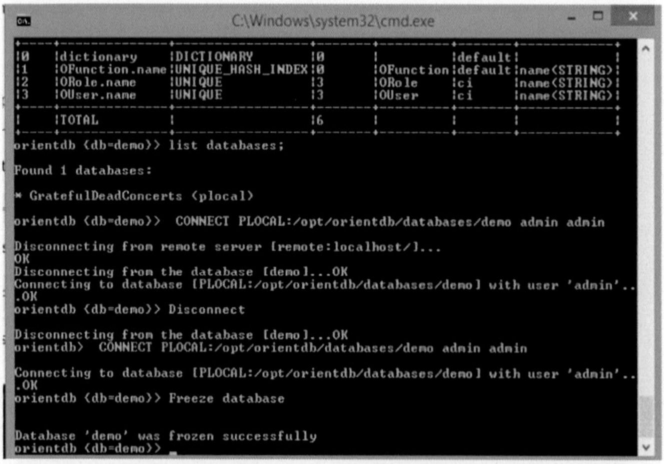

Fig. 72 Query to freeze the databases in OrientDB.

(h) The **release** command to release the database from the freeze state (Fig. 73).

(i) The **config** command to display the configuration of demo database as depicted in Figs. 74 and 75.

(j) The **config set** command to update the configuration variable value (Fig. 76).

(k) The **config get** command to display the configuration variable value (Fig. 77).

(l) The **create** command to create classes and property in the database (Fig. 78).

(m) The **insert** command to insert record into the table (Fig. 79).

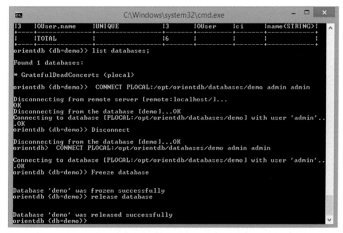

Fig. 73 Query to release the freeze database in OrientDB.

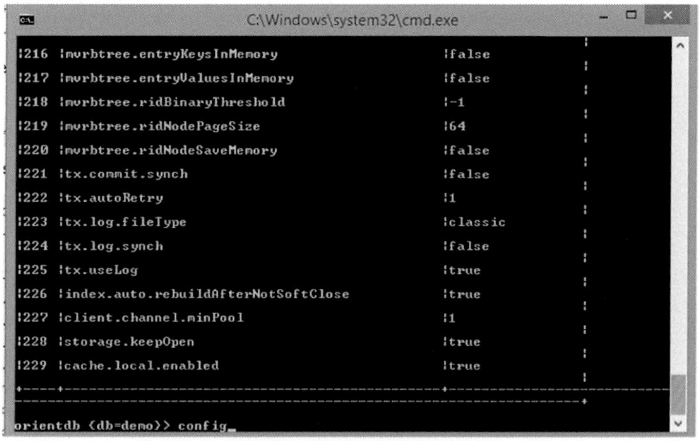

Fig. 74 Query to check the configuration in OrientDB.

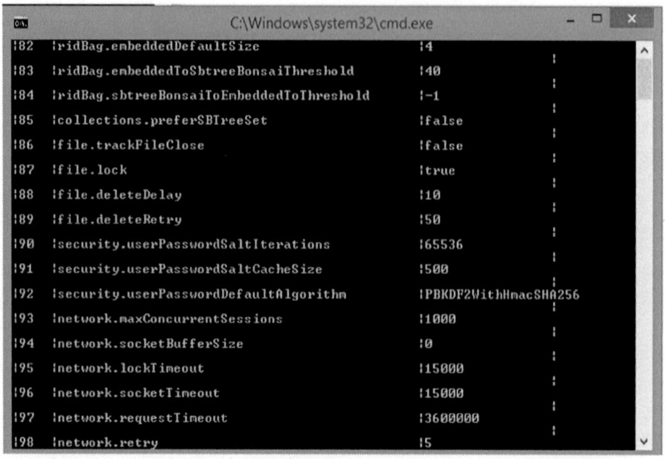

Fig. 75 Query displaying the configuration in OrientDB.

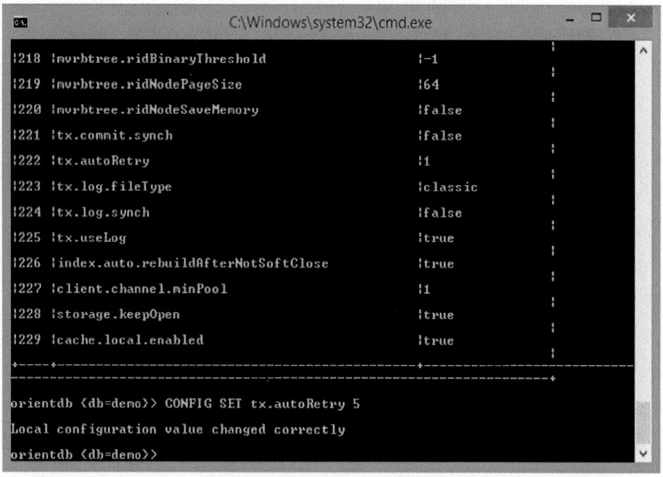

Fig. 76 Query to update the configuration variable's value in OrientDB.

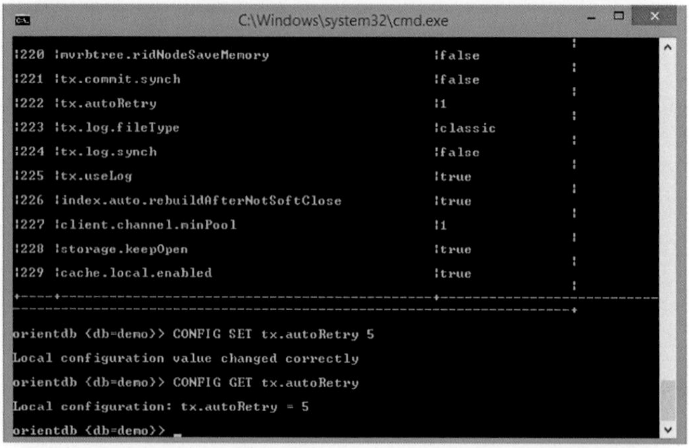

Fig. 77 Query to display the configuration variable's value in OrientDB.

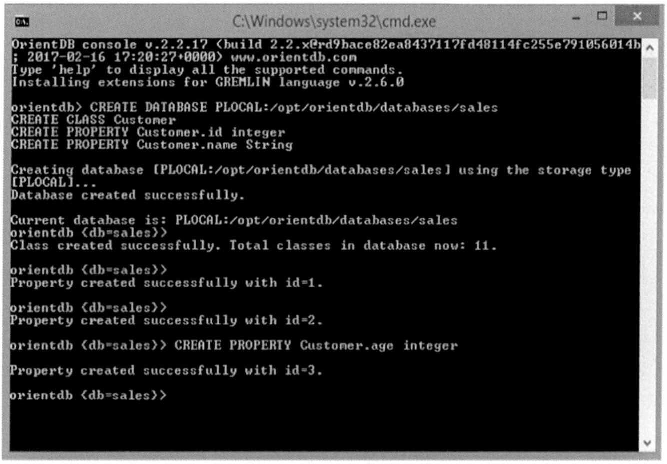

Fig. 78 Query to create class and property in OrientDB.

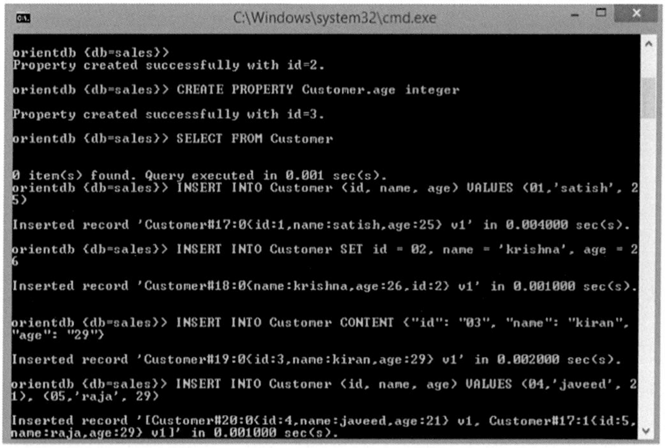

Fig. 79 Query to insert record in OrientDB.

(n) The **load** command to load particular record based on the record ID (Fig. 80).

(o) The **update** command to update record based on certain condition specified (Fig. 81).

(p) The **truncate record** command to delete the values of a particular record based on certain condition specified (Fig. 82).

(q) The **delete record** command to delete one or more records completely from the database (Fig. 83).

The various other commands that can be triggered are as shown in Figs. 84–89.

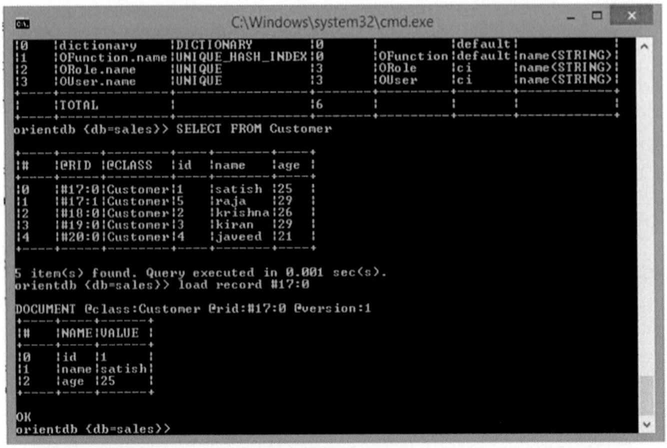

Fig. 80 Query to load record in OrientDB.

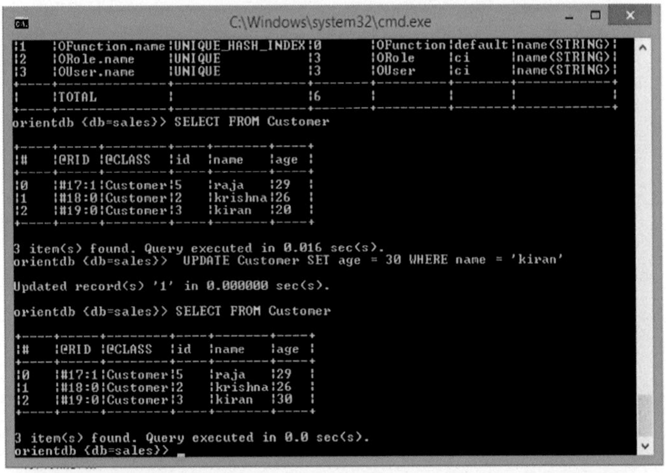

Fig. 81 Query to update record in OrientDB.

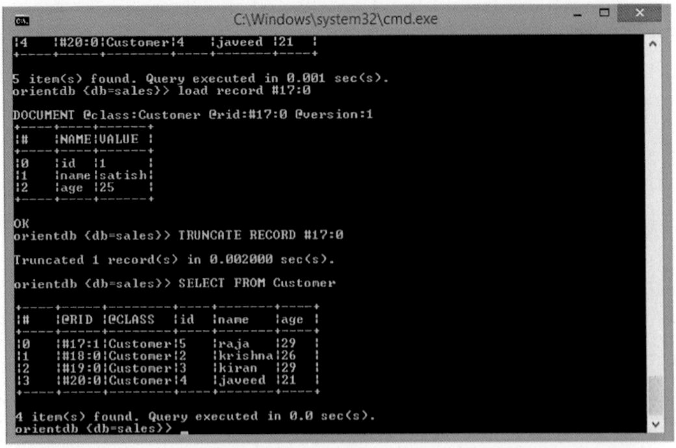

Fig. 82 Query to truncate record in OrientDB.

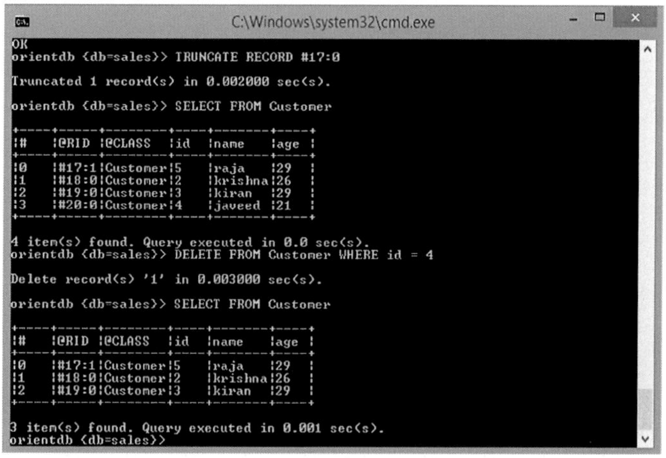

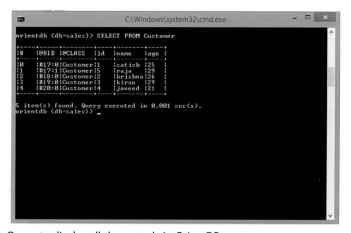

Fig. 83 Query to delete record in OrientDB.

Fig. 84 Query to display all the records in OrientDB.

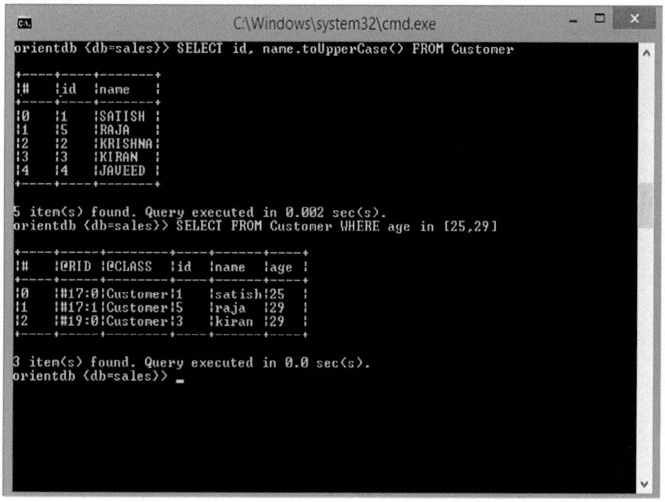

Fig. 85 Query using command to UpperCase( ) all the records in OrientDB.

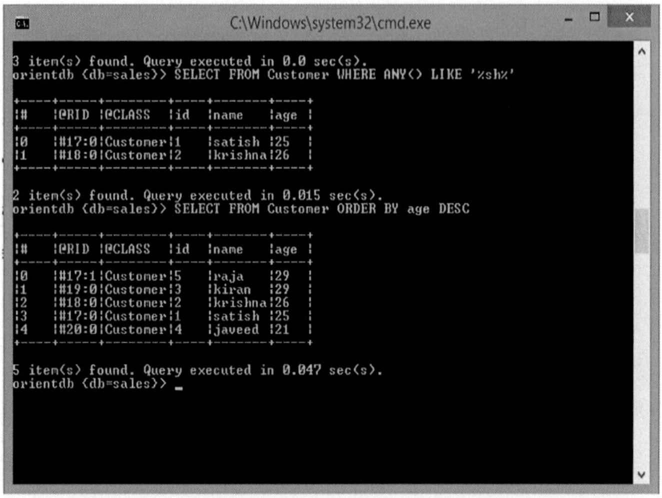

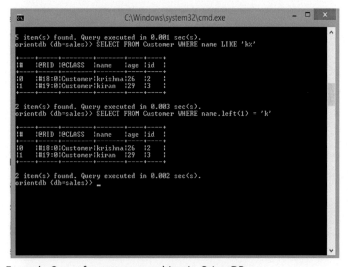

Fig. 86 Query for pattern matching in OrientDB.

Fig. 87 Example Query for pattern matching in OrientDB.

4. KEY–VALUE DATABASES

A key–value store, or key–value database, is a data storage paradigm designed for storing, retrieving, and managing associative arrays, a data structure more commonly known today as a dictionary or hash. Dictionaries contain a collection of objects, or records, which in turn have many different

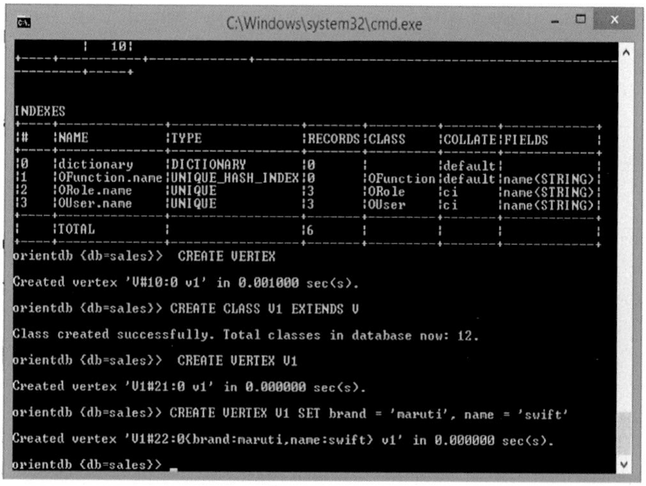

Fig. 88 Query to create vertex in OrientDB.

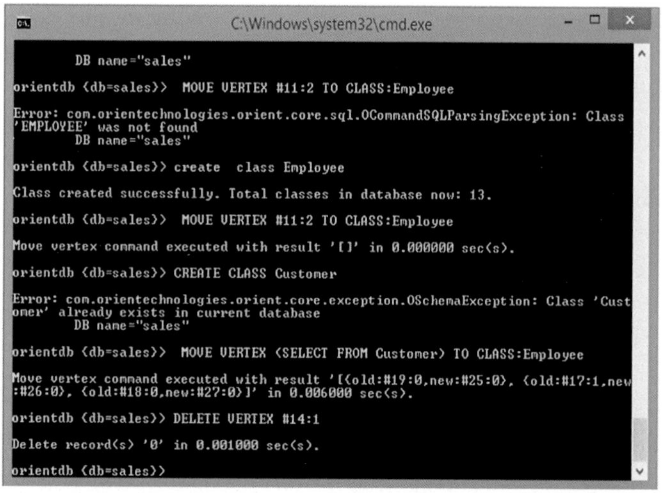

Fig. 89 Query to move and delete vertex in OrientDB.

fields within them, each containing data. These records are stored and retrieved using a key that uniquely identifies the record and is used to quickly find the data within the database. Key–value stores work in a very different fashion from the better known relational databases (RDB). RDBs predefine the data structure in the database as a series of tables containing fields with well-defined data types. Exposing the data types to the database program allows it to apply a number of optimizations. In contrast, key–value systems treat the data as a single opaque collection which may have different fields for every record. This offers considerable flexibility and more closely follows modern concepts like object-oriented programming. There are many key–value databases such as Aerospike, Redis [16], Riak, Oracle NoSQL, Voldemort, FoundationDB, InfluxDB, etc.

In this section, three key–value-based NoSQL databases have been taken into account, i.e., Redis, InfluxDB, and Aerospike.

4.1 Introduction to Redis

Redis stands for REmote DIctionary Server is an open source, flexible, BSD licensed, in-memory advanced key–value store written in ANSI C programming language by Salvatore Sanfilippo. It is used as database, cache and message broker developed in the year 2009 following the footfalls of other NoSQL databases such as MongoDB, Cassandra, CouchDB, etc. It is often referred to as a data structure server, since the keys can contain and support several data structures like strings, hashes, lists, sets, geospatial indices with radius queries, hyperlogs, and sorted sets allowing user to store massive data without the limits encountered in the case of relational database. Redis is written in C.

Redis can be compiled and used on Windows, Linux, OSX, OpenBSD, NetBSD, and FreeBSD. It supports big endian and little endian architectures, and both 32 bit and 64 bit systems. Redis has built-in replication, Lua scripting, LRU eviction, transactions and different levels of on-disk persistence, and provides high availability via Redis Sentinel and automatic partitioning with Redis Cluster.

User can easily build complex systems on top of Redis, here is a sample list:

(a) User-defined indexing schemes.
(b) Message queues with real-time new element notification.
(c) Directed and undirected graph stores for following or friending systems.
(d) Real-time publish/subscribe notification systems.
(e) Real-time analytics back ends.
(f) Bloom filters servers.
(g) Task queues and job systems.
(h) High score leaderboards.
(i) User ranking systems.
(j) Hierarchical/tree structured storage systems.
(k) Individual personalized news or data feeds for your users.

4.1.1 Features of Redis

The key features of Redis are:

(a) *Exceptionally fast*: Redis is very fast and can perform about 110,000 SETs per second, about 81,000 GETs per second. Redis supports pipelining of commands and getting and setting multiple values in a single command to speed up communication with the client libraries.

(b) *Supports rich data types*: Redis natively supports most of the data types that most developers already know like list, set, sorted set, and hashes. This makes it very easy to solve a variety of problems because we know which problem can be handled better by which data type.

(c) *Operations are atomic*: All the Redis operations are atomic, which ensures that if two clients concurrently access Redis server will get the updated value.

(d) *Multiutility tool*: Redis is a multiutility tool and can be used in a number of use cases like caching, messaging-queues (Redis natively supports Publish/Subscribe), any short lived data in your application like web application sessions, web page hit counts, etc.

(e) *Data structures*: It supports data structures such as strings, hashes, sets, lists, sorted sets with range queries, bitmaps, hyper logs, and geospatial indices with radius queries.

(f) *Supported languages*: Many languages have Redis bindings, including: ActionScript, C, C++, C#, Clojure, Common Lisp, D, Dart, Erlang, Go, Haskell, Haxe, Io, Java, JavaScript (Node.js), Julia, Lua, Objective-C, Perl, PHP, Pure Data, Python, R, Racket, Ruby, Rust, Scala, Smalltalk, and Tcl.

(g) *Master/Slave replication*: Redis supports a very simple and fast Master/Slave replication. Is so simple it takes only one line in the configuration file to set it up, and 21 s for a slave to complete the initial sync of 10 MM key set on an Amazon EC2 instance.

(h) *Sharding*: Distributing the dataset across multiple Redis instances is easy in Redis, as in any other key–value store. And this depends basically on the Languages client libraries being able to do so.

(i) *Portable*: Redis is written in ANSI C and works in most POSIX systems like Linux, BSD, Mac OS X, Solaris, and so on. Redis is reported to compile and work under WIN32 if compiled with Cygwin, but there is no official support for Windows currently.

(j) Redis is an in-memory but persistent on disk database, so it represents a different trade off where very high write and read speed is achieved with the limitation of data sets that cannot be larger than memory.

4.1.2 Installation Steps and Setting Environment for Redis

The installers for Redis are available in both the 32-bit and 64-bit formats. There is a desktop manager for Redis available for Windows, Mac OS X and Ubuntu, Other Inux such as Fedora, CentOS and OpenSUSE bearing Community package called AArch Linux. The various platform supported by Redis are Windows, Linux, OSX, OpenBSD, NetBSD, and FreeBSD. The various steps are as follows:

(a) Download precompiled version of Redis for 32-bit and 64-bit Windows from Dusan Majkic's Githhub page using https://github. com/dmajkic/redis/downloads. On clicking, this link will take the user to the page as depicted in Fig. 90. Click on the first package to download.

(b) After Redis is downloaded, executable from the zip file as shown in Fig. 91 needs to be extracted (depending on the platform and preferences).

(c) Start Redis by double clicking on the Redis-server executable as depicted in Fig. 92.

(d) The server starts and the window appear indicating that server is ready to accept the clients as depicted in Fig. 93.

(e) Furthermore, double click on redis-client.exe as depicted in Fig. 94 to start the client side.

(f) Further in the client side window, queries are triggered as depicted in Fig. 95. Further type the command "EXIT" to quit the client window.

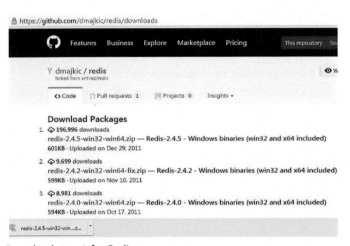

Fig. 90 Download step 1 for Redis.

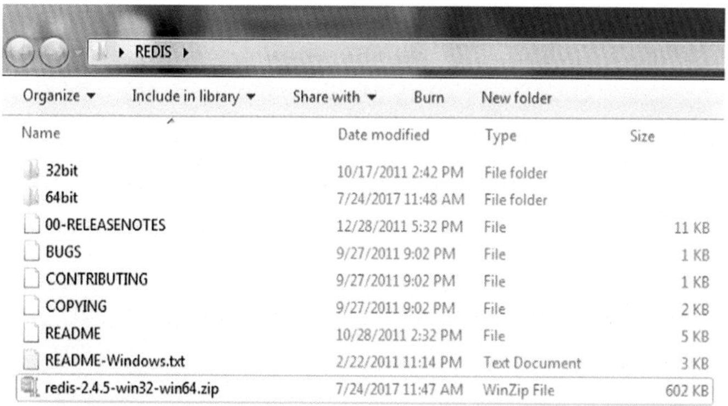

Fig. 91 Installation step 1 for Redis.

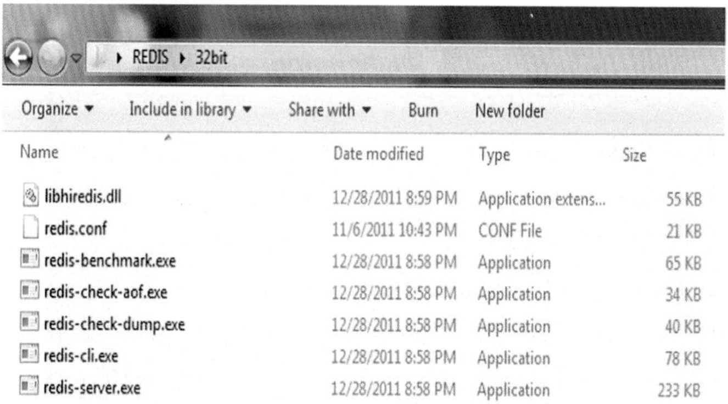

Fig. 92 Installation step 2 for Redis.

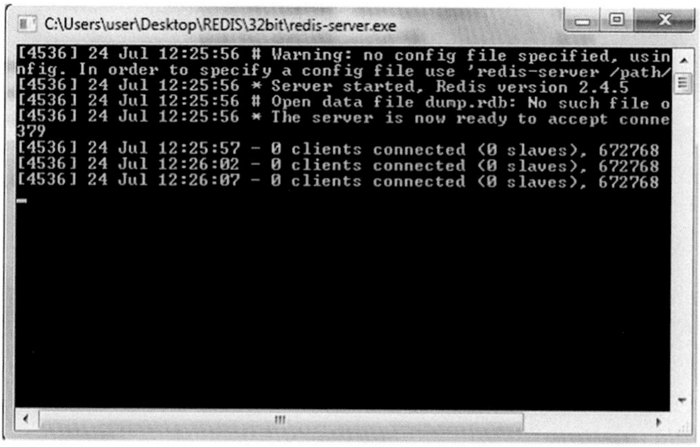

Fig. 93 Starting the server in Redis.

Fig. 94 Starting the client in Redis.

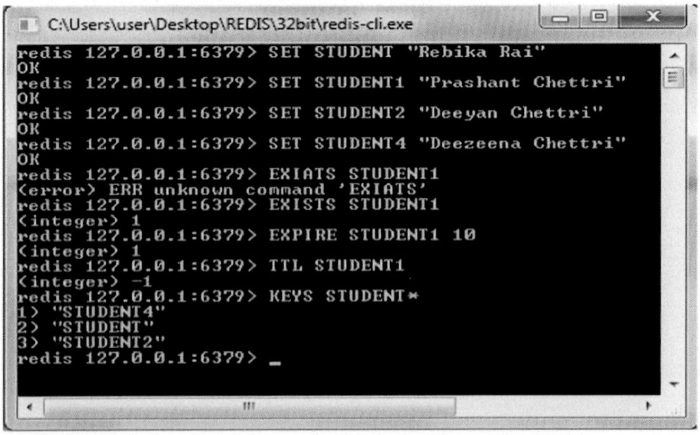

Fig. 95 Starting with the queries in client side window in Redis.

4.1.3 Various Operations on Redis

This section deals with the various operations that can be performed on Redis.

(a) The **SET** command to create keys.

This command is used to create keys in Redis and set certain values to it which may be further deleted if not required.

Syntax:

SET Key_Name "Values"

Example:

> SET Student "Prashant Chettri"

(b) The Redis **DEL** command to delete key.

This command is used to remove the specified keys. However, if a key does not exist, it is ignored.

Syntax:

DEL key Key_Name

Example:

>DEL key Student

(c) The **DUMP** command to serialize value stored at key.

This command permits the user to serialize the value stored at specified key in a Redis-specific format and return it to the user. The returned value can be synthesized back into a Redis key using the command **RESTORE**.

Syntax:

DUMP Key_Name

Example:

>DUMP Student

(d) The **EXISTS** command to check the existence of the key.

This command permits the user to check whether key exists in Redis or not thereby returning 1, if the key exists else returns 0 if the key does not exist.

Syntax:

EXISTS Key_Name

Example:

>EXIST Student

This query will return value 1 as the key Student exists. However, if the query would be **EXIST** Students; the return value would have been 0 as such key does not prevail in the database.

(e) The **EXPIRE** command to set timeout on key.

This command permits the user to set timeout on key and key gets automatically deleted after the specified timeout. The set timeout can only be cleared by two ways: First, if the key is removed or overwritten using commands such as DEL or SET commands, respectively. Second, by turning the key to persistent key using the command called as PERSIST. Also, the query will return 1 if the timeout is successfully set and key exists. However, if key does not exist or timeout cannot be set, the query will return value 0.

[*Note*: If the key in future is renamed, the associated time is transferred to the newly named key.]

Syntax:

EXPIRE Key_Name Time_In_Seconds

Example:

>EXPIRE Student 8

This query will return value 1 as the key Student exists and the timeout of 8 s is set. However, if the query would be **EXPIRE** Students, the return value would have been 0 as such key does not prevail in the database and timeout cannot be set.

(f) The **TTL** command to check the remaining time.

This command permits the user to check the remaining time to live for a key associated with timeout. This introspection capability allows a Redis client to check how many seconds a given key will continue to be part of the dataset. The query returns an integer value in milliseconds, i.e., the remaining time left if the expiry timeout is associated and further returns −1, if the key does not have expiry timeout associated else will return −2, if the key does not exists.

Syntax:

TTL Key_Name

Example:

>TTL Student

(g) The **PERSIST** command to remove timeout.

This command permits the user to remove the existing timeout specified on a key, turning the key from volatile to persistent which shall make the a key nonvolatile, that will never expire as no timeout is associated with it.

Syntax:

PERSIST Key_Name

Example:

>PERSIST Student

This query will remove the timeout that is associated with the key Student that was set otherwise using **EXPIRE** command.

(h) The **KEYS** command to match pattern.

This command permits the user to view the list of keys that matches the specified pattern.

Syntax:

KEYS pattern

Example:

>SET key1 Student1 "Rebika Rai"

> SET key1 Student2 "Deeyan Chettri"

> SET key1 Student3 "Deezeena Chettri"

>KEYS Student*

This query will return all the keys that match the pattern specified by the user.

[*Note*: * is used to return all the keys in Redis. For instance the query **KEYS** * will return all the created keys available in the Redis database irrespective of the pattern.]

(i) The **MOVE** key to move key from one to another database.

This command enables the user to move a key from the currently selected database to the specified database. The output of the command is nothing if the key does not exist in the source database or already available in the destination database. Furthermore, the query returns 1, if the key is moved else it returns 0.

Syntax:

MOVE Key_Name Destination_db

Example:

>MOVE Student1 RebikaRR

The query will return value 1 if it is moved to database RebikaRR. To check the existence of key the command **EXISTS** can be used.

(j) The **SADD** command to add specified members.

This command is used to add the specified members to the set stored at key. Specified members that are already a member of this set are ignored. If key does not exist, a new set is created before adding the specified members. An error is returned when the value stored at key is not a set. The query returns an integer value, i.e., the number of elements that were added to the set, not including all the elements already present into the set.

Syntax:

SADD Key_Name Member_1 Member_2 Member_3 ...... Member_n.

Example:

> SADD url Facebook.com Yahoo.com Youtube.com Gmail.com Hotmail.com

This query will return an integer value 5 as five members have been added.

(k) The **SORT** command to sort the elements.

This command is used to sort elements in the list or set based on the following modifier: Ascending, descending, lexicographically, placing a limit to sort only first few elements, etc. If no modifier is specified, by default the list is sorted in ascending order.

Syntax:

SORT My_List

OR

SORT My_List DESC / ALPHA / LIMIT 0 5

Example:
```
>SORT url ALPHA DESC
```

(l) The **RENAME** command to change the name of the key.
This command is used to change the name of a key to a new key name. The query returns an error when the source and destination names are the same, or when a key does not exist. If new key already exists it is overwritten, when this happens **RENAME** executes an implicit **DEL** operation, so if the deleted key contains a very big value.
Syntax:
```
RENAME Old_Key_Name New_Key_Name
```
Example:
```
>RENAME Student StudentDetails
```
This query returns a string "OK" or "Error."

4.2 Introduction to InfluxDB

InfluxDB is an open-source time series database developed by InfluxData, written in Go (Golang) programming language initially released on September 2013 and permissive free software license provided by MIT. It is optimized for fast, high-availability storage, and retrieval of time series data in fields such as operations monitoring, application metrics, Internet of Things sensor data, and real-time analytics. It also has support for processing data from Graphite. InfluxDB has no external dependencies and provides an SQL-like language with built-in time-centric functions for querying a data structure composed of measurements, series, and points. Each point consists of several key–value pairs called the field set and a timestamp. When grouped together by a set of key–value pairs called the tag set, these define a series. Finally, series are grouped together by a string identifier to form a measurement.

4.2.1 Features of InfluxDB
InfluxDB features a SQL-like query language, only used for querying data. The HTTP API has endpoints for writing data and performing other database administration tasks. The only exception to this is continuous queries, which perpetually write their results into one or more time series. Various features of InfluxDB are:
(a) SQL-like query language: InfluxDB has no external dependencies and provides an SQL-like language with built-in time-centric functions for querying a data structure composed of measurements, series, and points.

(b) InfluxDB accepts data via HTTP, TCP, and UDP. It defines a line protocol which is backward compatible with graphite's and stores metrics data such as response time, CPU load, and events data such as exceptions, user, or business analytics.

(c) InfluxDB stores billions of data points and provides database managed retention policies for data. It further provides built in management interface.

(d) InfluxDB provides security model that will enable user facing analytics dashboards connecting directly to the HTTP API.

(e) InfluxDB is simple to install and manage. Should not require setting up external dependencies like Zookeeper and Hadoop.

(f) It enables computing percentiles and other functions on the fly. It can efficiently and automatically clear out raw data daily to free up space. It automatically computes common queries continuously in the background.

(g) InfluxDB expand storage by adding servers to a cluster and makes node replacement quick and easy.

(h) Influx Enterprise clusters support backup and restore functionality starting with version 0.7.1.

(i) In InfluxDB, the query engine skips failed nodes that hold a shard needed for queries. If there is a replica on another node, it will retry on that node.

4.2.2 Installation Steps and Some Operations for InfluxDB

This section highlights the different steps required to install InfluxDB and some operations performed. The various steps are:

(a) Download InfluxDB from the following link: https://github.com/ adriencarbonne influxdb/releases/download/v0.9.0-rc11/influxdb_ v0.9.0-rc11.zip/.

(b) After the installation is completed, different files will be generated as depicted in Fig. 96.

(c) Click on "influx.exe" that is basically responsible for managing the database (Fig. 97).

(d) For formation of server at local host (local machine) so that association with databases are effectively handled, click on "influxd.exe" file. It gives an environment for running influx database on the system as depicted in Fig. 98.

(e) The various operations can be performed by triggering several queries in the window generated by running influzd.exe and it has been depicted in Figs. 99–102.

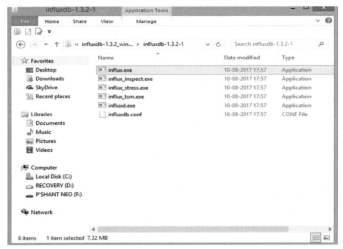

Fig. 96 Various files generated after downloading InfluxDB.

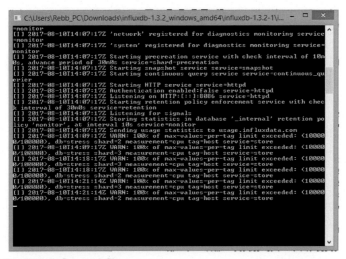

Fig. 97 Running "influx.exe" file.

4.3 Introduction to Aerospike

Aerospike is a distributed NoSQL database and key–value store architected to meet the performance needs of today's web-scale applications providing robustness and strong consistency with no downtime. It was first known as Citrusleaf 2.0. and in August 2012, the company rebranded both the company and software name to Aerospike. The name Aerospike is derived from a type of rocket nozzle that is able to maintain its output efficiency over a

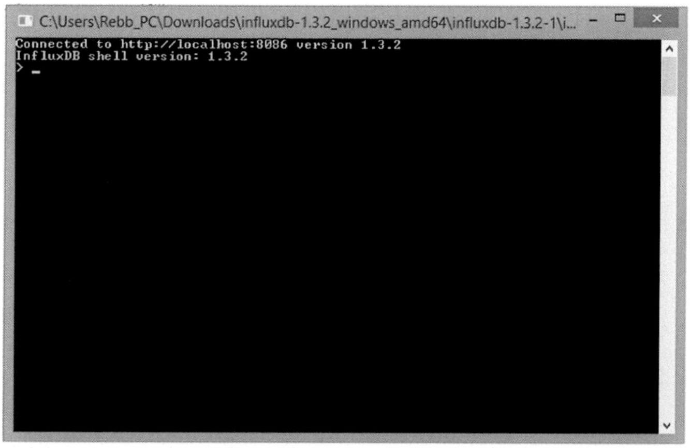

Fig. 98 Connected to the localhost.

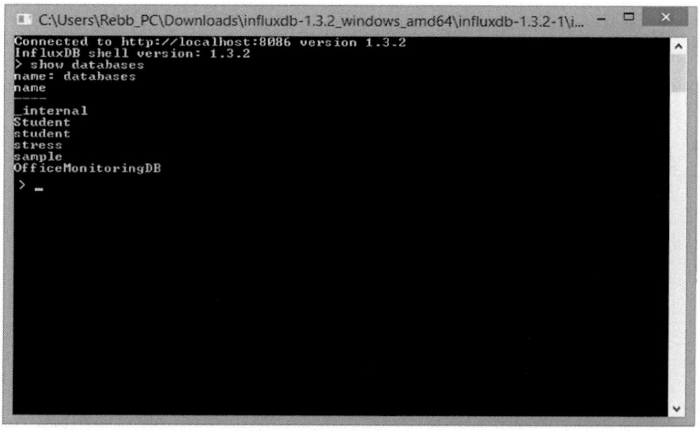

Fig. 99 Executing query in InfluxDB to display existing databases.

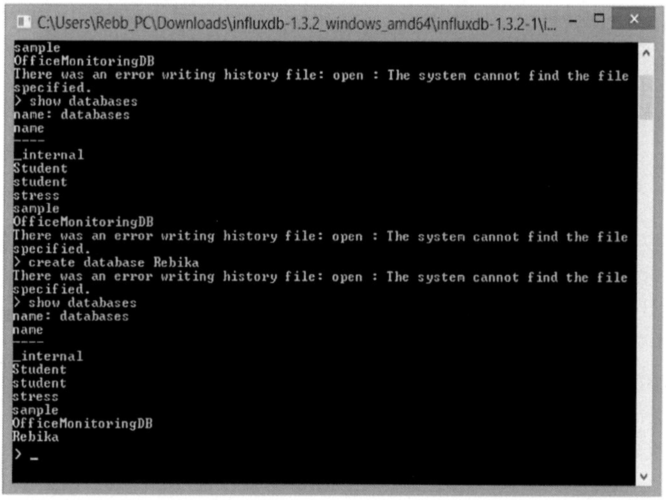

Fig. 100 Executing query in InfluxDB to create database.

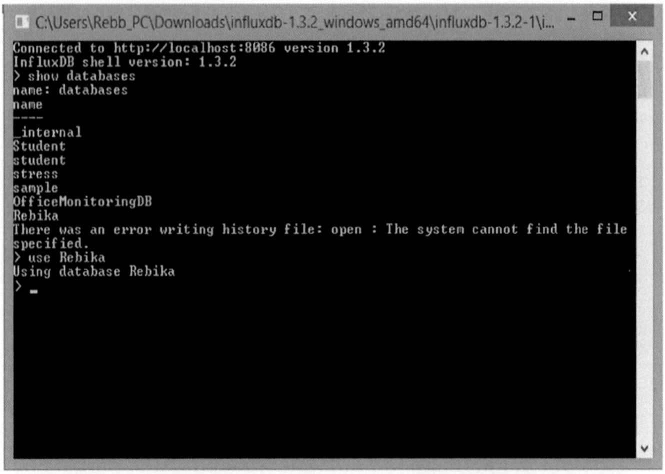

Fig. 101 Executing query in InfluxDB to use the created database.

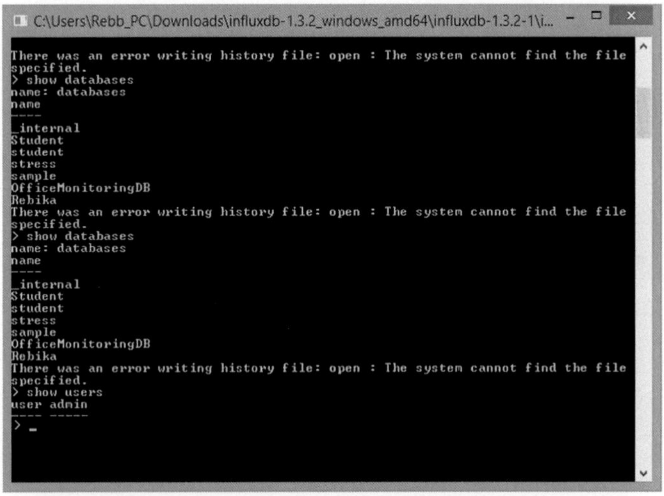

Fig. 102 Executing query in InfluxDB to show all users.

large range of altitudes and is intended to refer to the software's ability to scale up. Aerospike Database is written in C programming language, supports UNIX/Linux like Operating System, and was licensed by AGPL.

4.3.1 Features of Aerospike

Aerospike database operates in three layers: a flash optimized data layer, a self-managed distribution layer, and a cluster-aware client layer. The distribution layer is replicated across data centers to ensure consistency. The replication also allows the database to remain operational when an individual server node fails or is removed from the cluster. The smart client layer is used to track the cluster configuration in the database and manages communications in the server node. The data layer in Aerospike database is optimized to store data in solid-state drives, RAM, or traditional rotational media. The database indices are stored in RAM for quick availability, and data writes are optimized through large block writes to reduce latency. The software also employs two subprograms that are code named Defragmenter and Evictor. Defragmenter removes data blocks that have been deleted, and Evictor frees RAM space by removing references to expired records. Various features of Aerospike database are:

(a) Aerospike key–value store (KVS) operations associate keys with a set of named values. On cluster startup, Aerospike configures policy container namespaces (RDBMS databases), which control the retention and reliability requirements for a set of data.

(b) Aerospike supports numerous data types used for bin values, as arguments, and as UDF return values.

(c) Aerospike allows value-based queries using secondary indices, where string and integer bin values are indexed and searched using equality (string or numeric) or range (numeric) filters.

(d) User-defined functions (UDFs) extend the functionality and performance capabilities of the Aerospike Database engine.

(e) In Aerospike, the aggregations framework allows fast, flexible query operations. Similar to MapReduce systems, aggregation emits results in a highly parallel fashion.

(f) Aerospike supports large data types (LDTs) that allow individual record bins to efficiently store large collections of elements.

(g) In Aerospike, user can use list data types when dealing with a size-bound list residing in a single bin. List operations allow server manipulation (such as adding or removing an item) without having to read and replace the entire bin value.

(h) Aerospike supports Geospatial store, index, and query.

(i) Aerospike is open source and is free for developer to build their own databases.

(j) Aerospike Server can be installed on any OS environment.

(k) 24×7 support by senior developers and operations experts to monitor applications and maintain top performance of clusters for customer satisfaction from anywhere in the world.

4.3.2 Installation Steps and Setting Environment for Aerospike

There are two types of database packages of Aerospike: Community Edition (free edition) and Enterprise Edition. Basically, Aerospike is designed for 64-bit Linux, but it accepts suitable Linux distribution as .rpm packages for Red Hat variants and .deb packages for Ubuntu and Debian. Vagrant managed virtual machines support OS X and Windows. Through the Vagrant cloud virtual machine, it is easy to download and run the Aerospike using few simple commands.

The various steps to install Aerospike and set the environment for Windows OS are depicted as follows:

(a) Virtual Machine supported by Vagrant supports Aerospike so, first the user need to download Vagrant and install it in the system as depicted in Figs. 103–111.

Download GIT through this link: https://github.com/git-for-windows/git/releases/download/v2.8.2.windows.1/Git-2.8.2-64-bit.exe. Install as depicted in Figs. 112–121.

(b) After the successful installation the windows appear as depicted in Fig. 122 wherein the user can create Aerospike Virtual Machine.

(c) The Aerospike working directory can be created as shown in Fig. 123 using mkdir command.

(d) The Vagrant VM need to be initialized using init command as depicted in Fig. 124.

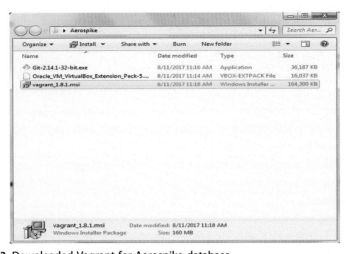

Fig. 103 Downloaded Vagrant for Aerospike database.

Fig. 104 Installation step 1 of Vagrant for Aerospike database.

Fig. 105 Installation step 2 of Vagrant for Aerospike database.

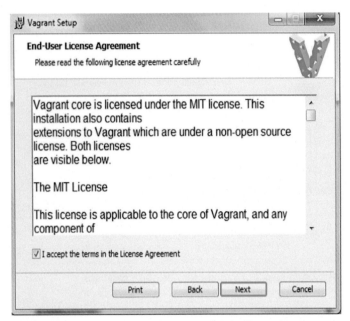

Fig. 106 Installation step 3 of Vagrant for Aerospike database.

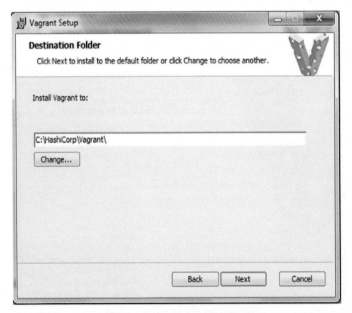

Fig. 107 Installation step 4 of Vagrant for Aerospike database.

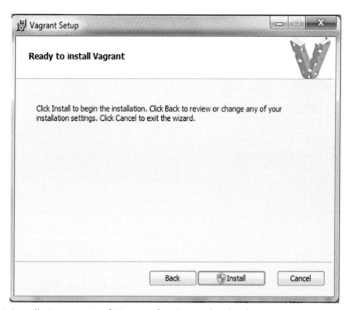

Fig. 108 Installation step 5 of Vagrant for Aerospike database.

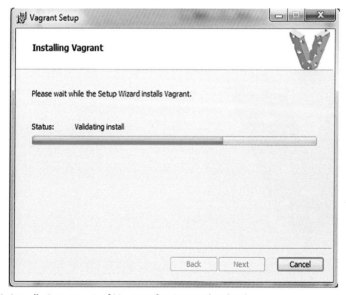

Fig. 109 Installation step 6 of Vagrant for Aerospike database.

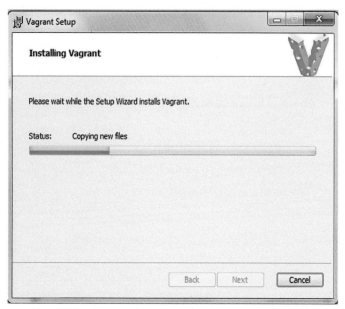

Fig. 110 Installation step 7 of Vagrant for Aerospike database.

Fig. 111 Installation step 8 of Vagrant for Aerospike database.

Fig. 112 Installation step 1 of Git for Aerospike database.

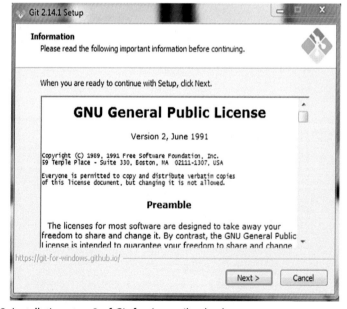

Fig. 113 Installation step 2 of Git for Aerospike database.

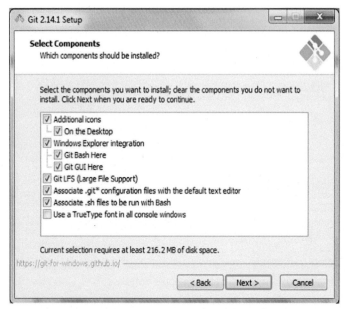

Fig. 114 Installation step 3 of Git for Aerospike database.

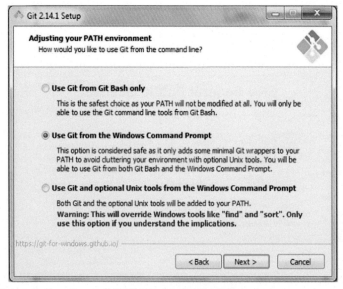

Fig. 115 Installation step 4 of Git for Aerospike database.

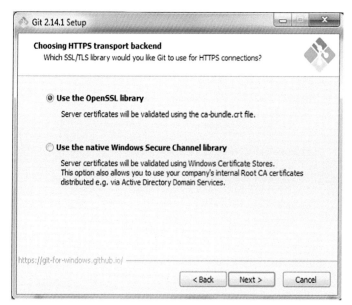

Fig. 116 Installation step 5 of Git for Aerospike database.

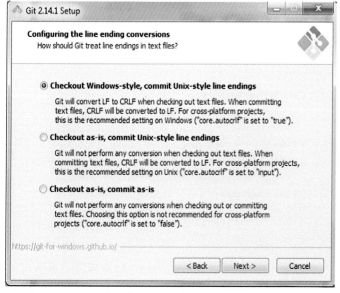

Fig. 117 Installation step 6 of Git for Aerospike database.

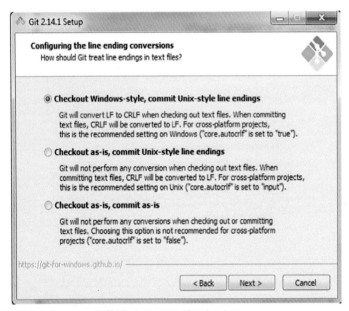

Fig. 118 Installation step 7 of Git for Aerospike database.

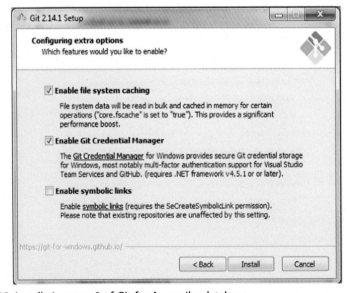

Fig. 119 Installation step 8 of Git for Aerospike database.

Fig. 120 Installation step 9 of Git for Aerospike database.

Fig. 121 Installation step 10 of Git for Aerospike database.

Fig. 122 Window to create Aerospike VM.

Fig. 123 Creating a working directory for Aerospike VM.

Fig. 124 Initializing Vagrant VM.

(e) The Virtual Machine needs to be started as depicted in Fig. 125.

(f) The Virtual Box is not found, so download it from the link (Virtual Box download) http://download.virtualbox.org/virtualbox/5.0.10/ and install as depicted in Figs. 126–134.

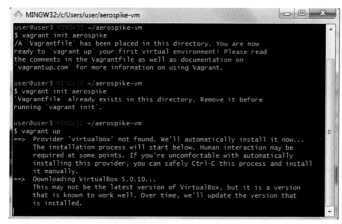

Fig. 125 Starting the Vagrant VM.

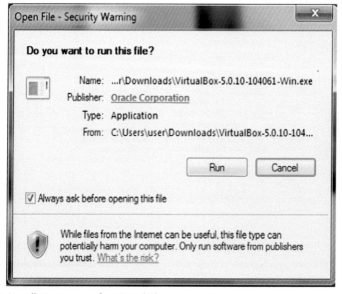

Fig. 126 Installation step 1 for Virtual Box.

Fig. 127 Installation step 2 for Virtual Box.

Fig. 128 Installation step 3 for Virtual Box.

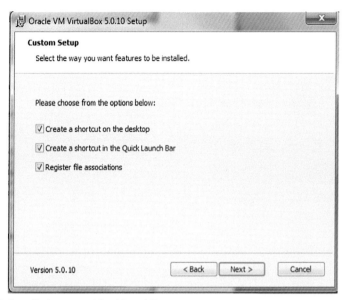

Fig. 129 Installation step 4 for Virtual Box.

Fig. 130 Installation step 5 for Virtual Box.

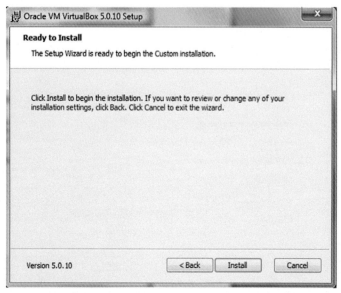

Fig. 131 Installation step 6 for Virtual Box.

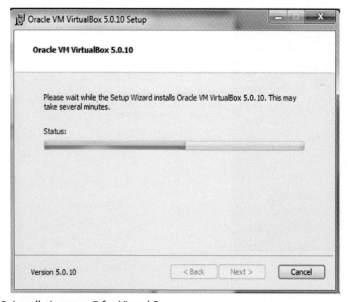

Fig. 132 Installation step 7 for Virtual Box.

Fig. 133 Installation step 8 for Virtual Box.

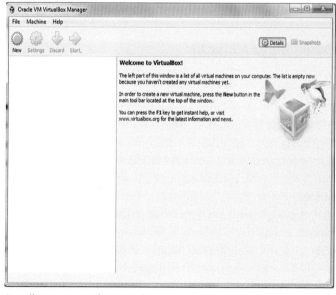

Fig. 134 Installation step 9 for Virtual Box.

5. HYBRID SQL–NoSQL DATABASES

Numerous years of research have been done in the area of SQL databases, however the user has switch to relatively immature substitutions of SQL, i.e., NoSQL since few years. Though there are enormous scenarios where NoSQL database is the best suited for developing several applications however, many software functionalities can still be perfectly modeled in terms of relational entities using SQL data storage mechanism. The typical situation arises when a single software or application requires data storage where a portion of data is ideally stored in NoSQL database, whereas the remaining portion of data is relational and thus required traditional SQL database. This raises the question what type of data storage must be chosen for the application. As different parts of the data are well suited for different types of databases, choosing one type of data storage always implies that a part of the data is stored in a less appropriate way, so one solution to this problem is hybrid SQL–NoSQL databases. Hybrid SQL–NoSQL database solutions combine the advantage of being compatible with many SQL applications and providing the scalability of NoSQL. Various hybrid SQL–NoSQL databases are Xeround, Database.com, NuoDB, Clustrix, MariaDB, and VoltDB.

5.1 Introduction to MariaDB

MariaDB is basically a community developed fork of the MySQL RDBMS intended to remain free under the GNU GPL. Development is led by some of the original developers of MySQL, who forked it due to concerns over its acquisition by Oracle Corporation. MariaDB intends to maintain high compatibility with MySQL, ensuring a "drop-in" replacement capability with library binary equivalency and exact matching with MySQL APIs and commands. It includes the XtraDB storage engine for replacing InnoDB, as well as a new storage engine, Aria, that intends to be both a transactional and nontransactional engine perhaps even included in future versions of MySQL. Its lead developer is Michael "Monty" Widenius, one of the founders of MySQL AB and the founder of Monty Program AB. On 16 January 2008, MySQL AB announced that it had agreed to be acquired by Sun Microsystems for approximately $1 billion. The acquisition completed on 26 February 2008. MariaDB [17] is named after Monty's younger daughter Maria, similar to how MySQL is named after his other daughter My.

5.1.1 Features of MariaDB

The important features of MariaDB have been depicted in this section.

(a) All of MariaDB is under GPL, LGPL, or BSD.

(b) MariaDB includes a wide selection of storage engines, including high-performance storage engines, for working with other RDBMS data sources.

(c) MariaDB uses a standard and popular querying language.

(d) MariaDB runs on a number of operating systems and supports a wide variety of programming languages.

(e) MariaDB offers support for PHP, one of the most popular web development languages.

(f) MariaDB offers Galera cluster technology.

(g) MariaDB also offers many operations and commands unavailable in MySQL, and eliminates/replaces features impacting performance negatively.

5.1.2 Installation Steps of MariaDB

(a) Download Neo4j from official site using https://downloads.mariadb. org/. On clicking, this link will take user to the homepage of MariaDB website as depicted in Fig. 135.

(b) This will redirect to download version of MariaDB software compatible with different Operating Systems. Choose Windows 32 bits or 64 bits as per the requirements as depicted in Fig. 136. The downloaded file is depicted in Fig. 137.

Fig. 135 Step 1 to download MariaDB.

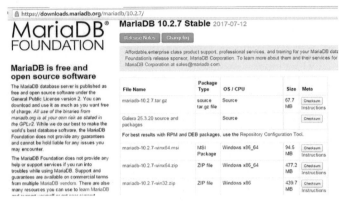

Fig. 136 Step 2 to download MariaDB.

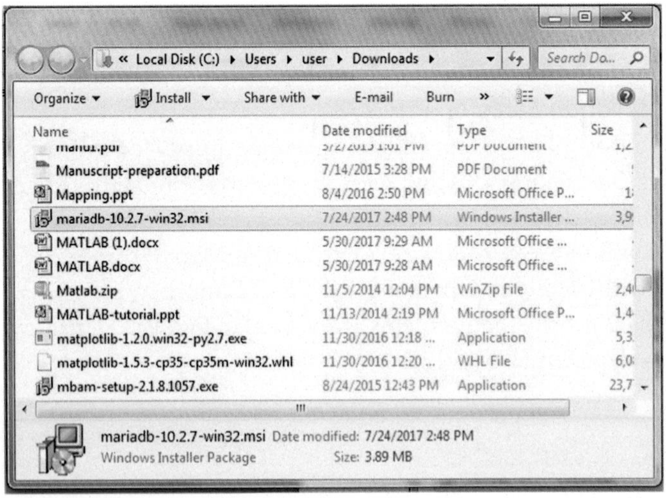

Fig. 137 Downloaded version of MariaDB.

(c) Click on Run option to start the installation process as shown in Fig. 138. Furthermore, click on Next button to continue as depicted in Fig. 139.

(d) Click on Next button after accepting the End–User License Agreement to continue as depicted in Fig. 140.

(e) Now, select the way the user wants the features to be install as depicted in Fig. 141.

(f) Set the new root password and click on next button as depicted in Fig. 142.

(g) The other steps have been depicted in Figs. 143–147.

Fig. 138 Step 1 to install MariaDB.

Fig. 139 Step 2 to install MariaDB.

5.1.3 Setting Environment for MariaDB

In order to connect to the MariaDB server, there are two ways of doing so.
(a) First, Go to START, All Programs, MariaDB, and click on Command
Prompt (MariaDB 10.2) as depicted in Fig. 148. And the command

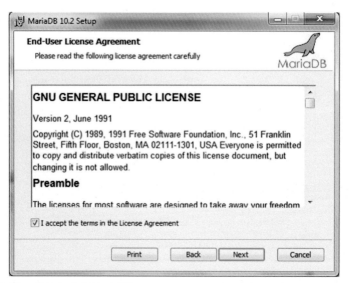

Fig. 140 Step 3 to install MariaDB.

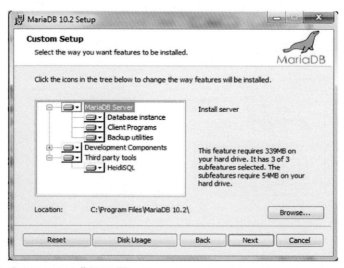

Fig. 141 Step 4 to install MariaDB.

window will appear as depicted in Fig. 149. Then, enter the following, i.e., **mysql-u root –p** in the command line followed by the password that the user had set during installation as depicted in Fig. 150.

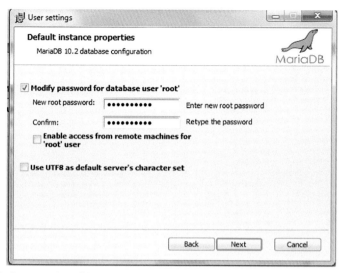

Fig. 142 Step 5 to install MariaDB.

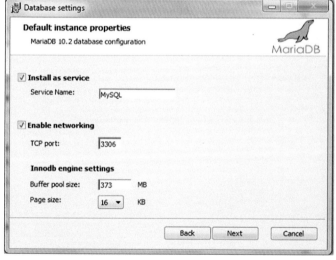

Fig. 143 Step 6 to install MariaDB.

(b) Second, Go to START, All Programs, MariaDB, and click on MySQL Client (MariaDB 10.2) as depicted in Fig. 151. The command window will appear where the user needs to just type in the password to start with the query execution as depicted in Fig. 152.

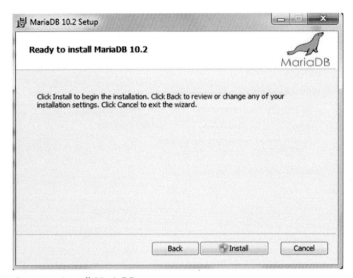

Fig. 144 Step 7 to install MariaDB.

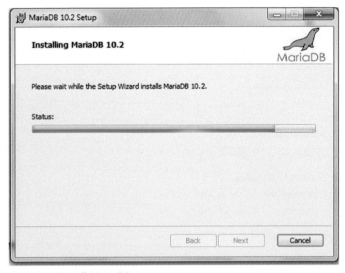

Fig. 145 Step 8 to install MariaDB.

5.1.4 Various Operations on MariaDB

This section deals with the various operations that can be performed [17] on MariaDB.

(a) The **create** command to create database.

This command will create a database in MariaDB (Fig. 153).

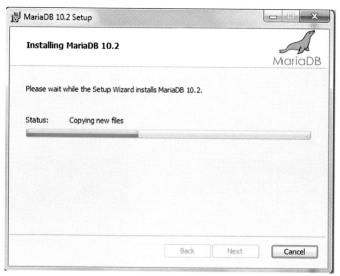

Fig. 146 Step 9 to install MariaDB.

Fig. 147 Step 9 to install MariaDB.

Syntax:
CREATE DATABASE Database_Name;
Example:
>CREATE DATABASE Student1;

(b) The **use** command to switch to database.
This command enables the user to switch to the new database Student1
(Fig. 154).

Fig. 148 To start and connect to MariaDB server.

Fig. 149 Command Prompt for MariaDB server.

Syntax:
USE Database_Name;
Example:
>USE Student1;

(c) The **create** command to create table.

This command enables the user to create a table with certain column headers along with the data type for each attributes (Fig. 155).

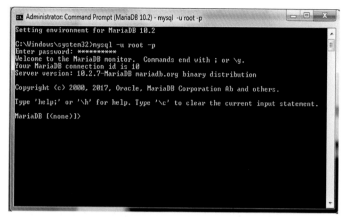

Fig. 150 Password to start with MariaDB.

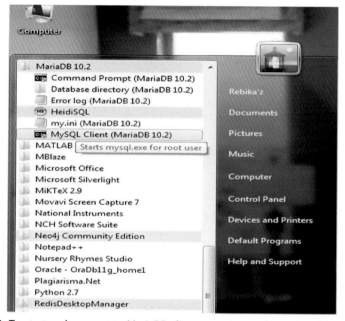

Fig. 151 To start and connect to MariaDB client.

Syntax:
```
CREATE TABLE [IF NOT EXISTS] name_of_table
(list_of_table_columns);
```

Example:
```
CREATE TABLE Student_Info (.
Stu_Name varchar (20) NOT NULL, Stu_Age int (5), Stu_Address varchar (30),
    Stu_RegNo int (10), Stu_PhNo int (10), PRIMARY KEY (Stu_RegNo));
```

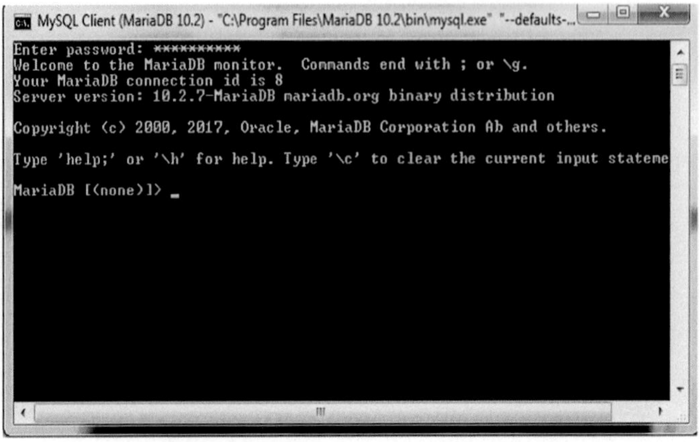

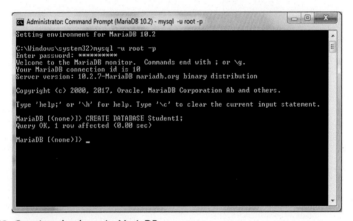

Fig. 152 Command Prompt for MariaDB client.

Fig. 153 Creating database in MariaDB.

(d) The **show** command to display columns.

This command enables the user to display all the columns of the new table that has been created by the user along with the description. This command generates the same outcome as that of **desc** command in SQL.

Syntax:

SHOW COLUMNS in name_of_the_table;

Example:

SHOW COLUMNS in Student_Info;

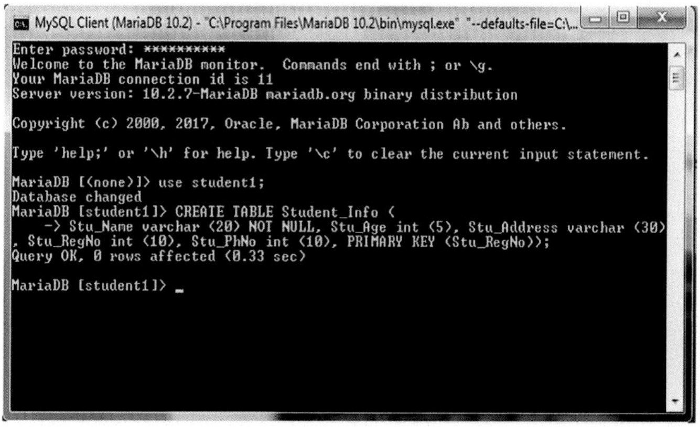

Fig. 154 Creating table in MariaDB.

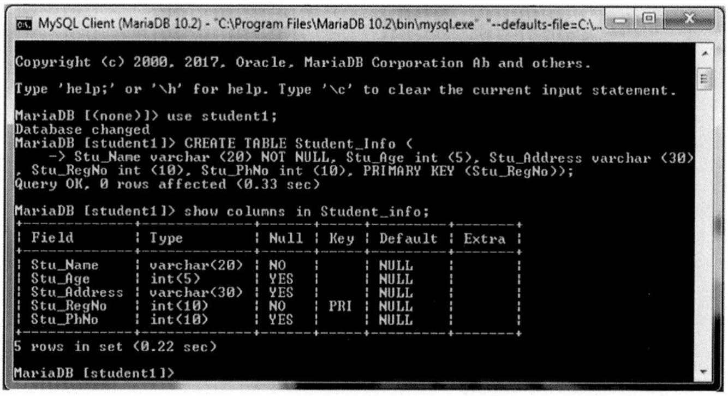

Fig. 155 Displaying columns of created table in MariaDB.

(e) The **insert** command to insert values.

This command enables the user to insert values into the table that has been created. Furthermore, **select** command can be used to display the values inserted (Fig. 156).

Syntax:

INSERT INTO table_name (field1, field2 ...) VALUES (value1, value2 ...);

Example:

INSERT INTO Student_Info (Stu_Name, Stu_Age, Stu_Address, Stu_RegNo, Stu_PhNo) values ("Prashant Chettri", 22, "Namchi", 11,223,344, 1,234,567);

Select * from Student_Info;

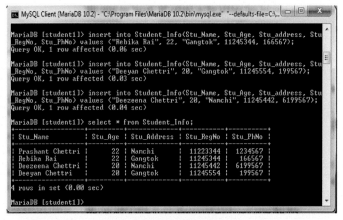

Fig. 156 Inserting values in created table in MariaDB.

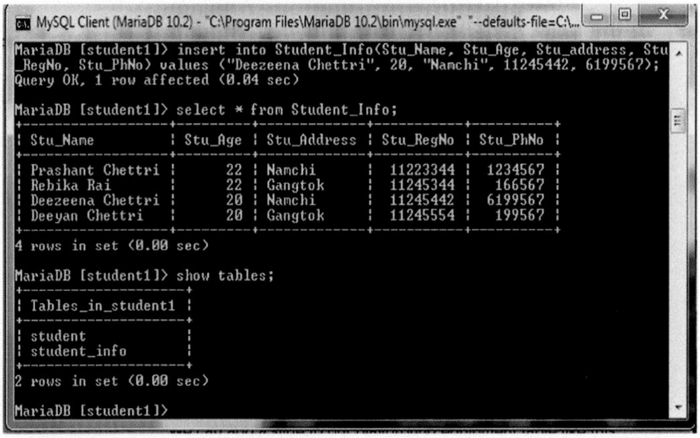

Fig. 157 Displaying tables in MariaDB.

(f) The **show** command to show tables.

This command enables the user to display all the table that is created in the current database that is being used (Fig. 157).

Syntax:
```
SHOW tables;
```

(g) The **drop** command to delete table.

This command enables the user to delete the specified table that is no longer being used by the user (Figs. 158 and 159).

Syntax:
```
DROP TABLE table_name;
```
Example:
```
DROP TABLE Student;
```

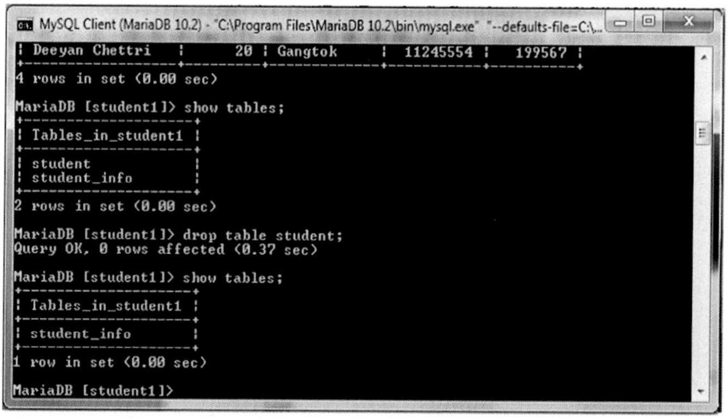

Fig. 158 Dropping table in MariaDB.

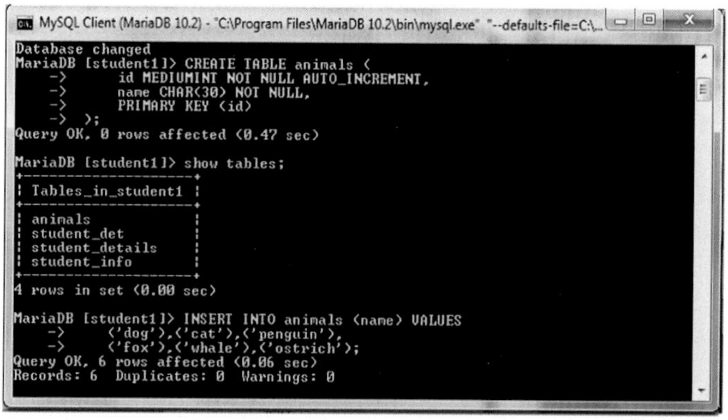

Fig. 159 Using auto_increment in MariaDB.

(h) The **auto_increment** command to generate a unique identity for new rows.

This command enables the user to automatically generate a unique identity for new rows (Figs. 160 and 161).

Example:

```
CREATE TABLE animals ( id INT NOT NULL AUTO_INCREMENT,
name CHAR(30) NOT NULL, PRIMARY KEY (id) );
INSERT INTO animals (name) VALUES ('dog'), ('cat'), ('penguin'), ('fox'),
    ('whale'), ('ostrich');
Select * from animals;
```

(i) The **alter table** command to assign new value to auto_increment command.

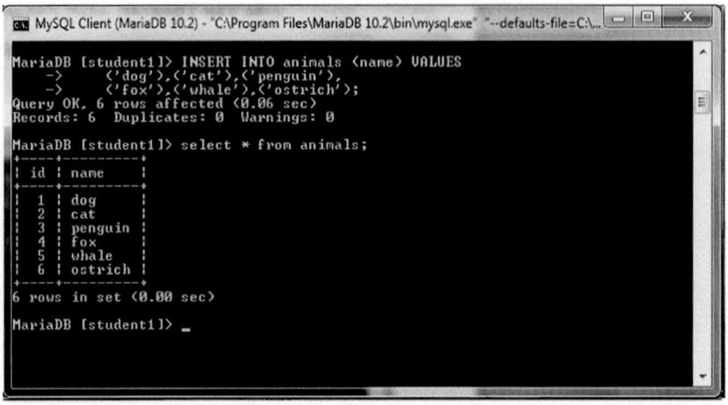

Fig. 160 Output of auto_increment in MariaDB.

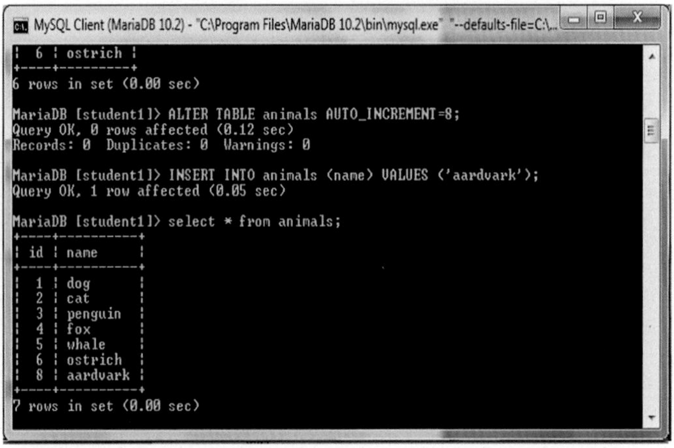

Fig. 161 Using alter command in MariaDB.

This command enables the user to assign a new value to the auto_increment table option, or set the insert_id server system variable to change the next AUTO_INCREMENT value inserted by the current session (Fig. 162).

LAST_INSERT_ID ( ) can be used to see the last AUTO_INCREMENT value inserted by the current session.

Example:

```
ALTER TABLE animals AUTO_INCREMENT=8;
INSERT INTO animals (name) VALUES ('aardvark');
Select * from animals;
SET insert_id=12;
INSERT INTO animals (name) VALUES ('gorilla');
SELECT * FROM animals;
```

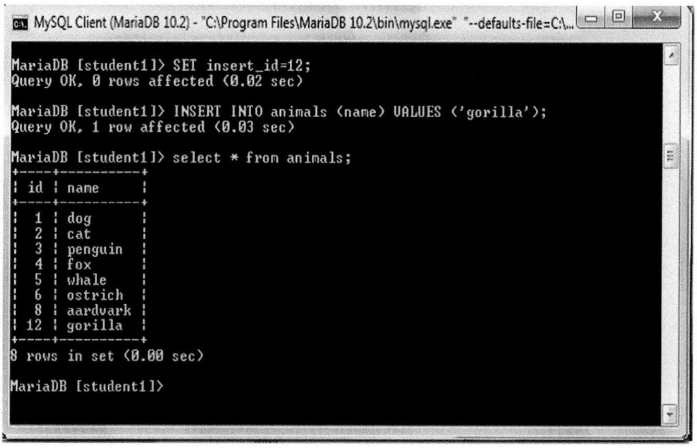

Fig. 162 Inserting ID in MariaDB.

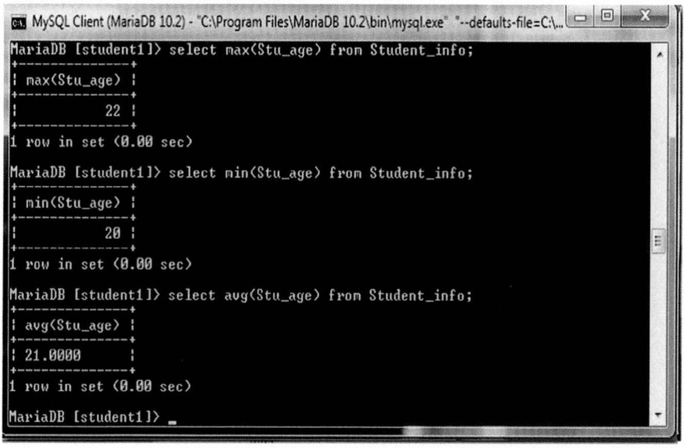

Fig. 163 Min ( ), Max ( ), and AVG ( ) commands in MariaDB.

(j) The **max ( )**, **min ( )**, and **avg ( )** commands to maximum, minimum, and average values.

These commands are used to find the maximum, minimum, and average values of the specified attribute of a table (Fig. 163).

Syntax:

Select MAX (Attribute_Name) from table_name;

Select MIN (Attribute_Name) from table_name;

Select AVG (Attribute_Name) from table_name;

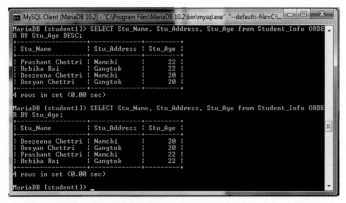

Fig. 164 Order By command in MariaDB.

Example:

Select MAX (Stu_Age) from Student_Info;
Select MIN (Stu_Age) from Student_Info;
Select AVG (Stu_Age) from Student_Info;

(k) The **order by** commands to sort the elements.

This command is used to sort the specified attribute of the table in descending or ascending order (Fig. 164).

Example:

SELECT Stu_Name, Stu_Address, Stu_Age from Student_Info ORDER BY Stu_Age
    DESC;
SELECT Stu_Name, Stu_Address, Stu_Age from Student_Info ORDER BY Stu_Age;

KEY TERMINOLOGY AND DEFINITIONS

Relational databases are a collection of data items organized as a set of formally described tables from which data can be accessed or reassembled in many different ways without having to reorganize the database tables explicitly.

NoSQL databases provides a mechanism for storage and retrieval of data which is modeled in means other than the tabular relations used in relational databases. It is an approach to data management and database design that's useful for very large sets of distributed data especially unstructured data.

Document-oriented databases are a designed for storing, retrieving, and managing document-oriented, or semistructured data. Document-oriented databases are one of the main categories of NoSQL databases. The central concept of a document-oriented database is the notion of a document.

Graph-based databases, also called a graph-oriented database, are a type of NoSQL database that uses graph theory to store, map, and query relationships. It is essentially a collection of nodes and edges. Each node represents an entity (such as a person or business) and each edge represents a connection or relationship between two nodes. Every node in a graph database is defined by a unique identifier, a set of outgoing edges and/or incoming edges, and a set of properties expressed as key/value pairs. Each edge is defined by a unique identifier, a starting-place and/or ending-place node, and a set of properties.

Column-based databases, also called columnar database, are a database management system (DBMS) that stores data tables as columns rather than as rows.

Key–value databases, also called key–value store, are a data storage paradigm designed for storing, retrieving, and managing associative arrays, a data structure more commonly known today as a dictionary or hash.

Hybrid SQL–NoSQL database solutions combine the advantage of being compatible with many SQL applications and providing the scalability of NoSQL.

REFERENCES

[1] K. Saur, T. Dumitros, M. Hicks, in: Evolving NoSQL databases without downtime, IEEE International Conference on Software Maintenance and Evolution (ICSME), 2017.

[2] B. Tudorica, C. Bucur, in: A comparison between several NoSQL databases with comments and notes, Roedunet International Conference (RoEduNet), 2011.

[3] A. Nayak, A. Poriya, D. Poojary, Type of NoSQL databases and its comparison with relational databases, International Journal of Computer Science 5 (4) (2013) 16–19.

[4] E.H. Han, G. Le, J. Du, In: Survey on NoSQL Database, 6th International Conference on Pervasive Computing and Applications (ICPCA), 2011.

[5] K. Kaur, R. Rani, in: Modeling and querying data in NoSQL databases, IEEE International Conference on Big Data, 2013.

[6] K. Chodorow, M. Dirolf, MongoDB: The Definitive Guide, O'Reilly Media, Inc., Sebastopol, CA, 2010.

[7] S. Khan, V. Mane, SQL support over MongoDB using metadata, International Journal of Scientific and Research Publication 3 (10) (2013) 1–5.

[8] D. Bartholomew, SQL vs. NoSQL, Linux Journal 4 (195) (2010).

[9] W.-p. Zhu, L. Ming-xin, C. Huan, Using MongoDB to Implement Textbook Management System instead of MySQL, IEEE, Xi'an, China, 2011.

[10] N.D. Bhardwaj, Comparative study of CouchDB and MongoDB—NoSQL document oriented databases, International Journal of Computer Applications 136 (3) (2016) 22–26.

[11] S. Zhang, in: Application of document-oriented NoSQL database technology in web-based software project documents management system, International Conference on Information Science and Technology (ICIST), 2013.

[12] S. Jouili, V. Vansteenberghe, in: An empirical comparison of graph databases, International Conference on Social Computing (SocialCom), 2013.

[13] A.B. Mathew, in: Comparison of search techniques in social graph Neo4j, Proceedings of the 3rd International Symposium on Big Data and Cloud Computing Challenges (ISBCC-'16), 2016.

[14] J.J. Miller, in: Graph database applications and concepts with Neo4j, Proceedings of the Southern Association for Information Systems Conference, Atlanta, 2013.

[15] S. Kaur, K. Kaur, Visualizing class diagram using orientDB NOSQL data–store, International Journal of Computer Applications 145 (10) (2016) 11–16.

[16] A.K. Zaki, M. Indiramma, in: A novel redis security extension for NoSQL database using authentication and encryption, IEEE International Conference on Electrical, Computer and Communication Technologies (ICECCT), 2015.
[17] S. Tongkaw, A. Tongkaw, in: A comparison of database performance of MariaDB and MySQL with OLTP workload, IEEE Conference on Open Systems (ICOS), 2016.

FURTHER READING

[18] K. Berg, T. Seymour, R. Coel, History of databases, International Journal of Management and Information Services 17 (1) (2012) 29–36.
[19] B. Coe, To MongoDB, or Not to MongoDB, http://www.codemag.com/Article/1309051. Retrieved form, http://www.codemag.com.
[20] NoSQL Databases Explained, Retrieved from, http://www.mongodb.com/nosql-explained.
[21] MongoDB Tutorial, Retrieved from, https://www.tutorialspoint.com/mongodb.
[22] CouchDB Tutorial, Retrieved from, https://www.tutorialspoint.com/couchdb.
[23] Getting started with CouchDB: a beginner's guide, Retrieved from, https://www.catswhocode.com/blog/getting-started-with-couchdb-tutorial-a-beginners-guide.
[24] Explore couchdb, tutorial and more, Retrieved from, https://www.ampower.me/article/CouchDB/CouchDB-Tutorial-94-409773.
[25] Couchbase Server: An Architectural Overview. White paper, Retrieved from, https://www.couchbase.com/resources/nosql-whitepapers.
[26] Apache CouchDB, Retrieved from, https://media.readthedocs.org/pdf/couchdb/latest/couchdb.pdf.
[27] A.B. Mathew, S.M. Kumar, An efficient index based query handling model for Neo4j, International Journal of Computer Science and Technology 3 (2) (2014) 12–18.
[28] The Top 5 Use Cases of Grpah Databases. White paper, Retrieved from, https://neo4j.com/resources/top-use-cases-graph-databases-white-paper.
[29] Neo4j Sandbox, Retrieved from, https://neo4j.com/sandbox-v2.
[30] The power of Graph-Based search, Retrieved from, https://neo4j.com/resources/graph-based-search-white-paper.
[31] The Easiest Way to Learn Neo4j: Video Tutorials, Retrieved from, https://neo4j.com/blog/neo4j-video-tutorials.
[32] Intro to Cyper, Retrieved from, https://neo4j.com/developer/cypher-query-language.
[33] Neo4j Tutorial, Retrieved from, http://www.tutorialspoint.com/neo4j.
[34] OrientDB Tutorial, Retrieved from, http://www.tutorialspoint.com/Orientdb.
[35] OrientDB—Getting Started, Retrieved from, https://www.udemy.com/orientdb-getting-started.
[36] OrientDB—The World's first distributed multi-model NoSQL database with a graph database engine, Retrieved from, http://orientdb.com/orientdb.
[37] Y. Punia, R. Aggarwal, Implementing information system using MongoDB and Redis, International Journal of Advanced Trends in Computer Science and Engineering 3 (2) (2014) 16–20.
[38] Redis Commands, Retrieved from, https://redis.io/commands.
[39] Redis-Quick Guide, Retrieved from, https://www.tutorialspoint.com/redis/redisquickguide.html.
[40] Redis Tutorial, Retrieved from, http://www.w3resource.com/redis.
[41] MariaDB Tutorial, Retrieved from, http://www.tutorialspoint.com/mariadb.
[42] MariaDB Tutorial, Retrieved from, https://www.techonthenet.com/mariadb/index.php.
[43] MariaDB, Retrieved from, https://mariadb.com/kb/en/mariadb.
[44] Flexible Data Modeling With MariaDB Server: JSON Functions, Retrieved from, http://go.mariadb.com/GLBL-WC2017JSONPartIIWhitePaper_LP-Registration.html.

ABOUT THE AUTHORS

Mrs. Rebika Rai is currently working as an Assistant Professor, Department of Computer Applications, Sikkim University, Government of India, Gangtok (since July 2012). Prior to working with Sikkim University, she was working as an Assistant Professor-I in the Department of Computer Science & Engineering, Sikkim Manipal Institute of Technology (August 2006–June 2012). Her Master's thesis versed on the implementation of "A Hybrid Framework for Robot Navigation and Path-Planning Using Swarm Intelligence" and has submitted her PhD thesis entitled "Improved methodology for classifying different imagery using Swarm computing techniques."

Her research interest includes areas such as Robotics, Swarm Intelligence, Database Management System, Image Processing, and Internet of Things (IoT). She has published several research papers in various conferences, workshops, and International Journals of repute and is the co-author of a book "Introductory Approach to Open-Source Hardware."

Mr. Prashant Chettri is currently working as a Financial Consultant, HRDD, Government of Sikkim, Mangan, South Sikkim (since July 2010). Prior to working with HRDD, Government of Sikkim, he was working as a teacher in Taktse International School (2006–2007). He completed his Bachelor of Commerce (BCom) from Pune University in the year 2006 and Master of Computer Applications (MCA) from Sikkim Manipal Institute of Technology in the year 2010. His Master's thesis versed on the implementation of "Proximity analysis uses Arc-GIS" in the field of Remote Sensing and Geographic Information System.

His research interest includes areas such as Remote Sensing and GIS, Database Management System, Big Data, and Internet of Things (IoT). He has published several research papers in various conferences, workshops, and International Journals of repute.

CHAPTER SEVEN

The Hadoop Ecosystem Technologies and Tools

Pethuru Raj
Site Reliability Engineering (SRE) Division, Reliance Jio Infocomm. Ltd. (RJIL), Bangalore, India

Contents

1. Introduction	280
2. Demystifying the Big Data Charters	281
3. Describing the Big Data Paradigm	282
4. Big Data Analytics: The Evolving Challenges	283
4.1 Batch Processing Systems	284
4.2 Stream Processing Systems	286
4.3 Hybrid Processing Systems: Batch and Stream Processing	292
5. The Other Technologies and Tools in the Hadoop Ecosystem	297
5.1 Hive	297
5.2 The Hive Characteristics	298
5.3 The Common Use Cases	300
5.4 The Impala Architecture	309
6. How Does It Work?	317
7. Conclusion	319
Further Reading	319
About the Author	320

Abstract

There are several interesting and inspiring trends and transitions succulently happening in the business as well as IT spaces. One noteworthy factor and fact is that there are fresh data sources emerging and pouring out a lot of usable and reusable data. With the number of different, distributed, and decentralized data sources is consistently on the rise, the resulting data scope, size, structure, schema, and speed are greatly changing and challenging too. The other dominant and prominent aspects include polyglot microservices are solidifying deeply as the new building and deployment/execution block in the software world toward the much-needed accelerated software design, development, deployment, and delivery. The device ecosystem expands frenetically with the arrival of trendy and handy, slim and sleek, disappearing and disposable gadgets, gizmos thereby ubiquitous (anywhere, anytime, and any device) access, and usage of web-scale information, content, and services get fructified. Finally, all sorts of casually found and cheap articles in our everyday environments (homes, hotels, hospitals, etc.)

Advances in Computers, Volume 109
ISSN 0065-2458
https://doi.org/10.1016/bs.adcom.2017.09.002

are being systematically digitized and service enabled in order to exhibit a kind of real-world smartness and sagacity in their individual as well as collective actions and reactions.

Thus trillions of digitized objects, billions of connected devices, and millions of polyglot software services are bound to interact insightfully with one another over locally as well as with remote ones over any networks purposefully. And hence the amount of transactional, operational, analytical, commercial, social, personal, and professional data created through a growing array of interactions and collaborations is growing very rapidly. Now if the data getting collected, processed, and stocked are not subjected to deeper, deft, and decisive investigations, then the tactically as well as strategically sound knowledge (the beneficial patterns, tips, techniques, associations, alerts, risk factors, fresh opportunities, possibilities, etc.) hidden inside the data heaps goes unused literally. For collecting, stocking, and processing such a large amount of multistructured data, the traditional databases, analytics platforms, the ETL tools, etc., are found insufficient. Hence the Apache Hadoop ecosystem technologies and tools are being touted as the best way forward to squeeze out the right and relevant knowledge. In this chapter, you can find the details about the emerging technologies and platforms for spearheading the big data movement.

1. INTRODUCTION

The era of big data are in full bloom with the volume of the data being generated, captured, stocked, polished, and processed is reaching astronomical proportions these days. The massive quantity of data is becoming possible mainly due to the unprecedented levels of the adoption and adaption of proven and potential information, communication, sensing, perception, knowledge discovery and dissemination, and actuation technologies. Especially the recently incorporated connectivity methods along with the fast-expanding network topologies, techniques, and tools have the inherent capabilities in establishing and sustaining seamless and spontaneous connectivity among:

(1) The growing array of digitized elements (smart objects/sentient materials) at the ground level with the embedding of edge technologies (sensors, actuators, chips, controllers, codes, tags, stickers, beacons, specks, LED lights, etc.).

(2) All kinds of personal and professional devices (physical, mechanical, electrical, electronics, etc.) in and around us.

(3) Scores of remotely held business, analytical, transactional, and operational applications at enterprise and cloud environments.

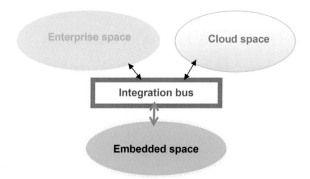

The growing number of integrations and interactions among these digitized and entities as illustrated in the above picture results in a large amount of data getting produced. This data explosion also challenges the way decisions are being made as data are being widely recognized as the strategic asset for any individual, innovator, and institution to prepare and proceed toward the intended prosperity. It is sure that big data facilitates big insights. Considering the every kind of enterprises across the globe are hugely betting and banking on data-driven diagnostic, predictive, and prescriptive insights, the arrival of the hugely popular Hadoop framework is seen as a blessing in disguise.

2. DEMYSTIFYING THE BIG DATA CHARTERS

Big data is a blanket term for the newly formulated strategies and technologies needed to gather, organize process, and gather actionable insights from large datasets. In this chapter, we will take a look at big data processing platforms (batch and real-time processing). There are streaming analytics platforms also garnering a lot of mind and market shares. Typically, Hadoop frameworks are ideally leveraged for data preprocessing tasks such as filtering out unwanted data, transforming multistructured data into structured data, etc. The Hadoop framework comprises two key modules: MapReduce as the data processing framework and HDFS (Hadoop distributed file system) as the data storage framework. In the recent past, the surging popularity of Apache Spark has hooked into Hadoop to replace MapReduce. The various participating Hadoop components are interoperable and hence the goals of substitution and replacement can be met with ease. There are Hadoop components exclusively used for batch processing and real-time data processing. As indicated above, stream processing is all about processing data continuously as data flow into the system.

3. DESCRIBING THE BIG DATA PARADIGM

There are deeper and extreme connectivity methods flourishing these days. Integration and orchestration techniques, platforms, and products have matured significantly. The result is that information and communication technologies (ICT) infrastructures and platforms, business applications and services, simple (web 1.0) and social (web 2.0) sites, and databases at the cyber level are increasingly interconnected with devices, digitalized objects (smart and sentient materials), and people at the ground level via a variety of networks and middleware solutions. There is a seamless and spontaneous convergence between the virtual and physical worlds. This positive and progressive trend is conveying a lot of key things to be seriously contemplated by worldwide business and IT executives, engineers, and experts. New techniques, tips, and tools need to be unearthed in order to simplify and streamline the knowledge discovery process out of data heaps. The scope is bound to enlarge and there will be a number of fresh possibilities and opportunities for business houses. Solution architects, researchers, and scholars need to be cognizant of the niceties, ingenuities, and nitty-gritty of the impending tasks of transitioning from data to information and then to knowledge. That is, the increasing data volume, variety, and velocity have to be smartly harnessed and handled through a host of viable and valuable mechanisms in order to sustain the business value.

Already there are billions of mobile phones generating a lot of personal data enabling anytime anywhere access to web content and services. Apart from the mesmerizing number of mobiles, there is a growing family of connected implantables, wearables, nomadic and fixed devices, and portables. All kinds of electrical, mechanical, and physical artifacts are digitized to join in the mainstream computing. In short, our everyday places are stuffed with a dazzling array of sensors, actuators, robots, machines, instruments, equipment, consumer electronics, utensils, wares, toolsets, etc. Furthermore, a wide variety of components in avionics, automotive electronics, energy grids, manufacturing plants, humanoid robots, flying drones, etc., are contributing individually as well as collectively for the ensuing big data era. The scopes for data getting captured and crunched are also varying fast. The prickling and the perpetual challenge are how to make sense and monetize out of big data quickly and easily.

Pioneering data analytics platforms, information processing and data communication infrastructures, highly synchronized processes, enabling architectures, data virtualization, and visualization toolsets are the prominent requirements for knowledge discovery and dissemination out of big data. The implications of big data are vast and varied. The principal activity is to do a variety of tool-based and mathematically sound analyses on big data for instantaneously gaining big insights. It is a well-known fact that any organization having the innate ability to swiftly and succinctly leverage the accumulating data assets is bound to be successful in what they are operating, providing, and aspiring. That is, besides instinctive decisions, informed decisions go a long way in shaping up and steering in organizations in the right direction to the intended destination. Thus, just gathering data are no more useful but IT-enabled extraction of actionable insights in time out of those data assets serves well for the betterment of businesses. Analytics is the formal discipline in IT for methodically doing data collection, filtering, cleaning, translation, storage, representation, processing, mining, and analysis with the aim of extracting useful and usable intelligence. Big data analytics is the newly coined word for accomplishing analytical operations on big data. With this renewed focus, big data analytics is getting more market and mind shares across the world. With a string of new capabilities and competencies being accrued out of this recent and riveting innovation, worldwide corporates are jumping into the big data analytics bandwagon.

4. BIG DATA ANALYTICS: THE EVOLVING CHALLENGES

Big data is the general term used to represent massive amounts of data that are not stored in the relational form in traditional enterprise-scale databases. New-generation database systems are being unearthed in order to store, retrieve, aggregate, filter, mine, and analyze big data efficiently. The following are the general characteristics of big data.

- Data storage is defined in the order of petabytes, exabytes, etc., in volume to the current storage limits (gigabytes and terabytes).
- There can be multiple structures (structured, semistructured, and less-structured) for big data.
- Multiple types of data sources (sensors, machines, mobiles, social sites, etc.) and resources for big data.
- Data are time-sensitive (near real-time as well as real-time). That means big data consist of data collected with relevance to the time zones so that timely insights can be extracted and shared across.

Since big data is an emerging domain, there can be some uncertainties, potential roadblocks, and landmines that could probably unsettle the expected progress. In short, big data applications, platforms, appliances, and infrastructures need to be designed in a way to facilitate their usage and leverage for everyday purposes. The awareness about the impending potentials needs to be propagated widely and professionals need to be trained in order to extract better business value out of big data. Competing for technologies, enabling methodologies, prescribing patterns, evaluating metrics, key guidelines, and best practices need to be unearthed and made as reusable assets.

A myriad of big data technologies and tools are emerging and evolving fast. This chapter is dedicated to describing how the glowing Hadoop ecosystem is to accelerate the process of emitting and spitting out usable insights out of big data. There are pioneering and path-breaking solutions and services for both batch and real-time processing of big data.

4.1 Batch Processing Systems

Batch processing has been there for a long time. Capturing, collecting, cleansing, and crunching aggregated data using compute clusters and grids is a common and conventional thing. Batch processing involves operating over a large amount of static data and the processing time takes minutes and even hours of time. Batch data are bounded in the sense that it represents a finite collection of data. Batch data get stocked in a kind of permanent and persistent storage system. Batch processing is well suited for calculations where access to a complete set of records is required. For instance, when calculating totals and averages, datasets must be treated holistically instead of as a collection of individual records. These operations also mandate that the state has to be maintained for the duration of the calculations. Batch operations are ideal for handling a large amount of data. The trade-off for handling large quantities of data is longer computation time.

4.1.1 Apache Hadoop Framework

Apache Hadoop is a big data processing framework that exclusively provides batch processing. The latest versions of Hadoop have been empowered with a number of several powerful components or layers that work together to process batched big data:

- HDFS: This is the distributed file system layer that coordinates storage and replication across the cluster nodes. HDFS is fault tolerant and highly available. It is used as the source of data, to store intermediate processed results, and to persist the final calculated results.
- YARN (Yet another resource negotiator) is the cluster coordinating component of the Hadoop stack. It is responsible for coordinating and managing the underlying resources and scheduling jobs to be run.
- MapReduce is the Hadoop's native batch processing engine.

A MapReduce job splits a large dataset into independent chunks and organizes them into key and value pairs for parallel processing. The mapping and reducing functions receive not just values, but (key, value) pairs. This parallel processing improves the speed and reliability of the cluster, returning solutions more quickly and with greater reliability.

4.1.2 Mapping Phase

A list of data elements are provided, one at a time, to a function called the Mapper, which transforms each element individually to an output data element. The Map function divides the input into ranges by the InputFormat and creates a map task for each range in the input. The JobTracker distributes those tasks to the worker nodes. The output of each map task is partitioned into a group of key–value pairs for each reduce.

4.1.3 Reducing Phase

A reducer function receives an iterator of input values from an input list. It then combines these values together to bring forth a single output value. The Reduce function then collects the various results and combines them to answer the larger problem that the master node needs to solve. Each "reduce" task pulls the relevant partition from the machines where the maps executed, then writes its output back into HDFS. Thus, the "reduce" function is able to collect the data from all of the maps for the keys and combine them to solve the problem. The figure below vividly illustrates the MapReduce process.

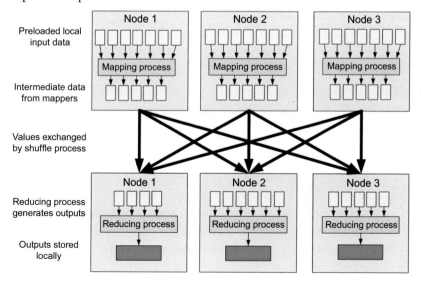

The Hadoop platform has to handle a large amount of data that are getting stored in persistent storages. There are multiple reads and writes and hence there is slowness and performance degradation in this setup. Also, the Hadoop tasks are being accomplished using inexpensive commodity servers. There are Hadoop experimentations leveraging thousands of commodity nodes. Lately, there are many new processing frameworks and engines that have the innate Hadoop integrations to utilize HDFS and the YARN resource manager. There are requirements and applications in need of efficient and cost–effective batch processing of big data. In the recent past, Apache Spark emerges as the next-generation batch processing platform for big data. The legacy MapReduce approach may make a way for the intrusion of Spark inside the Hadoop framework to tackle the big data needs.

4.2 Stream Processing Systems

It has been the client–server application architecture and then it evolved to become multitier architecture. The presentation tier (desktop applications (thick clients), web browsers (thin clients)) processed only mandatory requests before transmitting the rest to a high-throughput middle tier/ business logic tier's application server. The SQL databases at the final tier would then act on this "stream of events" and update backend databases. There are several I/O and IoT devices coming into the picture at the presentation layer and hence the number of events being generated by the presentation tier devices is growing at rapid pace. The applications are increasingly becoming enterprise-class and web scale. There are distributed file systems, clustered SQL databases, NoSQL and NewSQL databases to tackle big data. Therefore, the need is to bring forth powerful servers/ queues/brokers at the middle tier. Apache Kafka and other distributed messaging products started to occupy the middle slot supporting millions of messages/s by spreading queue partitions over clusters of machines to scale message throughput linearly. With the number of networked, resource-constrained, and intensive embedded systems grows at an exponential rate, the streams of data, and messages.

A stream is composed of immutable sequences of messages of a similar type or category. Stream processing systems compute over data as it enters the system. The data flowing frequency is continuous and there is no gap. For the big data, the volume matters. But for stream processing, the data velocity/speed matters. This is a kind of fast data and the appropriate data handling is real-time processing. Instead of batching up and processing multiple datasets using parallel, stream processing immediately crunches data whenever there is a new data. There is no waiting for processing and stream processors define operations that will be applied to each individual data item as it passes through the system.

The datasets in stream processing are typically considered "unbounded." That is, the total dataset is only defined as the amount of data that have entered the system so far. The working dataset is limited to a single item or a few items at a time and the results are immediately made available to the subscribers and users. Stream processing systems can handle a nearly unlimited amount of data.

There are specific workloads and requirements for stream processing. Stream analytics is the key task on streaming data. Event messages are streamed into analytics engine mandating real-time analytics for extracting predictive, prescriptive, and personalized insights in time. Operational, log, security, and performance insights are being extracted out of streaming data and messages instantaneously. The knowledge discovered and disseminated helps to plan out several countermeasures proactively and preemptively. Predictive and preventive maintenance capabilities are being realized through real-time analytics on streaming data. Any kind of abnormal deviations, spikes, changes, and anomalies are being identified and shared across quickly to empower decision makers and other key players to embark on arresting any kind of disasters in a timely manner.

4.2.1 Apache Storm

Apache Storm is a newly introduced stream processing framework that can instantly work on streaming data (record-by-record) to emit out actionable insights. It can handle very large quantities of data and deliver results with less latency. Storm makes it easy to reliably process unbounded streams of data. It supports real-time processing as the Hadoop's MapReduce does batch processing. Storm is simple and can be implemented using any programming language.

Storm is a distributed real-time computation system for processing large volumes of high-velocity data. Storm is extremely fast and has the ability to process over a million records per second per node on a cluster of the modest size. The key use cases include real-time customer service management, data monetization, operational dashboards, or cyber security analytics and threat detection. It is scalable, fault tolerant, and is easy to set up and operate. Storm integrates with the queuing and database technologies. A Storm topology consumes streams of data and processes those streams in arbitrarily complex ways, repartitioning the streams between each stage of the computation however needed. Storm works by orchestrating DAGs (directed acyclic graphs) in a framework (termed as topologies). These topologies describe the various translative and transformative steps that will be performed on each incoming data. The topologies are mainly composed of:

- Streams.
- Spouts: These can be APIs, queues, etc., for producing data to be operated on.
- Bolts: Bolts typically represent a processing step that consumes streams, applies an operation to them, and outputs the result as a stream. Bolts are connected to each of the spouts as well as connected with each other to arrange all of the data processing as per brewing needs. At the end of the topology, final bolt output may be used as an input for a connected system.

The idea behind Storm is to define small and discrete operations using the above components and then compose them into a process-aware topology. By default, Storm offers at-least-once processing guarantees. That is, Storm guarantees each message is processed at least once, but there may be duplicates in case of any failure. Storm does not guarantee that messages will be processed in order. In order to achieve exactly once and stateful processing, a new abstraction called Trident is made available. Storm without Trident is often referred to as Core Storm. Trident significantly alters the processing dynamics of Storm by increasing latency, adding a state to the processing, and implementing a microbatching model, which is different from the record-by-record data processing.

Trident's guarantee to process items exactly once comes handy in cases where the system cannot intelligently handle duplicate messages. Trident is the only way forward to maintain state between items. The Trident topologies are composed of:

- Stream batches: These are microbatches of stream data that are chunked in order to provide batch processing semantics.
- Operations: These are batch procedures that can be performed on the data.

Storm is very popular for real-time streaming analytics. For all kinds of real-time applications, the real-time insights are being supplied by Storm. Storm is affordable in the sense that the commodity servers are the key IT infrastructure for running Storm platform to perform real-time streaming analytics. Storm with Trident gives the option for using microbatches instead of pure stream processing. Precisely speaking, Apache Storm is emerging as the success factor in fulfilling the stream processing needs. There are viable alternatives as accentuated below.

4.2.2 Apache Samza

Apache Samza is a recently incorporated stream processing framework that is closely associated with the fully matured and stabilized Apache Kafka message-queuing system. Kafka is the widely used messaging system for processing streams of messages. Samza is designed to run on Kafka to leverage the distinct advantages of Kafka's unique capabilities. Especially Samza uses Kafka to guarantee the much-needed fault tolerance, buffering, and state storage. Samza uses YARN for resource negotiation and relies on Kafka's semantics to define the way that data streams are being handled. Kafka uses the following concepts when dealing with data. In short, Samza's goal is to provide a lightweight framework for processing continuous data to extract predictive insights proactively. The well-known Kafka semantics include:

- Topic is a stream of data entering into a Kafka system. A topic is basically a stream of useful information that consumers can subscribe to and use.
- Partitions—In order to enhance the message availability through distribution, a topic is sent to different nodes. For this, Kafka divides the incoming messages into multiple partitions based on a key. The idea is that each message with the same key is transmitted to the same partition in an orderly manner.
- Brokers are the individual nodes that make up a Kafka cluster.
- Producer is a component writing to a Kafka topic. The producer provides the key that is used to partition a well-defined topic.
- Consumers are any component that reads from a Kafka topic. Consumers are responsible for maintaining information about their own offset so that they are aware of which records have been processed if a failure occurs.

A job is a code that consumes and processes a set of input streams. In order to scale the throughput of the stream processor, jobs are broken into smaller units of execution called Tasks. Each task consumes data from one or more partitions for each of the job's input streams.

Samza's reliance on a Kafka-like queuing system opens up a variety of possibilities and opportunities. For example, Kafka offers replicated storage of data that can be accessed and processed with low latency. It also provides a very easy and inexpensive multisubscriber model to each individual data partition. All outputs, including intermediate results, are also written to Kafka and can be independently consumed by downstream stages.

Samza's strong relationship to Kafka allows the processing steps themselves to be very loosely and lightly coupled. An arbitrary number of subscribers can be added to the output of any step without any prior coordination. This capability comes handy for organizations where multiple teams might need to access similar data. Teams can subscribe to the topic entering the system or can subscribe to the topics created by other teams that have already undergone some processing. Kafka is designed to hold data for very long periods of time and this enables components to process at their convenience and can be restarted without consequence. The Samza's architecture is composed of three key components:

(1) A streaming layer is responsible for providing partitioned streams that are replicated and durable.

(2) An execution layer is responsible for scheduling and coordinating tasks across the machines.

(3) A processing layer is responsible for processing the input stream and applying transformations.

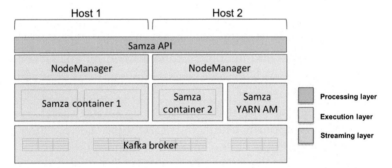

The actual implementation of the streaming layer and execution layer can be plugged on a need basis. Kafka does the streaming implementation whereas Apache YAWN and Mesos can be plugged-in for job execution systems.

4.2.3 Fault Tolerance and Isolation

As described below, Samza is able to store state using a fault tolerant checkpointing system implemented as a local key–value store. This allows Samza to offer an at-least-once delivery guarantee. However, it does not ensure the accurate recovery of aggregated state (like counts) in the event of a failure since data might be delivered more than once. Samza might help to update databases, compute counts or other aggregations, transform messages, or any number of other operations.

Apache Samza guarantees fault tolerance by restarting containers that fail potentially on another machine and resuming processing of the stream. Samza resumes from the same offset by using "checkpoints." The Samza container periodically checkpoints the current offset for each input stream partition that a task consumes. When the container starts up again after a failure, it looks for the most recent checkpoint and starts consuming messages from the checkpointed offsets. This guarantees at-least-once processing of messages.

Another key factor is that a failure in a downstream job does not cause back pressure on upstream jobs. Generally, when a job fails, the job producing an input for the failed job has to decide what to do. That is, it has the messages that need to be sent, but the downstream job is not available to process them. Also, it can drop the messages, block until the downstream processing resumes, or store them locally until the job returns. Samza artistically avoids this problem by sending all messages between jobs to Kafka. This enables upstream jobs to continue the processing at full speed without worrying about losing its output, even in scenarios where the jobs that are processing might be down. Messages are stored in Kafka brokers when a job is even unavailable.

By using independent tasks processing different partitions of the input streams and isolating the task execution by means of containers, Samza achieves the much-needed process isolation and fault tolerance. Since each container is a separate UNIX process, the execution framework that Samza integrates with can easily migrate a process from one machine to another if any of the containers starts hogging the resources of a machine. Process isolation means that when one job fails, it does not impact other jobs in the cluster.

4.2.4 Stateful Stream Processing

Some stream processing tasks are stateless and operate on one record at a time, but other uses such as counts, aggregation, or joins over a window in the stream mandate for state information to be buffered in the system. It is possible to use a remote data store to maintain state information. However, this model does not scale well. The message-processing rate of a streaming task is higher than the rate at which a DB processes requests. Furthermore, if a task accidentally fails, it is not possible to roll back the mutations to the DB. This means that a task cannot recover with the correct state. Thus stateful stream processing is required.

Samza solves this heady problem by bringing the data closer to the stream processor. Each Samza task has its own local data store and this improves the read/write performance by many orders of magnitude when compared with having a remote data store. All writes to the local data store are replicated to a durable change-log stream. When a machine fails, the task can consume the change-log stream to restore the contents of the local data store to be consistent.

As Samza stores the state for each task locally on disk on the same machine that the Samza container is running on, its stream processing performance is good. If the streaming applications mandate for stateful processing with high performance, Apache Samza is the way to go. Samza ensures that message streams must be ordered, highly available, partitioned, and durable. The tight linkage with the powerful Kafka elevates Samza to the next level in realizing high-performance stream processing. Samza is a good fit for organizations with multiple teams using data streams at various stages of processing. Samza greatly simplifies many parts of stream processing and offers low-latency performance.

4.3 Hybrid Processing Systems: Batch and Stream Processing

Some processing frameworks can aptly handle both batch and stream workloads. These frameworks simplify diverse processing requirements by allowing the same or related components and APIs to be used for both types of data. Apache Spark and Flink can provide the hybrid processing.

4.3.1 Apache Spark

Apache Spark is a next-generation batch processing framework with stream processing capabilities. Spark focuses primarily on speeding up batch processing workloads by offering full in-memory computation and processing optimization. Spark can be deployed as a standalone cluster by pairing with a capable storage layer or can hook into Hadoop's HDFS. Spark, in the beginning, loads the data into memory, processes all the data in memory, and at the end, persists the final results in the disk storage. All intermediate results are fully kept and managed in memory.

We all know that in-memory computing accelerates data processing drastically. That is, when data get stored in system memory rather on disk storages, the processing happens at 3000 times speedier. Spark is relatively fast on disk-related tasks because Spark brings forth a series of optimizations by analyzing the complete set of tasks ahead of time. It achieves this by creating DAGs, which represent all of the operations that must be performed, the data to be operated on, as well as the relationships between them, giving the processor a greater ability to intelligently coordinate work.

Resilient distributed datasets (RDDs)—To implement an in-memory batch computation, Spark uses this proven RDD model to work with data. These are immutable structures that exist within memory that represent collections of data. Operations on RDDs can produce new RDDs and each RDD can trace its lineage back through its parent RDDs and ultimately to the data on disk. Through the concept of RDDs, Spark is able to maintain the much-needed fault tolerance without needing to write back to disk after each operation. Precisely speaking, Spark started its golden innings by performing batch processing.

Spark Streaming is a newly introduced API in the Apache Spark family in order to simplify and speed upstream processing. Spark implements an original concept of microbatches to facilitate stream processing. The idea is to treat streams of data as a series of very small batches that can be handled using the native semantics of the batch engine. Spark Streaming works by buffering the stream in subsecond increments and they are sent as small fixed datasets for batch processing. This method can lead to different performance guarantees. Spark through its in-memory computing capability is able to do justice for both batch as well as streaming analytics. Adapting the batch methodology for stream processing can lead to buffering the data as it enters the system. The buffer helps to handle a high volume of incoming data and increasing the overall throughput. The problem here is that the waiting period to flush the buffer leads to high latency and hence for real-time processing, Spark is not a good fit. Ultimately Spark will replace the Hadoop's MapReduce module.

The Spark deployment and operational model are quite unique and versatile. That is, Spark can be deployed as a standalone cluster or integrated with an existing Hadoop cluster. That is, a single cluster can do both batch and stream processing. Because of its innate strength, Spark is on the right track by adding additional libraries such as machine learning (ML), etc. GraphX is the Apache Spark's API for graphs and graph-parallel computation. GraphX is capable of unifying ETL, exploratory analysis, and iterative graph computation within a single system. We can view the same data as both graphs and collections, transform and join graphs with RDDs efficiently, and write custom iterative graph algorithms using the Pregel API.

The principal advantages of Spark—There are many benefits being accrued out of the advancements happening in the Spark domain.

- Faster processing—Apache Spark essentially takes MapReduce to the next level with a performance that is significantly faster. Spark has the ability to hold intermediate results in memory itself instead of writing it back to disk and reading it again.
- Speed—Spark can execute batch processing jobs 10–100 times faster than MapReduce. That does not mean it lags behind when data have to be written to and fetched from disk.
- Ease of use—Apache Spark has easy-to-use APIs for easily operating on large datasets.
- Unified engine—Spark can run on top of Hadoop making use of its cluster manager (YARN) and underlying storage (HDFS, HBase, etc.). Also, it can run independently of Hadoop by joining hands with other cluster managers and storage platforms such as Cassandra and Amazon S3.
- Choose from Java, Scala, or Python—Spark supports all the prominent and dominant programming languages.
- In-memory data sharing—Different jobs can share data within the memory and this makes an ideal choice for iterative, interactive, and event stream processing tasks.

As the relatively expensive memory is being used for computation, Spark is to cost more. However, the increased processing speed means that tasks can be completed faster and resultingly the cost of computation is on the lower side. Precisely speaking, Spark emerges as the one-stop solution for big data analytics.

4.3.2 Apache Flink

Apache Flink is a new stream processing framework that can also handle batch tasks. It considers batches as data streams with finite boundaries and hence can perform batch processing as a subset of stream processing. This stream-first approach, touted as the Kappa architecture, to all the processing needs has a number of advantages. We are well aware of the Lambda architecture, wherein the batching is the primary processing method and the streaming comes as a supplement.

Flink's stream processing model handles incoming data on an item-by-item basis as a true stream. Flink provides its DataStream API to work with unbounded streams of data. The basic Flink components are:

- Streams.
- Operators are the functions that work on data streams to produce other streams.
- Sources are the entry point for streams entering the system.
- Sinks are the place where streams flow out of the Flink system. Sinks might represent a database or a connector to another system.

Stream processing tasks take snapshots at set points during their computation in case of any need for recovery. For storing state, Flink can work with a number of state-of-the-art backend storage mechanisms. Flink's batch processing model is just an extension of the stream processing model. Instead of reading from a continuous stream, it reads a bounded dataset off of persistent storage as a stream. Flink uses the same runtime for both stream and batch processing needs. Flink offers some crucial optimizations for batch workloads. Since batch operations are backed up by persistent storage, Flink removes the snapshotting from batch loads. Yet data are still recoverable and faster with normal processing.

In summary, while Spark performs batch and stream processing, its streaming mechanism is not appropriate for many use cases as it uses the concept of microbatching. On the other hand, the Flink's stream-first approach guarantees low-latency, high-throughput, and real entry-by-entry processing. Flink manages its own memory instead of relying on the native Java garbage collection mechanisms for performance reasons. Unlike Spark, Flink handles data partitioning and caching automatically. Flink analyzes its work and optimizes tasks in a number of ways.

Flink maps out the most effective way to implement a given task. And it is able to parallelize stages that can be completed in parallel while bringing data together for blocking tasks. For iterative tasks, Flink attempts to do computation on the nodes where the data are stored for performance reasons. It can also do "delta iteration" (i.e., it does the iteration on only the portions of data that have gone through any changes). Link simplifies several tasks through a host of automated tools. Flink offers a web-based scheduling view to easily manage tasks and view the system. Users can also display the optimization plan for submitted tasks to see how it will actually be implemented on the cluster. For analysis tasks, Flink offers SQL-style querying, graph processing and ML libraries, and in-memory computation. Flink integrates well with YARN, HDFS, and Kafka. The Flink's reference architecture is given below.

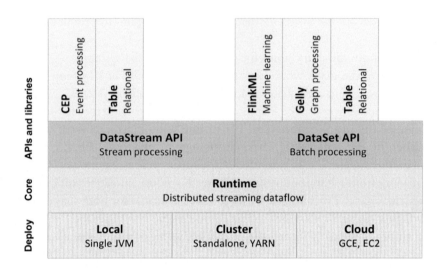

The defining hallmark of Apache Flink is the ability to process streaming data in real time. Flink was built around a streaming model and Flink can the same model to batch and SQL processing jobs as well. It has added newer libraries

- Table API enables the usage of SQL queries over the data. They are being easily embedded on both the DataStream and DataSets APIs and support the use of relational operators like selection, aggregations, and joins.
- Flink ML can be used for performing ML tasks over the DataSet API. It enables users to write ML pipelines, which make it easier to handle the ML workflow. ML pipelines bind the different steps of an ML flow together making it efficient to prepare and deploy the models in a production environment.
- Gelly for graph processing. It provides set of operators to create and modify graphs. A graph is typically represented by a DataSet of edges and DataSet of vertices. Gelly is only available for DataSet API and can only be used for batch processing.
- Flink CEP is the complex event processing library for Flink. It allows quickly detecting complex event patterns in a stream of endless data.

The streaming model benefits iterative processing. That is, repeating the processing of the same data to achieve better accuracy. ML is one shining example of iterative processing. Flink can be instructed to only process the parts of the data that have actually changed and this optimization significantly speeds up the job.

4.3.3 The Major Advantages of Flink

This has some definite and decisive advantages compared with other stream processing technologies as listed below.

- Flink is primarily for stream processing that can be leveraged for batch processing.
- Explicit memory management gets rid of the occasional spikes found in the popular in-memory processing framework (Spark).
- It achieves accelerated processing by allowing iterative processing to take place on the same node rather than having the cluster runs them independently. Its performance can be further tuned by tweaking it to reprocess only that part of data that have changed rather than the entire set.

For mixed workloads, Spark provides high-speed batch processing and microbatch processing for streaming. Flink provides true stream processing with batch processing support. It is heavily optimized, can run tasks written for other platforms, and provides low-latency processing.

5. THE OTHER TECHNOLOGIES AND TOOLS IN THE HADOOP ECOSYSTEM

Apache Hadoop has become a de facto software framework for reliable, scalable, distributed, and large-scale computing. The key difference with other accomplished computing models is that Hadoop brings the computation to data rather than sending data to computation. As Hadoop has matured and stabilized over the years, many new components and tools are being added to its ecosystem to enhance its usability, viability, versatility, and functionality. Hadoop Hive, HBase, ZooKeeper, Oozie, Pig, Sqoop, etc., are some of the new capabilities to make Hadoop as the one-stop solution for big data management.

5.1 Hive

This is a kind of data warehousing software that addresses how data are structured and queried in distributed Hadoop clusters. Hive is also a popular development environment that is used to write queries for data in the Hadoop environment. It provides tools for ETL operations and brings some SQL-like capabilities to the environment. Hive is a declarative language used to develop applications for the Hadoop environment. Hive has several contributing components as enlisted below.

- HCatalog—This enables data processing tools to read and write data on the grid. It supports MapReduce and Pig.

- WebHCat—This is an HTTP/REST interface to run MapReduce, YARN, Pig, and Hive jobs.
- HiveQL—This is the Hive's query language for SQL developers to easily work with Hadoop. It is similar to SQL and helps both structure and query data in distributed Hadoop clusters. Hive queries can run from the Hive shell, JDBC, or ODBC. MapReduce breaks down HiveQL statements for parallel execution across the cluster.

5.2 The Hive Characteristics

- Hive supports the familiar SQL-like interface. This helps to use existing SQL skills to run batch queries on data stored in Hadoop. Queries are written using HiveQL (SQL-like query language) and are executed through either MapReduce or Apache Spark.
- Hive simplifies any data processing tools such as Pig and MapReduce to access Hive's metadata using the HCatalog, a table and storage management layer for Hadoop. RESTful API is another popular way to access the metadata. This setup allows users to easily read and write data without worrying about where the data is stored, what format it is, etc. This also facilitates redefining the structure for each tool.
- Hive-on-Spark features the next generation of batch processing for Hive. With queries executed through Apache Spark, users will see dramatic performance improvements compared to MapReduce.
- With its familiar interface, Hive is the tool-of-choice for a variety of batch processing workloads such as
  - Data preparation
  - ETL
  - Data mining
  - Ad optimization

Hive on LLAP (live long and process) makes use of persistent query servers with intelligent in-memory caching to avoid Hadoop's batch-oriented latency and provide as fast as subsecond query response times against smaller data volumes, while Hive on Tez continues to provide excellent batch query performance against petabyte-scale datasets. The tables in Hive are similar to tables in a relational database, and data units are organized in a taxonomy from larger to more granular units. Databases are comprised of tables, which are made up of partitions. Data can be accessed via a simple query language and Hive supports overwriting or appending data.

Within a particular database, data in the tables are serialized and each table has a corresponding HDFS directory. Each table can be subdivided into partitions that determine how data are distributed within subdirectories of the table directory. Data within partitions can be further broken down into buckets. Hive supports all the common primitive data formats such as BIGINT, BINARY, BOOLEAN, CHAR, DECIMAL, DOUBLE, FLOAT, INT, SMALLINT, STRING, TIMESTAMP, and TINYINT. In addition, analysts can combine primitive data types to form complex data types, such as structs, map, and arrays.

Hive allows any MapReduce-like software to perform more sophisticated functions. Hive does not allow row-level updates and it does not support for real-time queries and hence Hive is not suitable for OLTP workloads. Hive is more effective for processing structured data whereas Pig is famously suited for processing unstructured data.

5.2.1 Pig

This is a procedural language for developing parallel processing applications for large datasets in the Hadoop environment. With YARN emerging as the architectural center of Apache Hadoop, multiple data access engines such as Apache Pig interact with data stored in the cluster. Apache Pig allows Apache Hadoop users to write complex MapReduce transformations using a simple scripting language called Pig Latin, which is a component of the Pig platform. Pig translates the Pig Latin script into MapReduce so that it can be executed within YARN for access to a single dataset stored in the HDFS. Pig is commonly used for complex use cases that require multiple data operations. Pig helps developing applications that aggregate and sort data and supports several inputs and exports. It is highly customizable in the sense that users can build their own functions using any scripting language (Ruby, Python, and Java). In short, Apache Pig is an abstraction over MapReduce. It is a scripting platform used to analyze larger sets of data representing them as data flows.

Pig can perform a long series of data operations and is emerging as the ideal platform for three big data tasks:
- extract-transform-load (ETL) data pipelines,
- research on raw data, and
- iterative data processing.

5.2.2 HBase

This is a scalable and distributed NoSQL database that sits atop the HFDS. It was designed to store structured data in tables that can have billions of rows and millions of columns. It has been used to power historical searches through large datasets, especially when the desired data are contained within a large amount of redundant, routine, and repetitive data. HBase is not a relational database and was not designed to support transactional and other real-time applications. It is accessible through a Java API and has ODBC and JDBC drivers. HBase does not support SQL queries. However, there are several SQL support tools available. Hive can be used to run SQL-like queries in HBase.

Apache HBase enables random, real-time read/write access to big data. This has the capability and capacity of accommodating billions of rows × millions of columns atop clusters of commodity servers. Apache HBase is an open-source, distributed, versioned, nonrelational database modeled after Google's Bigtable: Just as Bigtable leverages the distributed data storage provided by the Google file system, Apache HBase provides Bigtable-like capabilities on top of Hadoop and HDFS.

- Real-time speed—HBase performs fast and random reads and writes to all data stored and integrate with other components, like Apache Kafka or Apache Spark Streaming, to build complete end-to-end workflows all within the single platform.
- Hadoop scalable—HBase is designed for massive scalability to store unlimited amounts of data in a single platform and handle growing demands for serving data to more users and applications. HBase supports linear scalability.
- Flexibility—HBase stores poly-structured data (structured, semistructured, and unstructured) without any up-front modeling. Flexible storage facilitates complete access to full-fidelity data for a wide range of analytics and uses cases, with direct access to the frameworks such as Impala and Apache Solr.
- Reliability—HBase ensures automatic and tunable replication to make multiple copies of data to guarantee high availability. Commodity servers are bound to fail frequently but this replication comes handy in facilitating data availability, recovery, and business continuity through fault tolerance.

5.3 The Common Use Cases

HBase enhances the benefits of HDFS with the ability to serve random reads and writes to many users or applications in real-time and is being proclaimed as the ideal one for a variety of critical use cases all within a single platform.

- Messaging service
- Sessionization

- Real-time metrics and analytics (advertising, auction, etc.)
- Graph data
- Internet of Things (IoT) applications

5.3.1 Apache Oozie

This is a Java web application used to schedule Apache Hadoop jobs. Oozie combines multiple jobs sequentially into one workflow. Oozie is integrated with the other components of the Hadoop stack as well as system-specific jobs (such as Java programs and shell scripts). Apache Oozie is a tool for enabling cluster administrators to build complex data transformations out of multiple component tasks. This provides greater control over jobs and also makes it easier to repeat those jobs at predetermined intervals. There are two basic types of Oozie jobs.

- Oozie workflow jobs are DAGs, specifying a sequence of actions to execute. The workflow job has to wait.
- Oozie coordinator jobs are recurrent Oozie workflow jobs that are triggered by time and data availability.
- Oozie bundle provides a way to package multiple coordinators and workflow jobs and to manage the lifecycle of those jobs.

Oozie runs as a service in the cluster and clients submit workflow definitions for immediate or later processing. Oozie workflow consists of action nodes and control-flow nodes.

1. An action node represents a workflow task, e.g., moving files into HDFS, running a MapReduce, Pig or Hive jobs, importing data using Sqoop or running a shell script of a program written in Java.
2. A control-flow node controls the workflow execution between actions by allowing constructs like conditional logic wherein different branches may be followed depending on the result of earlier action node.
   - Start Node, End Node, and Error Node fall under this category of nodes.
   - Start Node designates the start of the workflow job.
   - End Node signals end of the job.
   - Error Node designates an occurrence of an error and corresponding error message to be printed.

At the end of execution of the workflow, HTTP call back is used by Oozie to update the client with the workflow status. Entry-to or exit-from an action node may also trigger the callback. A sample workflow diagram is:

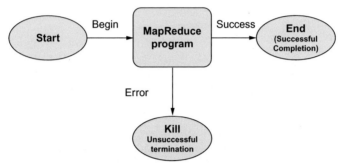

Oozie is a scalable, reliable, and extensible system. Oozie is the workflow scheduler managing how workflows start and execute, and also controls the execution path. Oozie is a server-based Java web application that uses workflow definitions written in hPDL, which is an XML process definition language similar to JBOSS JBPM jPDL.

5.3.2 Apache Sqoop

This is a tool for efficiently transferring bulk data between Apache Hadoop and external data stores such as relational databases and enterprises data warehouses. Sqoop is used to import data from external data stores into HDFS or related Hadoop eco-systems like Hive and HBase. Similarly, Sqoop can also be used to extract data from Hadoop or its ecosystem and export it to external data stores. The functionality of Sqoop (the short form of SQL to Hadoop) is vividly illustrated in the picture below. Sqoop works with relational databases such as Teradata, IBM Netezza, Oracle, MySQL, Postgres, etc. YARN coordinates data ingest from Apache Sqoop and other services that deliver data into the Enterprise Hadoop.

Sqoop uses MapReduce framework to import and export the data. MapReduce, as indicated in the beginning, guarantees parallel execution and fault tolerance. Sqoop provides command line interface and developers just need to provide basic information like source, destination, and database authentication details in the Sqoop command. Sqoop takes care of everything else. Sqoop provides many salient features like:

1. Full load
2. Incremental load
3. Parallel import/export
4. Import results of SQL query
5. Compression
6. Connectors for all major RDBMS databases

7. Kerberos security integration
8. Load data directly into Hive/HBase
9. Support for Accumulo

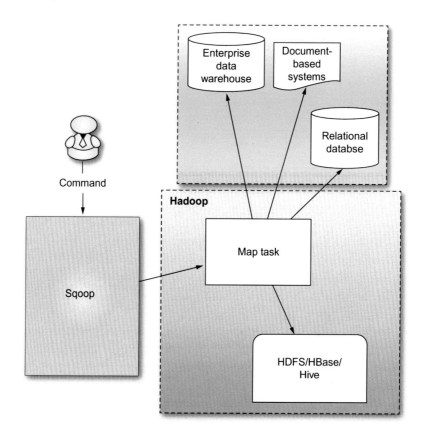

Sqoop provides a command line interface to the end users. Sqoop can also be accessed using Java APIs. Sqoop command submitted by the end user is parsed by Sqoop and launches Hadoop Map only job to import or export data Sqoop just imports and exports the data. The Map job launches multiple mappers depending on the number defined by the user in the command line. For Sqoop import, each mapper task will be assigned with part of data to be imported based on key defined on the command line. Sqoop distributes the input data among the mappers equally to get high performance. Then each mapper creates a connection with the database using JDBC and fetches the part of data assigned by Sqoop and writes it into HDFS or Hive or HBase based on the option provided in the command line. Sqoop is an important component for the ensuing era of big data in data movement.

5.3.3 *Ambari*

Many Hadoop distributions offer preinstalled sandbox of VM to try out things but it does not give the feel of distributed computing. However, installing a multinode Hadoop cluster is not an easy task and with growing number of components, it is very tricky to handle so many configuration parameters. Apache Ambari comes here to our rescue!

This is a web-based tool for provisioning, managing, and monitoring Apache Hadoop clusters. Ambari extends support for Hadoop HDFS, MapReduce, Hive, HCatalog, HBase, ZooKeeper, Oozie, Pig, and Sqoop. Ambari provides an intuitive and easy-to-use Hadoop management web UI backed by its RESTful APIs. Ambari enables system administrators to:

- Provision a Hadoop cluster
  - Ambari provides a step-by-step wizard for installing Hadoop services across any number of hosts.
  - Ambari handles configuration of Hadoop services for the cluster.
- Manage a Hadoop cluster
  - Ambari provides central management for starting, stopping, and reconfiguring Hadoop services across the entire cluster.
- Monitor a Hadoop cluster
  - Ambari provides a dashboard for monitoring health and status of the Hadoop cluster.
  - Ambari leverages Ambari Metrics System for metrics collection.
  - Ambari leverages Ambari alert framework for system alerting and will notify you when your attention is needed (e.g., a node goes down, the remaining disk space is low, etc.).

Ambari enables application developers and system integrators (SIs) to

- Easily integrate Hadoop provisioning, management, and monitoring capabilities to their own applications with the Ambari REST APIs.

It provides a highly interactive dashboard which allows the administrators to visualize the progress and status of every application running on the Hadoop cluster. Some of the key highlights are:

- Instantaneous insight into the health of Hadoop cluster using preconfigured operational metrics.
- User-friendly configuration providing an easy step-by-step guide for installation.
- Dependencies and performances monitored by visualizing and analyzing jobs and tasks.
- Authentication, authorization, and auditing by installing Kerberos-based Hadoop clusters.

Ambari has two main components:

(1) Ambari server—This is the master process which communicates with Ambari agents installed on each node participating in the cluster. This has Postgres database instance which is used to maintain all cluster-related metadata.

(2) Ambari Agent—These are acting agents for Ambari on each node. Each agent periodically sends his own health status along with different metrics, installed services status, and many more things. Accordingly master decides on next action and conveys back to the agent to act. The Ambari architecture is given below.

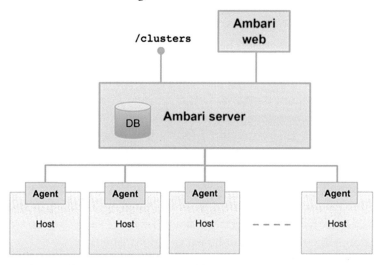

Apache Ambari comes as an enabling and elegant tool to automate Hadoop software installation, deployment, and delivery.

5.3.4 Avro

Data serialization is an important mechanism to translate data in a computer environment (like memory buffer, data structures, or object state) into a binary or textual form that can be transported over a network or stored in some persistent storage media. Apache Avro is a language-agnostic data serialization system and provides data structures, binary data format, and a container file format to store persistent data.

Since Hadoop writable classes lack language portability, Avro occupies an important slot as it deals with data formats that can be processed by multiple languages. Avro has emerged as the preferred tool to serialize data in Hadoop. Avro has a schema-based system. A language-independent schema is being associated with its "read and write" operations. Avro serializes the data, which has a built-in schema. Avro serializes the data into a compact binary format, which can be deserialized by any application. Avro uses JSON format to declare the data structures. Presently, it supports languages such as Java, C, C++, C#, Python, and Ruby.

Avro heavily depends on its schema. It allows every data to be written with no prior knowledge of the schema. It serializes fast and the resulting serialized data are lesser in size. The schema is stored along with the Avro data in a file for any further processing. In RPC, the client and the server exchange schemas during the connection. This exchange helps in the communication between same named fields, missing fields, extra fields, etc. Avro schemas are defined with JSON that simplifies its implementation in languages with JSON libraries. The distinct features of Avro:

- Avro is a language-neutral data serialization system.
- Avro does not require code generation to use and integrates well with JavaScript, Python, Ruby, C, C#, C++, and Java.
- The Hadoop ecosystem, as well as Kafka, uses Avro.
- Avro creates a binary structured format that is both compressible and splittable. Hence it can be efficiently used as the input to Hadoop MapReduce jobs.
- Avro provides rich data structures. For example, it is possible to create a record that contains an array, an enumerated type, and a subrecord. These data types can be created in any language, can be processed in Hadoop, and the results can be fed to a third language.
- Avro schemas defined in JSON facilitate the implementation in the languages that already have JSON libraries.
- Avro creates a self-describing file named Avro data File, in which it stores data along with its schema in the metadata section.

Avro supports polyglot bindings to many programming languages and a code generation for static languages. Avro supports evolutionary schemas, which support compatibility checks, and allows evolving data over time. Avro supports platforms like Kafka that has multiple producers and consumers, which evolve over time. Avro schemas help keep data clean and robust. Avro has a lot of potentials and can be used outside of Hadoop ecosystem and is a good solution for language interoperability. Avro's compact binary size and the availability of the schema at the top of the binary file are a perfect solution for message payloads that can be easily transported over different types of message queues. It is widely used as a message payload in Apache Kafka. Avro has basic RPC functionality with a higher level of Schema abstraction.

5.3.5 Chukwa

Logs are being generated incrementally across many machines, but Apache Hadoop MapReduce works best on a small number of large files. Merging the reduced output of multiple runs may require additional MapReduce jobs. This creates some overhead for data management on Apache Hadoop. Apache Chukwa is a new project devoted to bridging that gap between logs processing and Hadoop ecosystem. Apache Chukwa is a scalable distributed monitoring and analysis system, particular logs from Apache Hadoop and other distributed systems.

Chukwa is a data collection system for managing large distributed systems. Apache Chukwa aims to provide a flexible and powerful platform for distributed data collection and rapid data processing. This is competent enough to take advantage of newer storage technologies (HDFS appends, HBase, etc.). In order to maintain this competency, Apache Chukwa is structured as a pipeline of collection and processing stages, with clean and narrow interfaces between stages. This will facilitate the future innovations without breaking the existing code. Apache Chukwa has five primary components.

- Adaptors that collect data from various data source.
- Agents that run on each machine and emit data.
- ETL processes for parsing and archiving the data.
- Data analytics scripts for aggregate Hadoop cluster health.
- The Hadoop Infrastructure Care Center (HICC)—This is a web-portal style interface for displaying data.

The high-level architecture is given below.

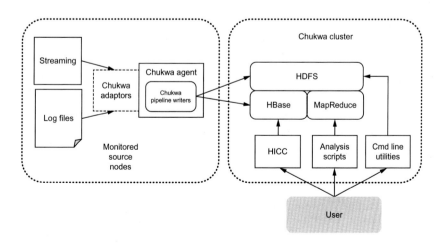

In summary, Chukwa is an open-source data collection system for monitoring large distributed systems. Chukwa is built on top of the HDFS and MapReduce framework and inherits Hadoop's scalability and robustness. Chukwa also includes a flexible and powerful toolkit for displaying, monitoring, and analyzing results to make the best use of the collected data.

5.3.6 Impala

This is an MPP (massive parallel processing) SQL query engine for processing huge volumes of data that are getting stored in Hadoop cluster. It provides high-performance and low-latency compared with the other SQL engines for Hadoop. Impala combines the SQL support and multiuser performance of a traditional analytic database with the scalability and flexibility of Apache Hadoop by utilizing standard components such as HDFS, HBase, Metastore, YARN, and Sentry.

- Hive MapReduce job will take some minimum time in launching and processing queries whereas Impala gives results in seconds.
- Impala can read almost all the file formats such as Parquet, Avro, RCFile used by Hadoop.
- Impala uses the same metadata, SQL syntax (Hive SQL), ODBC driver, and user interface (Hue Beeswax) as Apache Hive, providing a familiar and unified platform for batch-oriented or real-time queries.

Unlike Apache Hive, Impala is not based on the slow MapReduce algorithms. It implements a distributed architecture based on daemon processes that are responsible for all the aspects of query execution that run on the same machines. This makes Impala faster than Apache Hive. Here is a list of some noted advantages of Apache Impala:

- Using Impala, it is possible to process data that is stored in HDFS at lightning-fast speed with the traditional SQL knowledge. Since the data processing is carried where the data resides on Hadoop cluster, data transformation, and data movement are not required for data stored on Hadoop. Precisely speaking, Impala gives parallel processing database technology on top of Hadoop ecosystem. It allows users to perform low-latency queries interactively.
- Using Impala, it is possible to access the data that are stored in HDFS, HBase, and Amazon s3 without the knowledge of Java MapReduce jobs. Administrators can access them with a basic idea of SQL queries.
- To write queries in business tools, the data have to be gone through a complicated extract-transform-load (ETL) cycle. But, with Impala, this long procedure is significantly shortened. The time-consuming

stages of loading and reorganizing are overcome with the new techniques such as exploratory data analysis and data discovery making the process faster.

- Impala being real-time query engine is best suited for analytics and for data scientists to perform analytics on data stored in HDFS.
- As Impala gives results in real time, it is best fit in reporting tools or visualization tools like Pentaho and Tableau which already comes with connectors that allow to query and perform visualizations directly from GUI.
- Impala is pioneering the use of the Parquet file format, a columnar storage layout that is optimized for large-scale queries typical in data warehouse scenarios.
- Impala supports in-memory data processing, i.e., it accesses/analyzes data that are stored on Hadoop data nodes without data movement.

The major constrictions:
- Impala does not provide any support for serialization and deserialization.
- Impala can only read text files, not custom binary files.
- Whenever new records/files are added to the data directory in HDFS, the table needs to be refreshed.

5.4 The Impala Architecture

There are three vital daemons of Impala playing a major role in its architecture.

- Impalad
- Statestored
- Catalogd

The core part of Impala is a daemon that runs on each node of the cluster called impalad. It reads and writes to data files, accepts queries transmitted from Hue or any other connection requests from Data Analytic tools with JDBC and ODBC connections. It distributes the work to other nodes in the Impala cluster and transmits the intermediate results back to the coordinator node. The node in which job launched is known as the coordinator node. One can submit the query to Impala daemon running on any node of your cluster and that node serves as the coordinator node for that query. Coordinator node distributes the query for processing among the cluster of impalad daemons of other nodes. The other nodes transmit the partial data back to the coordinator, which then constructs the final result set for that query.

Statestore checks for Impala daemons availability on all the nodes of the cluster. A daemon process called state stored physically represents it. Only one node in the cluster needs to have this process. Impala daemons are in continuous communication with the statestore, to confirm which nodes are healthy and accept new work for processing. If an impalad daemon goes down because of any of the reason like hardware failure, network connection issue, then the statestored daemon informs all the other impalad daemons running in the cluster, so that future queries can avoid assigning tasks/queries to process. If statestore is not available, the other nodes continue running and distribute work among the other impalad daemons as usual with the assumption that all impalad daemons are running fine.

Catalog service relays the metadata changes from Impala DDL (data definition language) queries or DML (data manipulation language) queries to all nodes in the cluster. This process is represented as catalogd daemon and we need such process in one of the hosts in the cluster.

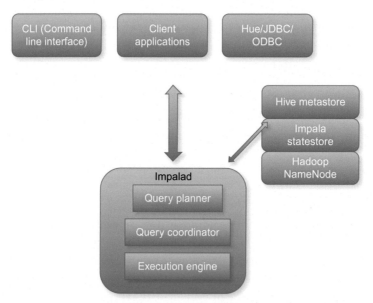

Third-party applications or any query requests will be taken by impalad then execution plan will be prepared and submit the query plan to rest of the other impalad's in the cluster. Before the execution plan, impalad will communicate with statestore and hive metastore and namenode for live impalad nodes and table metadata and file information. Impala provides command line interface, which gives interactive query results.

5.4.1 Flume

This is a distributed, reliable, and available service for efficiently collecting, aggregating, and moving large amounts of log data. It has a simple and flexible architecture based on streaming data flows. It is robust and fault tolerant with tunable reliability mechanisms. Furthermore, Flume is blessed with many failover and recovery mechanisms. It uses a simple extensible data model that allows for the online analytic application. The macrolevel architecture is given below.

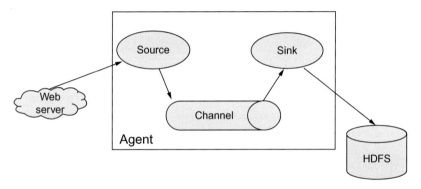

Enterprises extensively use Flume's powerful streaming capabilities to land data from high-throughput streams into the HDFS. The common streaming sources are application logs, infrastructure operational data, sensor and device data, geolocation data, and social media. These different types of data can be landed in Hadoop for future analysis using interactive queries in Apache Hive.

Apache Flume is composed of six important components:

- Events—The data units that are transferred over a channel from source to sink. The size of events is usually 4 KB.
- Sources—They accept data from a server or an application. Sources listen for events and write events to a channel.
- Sinks receive data and store it in HDFS repository or transmit the data to another source. It writes event data to a chosen target and removes the event from the queue.
- Channels connect between sources and sink by queuing event data for transactions.
- Interceptors drop data or transfer data as it flows into the system.
- Agents are used to running the sinks and sources in the flume.
- **(1)** Efficiently stream data—Flume easily collects, aggregates, and moves streaming logs or event data from multiple sources into Hadoop. Flume

is designed to ingest this data as it is generated for near real-time analytics. Flume is a vital module for streaming analytics and emerging as a silver bullet for sensor data aggregation or "Internet of Things (IoT)" use cases.

(2) Built for Hadoop scale—With the growing data volumes, Flume can simply scale horizontally to handle the increased load. Furthermore, Flume can efficiently gather logs from new data sources and multiple systems. On the other hand, connectors are made available to enable Flume to stream data into multiple systems.

(3) Always-on reliability—Flume protects against data loss and ensures that streaming data will continue to be delivered, even in the event of failure, with the built-in fault tolerance.

Here is a specific example. Flume is used to log manufacturing operations. When one run of product comes off the assembly line, it generates a log file about that run. Even if this occurs hundreds or thousands of times per day, the large volume log file data can stream through Flume into a tool for immediate analysis with Apache Storm. The other option is to aggregate months or years of production runs in HDFS and the batched data get analyzed using Apache Hive.

5.4.2 Apache Kafka

A messaging system is responsible for transferring data from one application to another so the applications can focus on data without getting bogged down on data transmission and sharing. Distributed messaging is based on the concept of reliable message queuing. Messages are queued asynchronously between client applications and messaging system. There are two types of messaging patterns. The first one is a point-to-point and the other one is "publish–subscribe" (pub-sub) messaging system. Most of the messaging systems follow the pub-sub pattern.

5.4.3 Point-to-Point Messaging System

In this type, messages are persisted in a queue. One or more consumers can consume the messages in the queue, but a particular message can be consumed by a maximum of one consumer only. Once a consumer reads a message in the queue, it disappears from that queue. The strength of queuing is that it allows dividing up the processing of data over multiple consumer instances, which lets scaling out the data processing. Unfortunately, queues are not multisubscriber.

5.4.4 Publish–Subscribe Messaging System

In this category, messages are persisted in a topic. Unlike point-to-point system, consumers can subscribe to one or more topic and consume all the messages in that topic. In the Publish–Subscribe system, message producers are called publishers and message consumers are called subscribers. The drawback with this is that there is no way of scaling processing since every message goes to every subscriber.

5.4.5 The Key Differentiators

The distinct advantage of Kafka's model is that every topic has both the properties. It can scale processing and is also multisubscriber. Kafka has stronger ordering guarantees than a traditional messaging system. A traditional queue retains records in-order on the server, and if multiple consumers consume from the queue, then the server hands out records in the order they are stored. However, although the server hands out records in order, the records are delivered asynchronously to consumers, so the records may arrive out of order to different consumers. This effectively means the ordering of the records is lost in the presence of parallel consumption.

Kafka solves this problem. By having a notion of parallelism—the partition—within the topics, Kafka is able to provide both ordering guarantees and load balancing over a pool of consumer processes. This is achieved by assigning the partitions in the topic to the consumers in the consumer group so that exactly one consumer in the group consumes each partition. Since there are many partitions, this balances the load over many consumer instances. Any message queue that allows publishing messages decoupled from consuming them is effectively acting as a storage system for the in-flight messages. Data written to Kafka is written to disk and replicated for fault tolerance. Kafka allows producers to wait on acknowledgment so that a write is not considered complete until it is fully replicated and guaranteed to persist even if the server written to fails. This enables Kafka to scale well and hence Kafka will perform the same whether there are 50 KB or 50 TB of persistent data on the server.

In Kafka, a stream processor is anything that takes continual streams of data from input topics, performs some processing on this input, and produces continual streams of data to output topics. It is possible to do simple processing directly using the producer and consumer APIs. However, for more complex transformations, Kafka provides a fully integrated streams API. This allows building applications that do nontrivial processing that compute aggregations off of streams or join streams together. This facility helps solve the hard problems such as handling out-of-order data, reprocessing input as code changes, performing stateful computations, etc.

The streams API builds on the core primitives Kafka provides: it uses the producer and consumer APIs for input, uses Kafka for stateful storage, and uses the same group mechanism for fault tolerance among the stream processor instances. A distributed file system like HDFS allows storing static files for batch processing. Effectively a system like this allows storing and processing historical data from the past. A traditional enterprise messaging system allows processing future messages that will arrive after one subscribes. Applications built in this way process future data as it arrives. Kafka combines both of these capabilities and this combination elevates Kafka as a key streaming application platform and data pipelines. By combining storage and low-latency subscriptions, streaming applications can treat both past and future data the same way. That is, a single application can process historical and stored data but rather than ending when it reaches the last record it can keep processing as future data arrives.

Apache Kafka is a distributed "publish–subscribe" messaging system and a robust queue that can handle a high volume of data and enables in passing messages from one endpoint to another. Kafka is suitable for both offline and online message consumption. Kafka messages are persisted on the disk and replicated within the cluster to prevent data loss. Kafka is built on top of the ZooKeeper synchronization service. It integrates very well with Apache Storm and Spark for real-time streaming data analysis. The advantages of Kafka are:

- Reliability—Kafka is distributed, partitioned, replicated, and fault tolerance.
- Scalability—Kafka messaging system scales easily without down time.
- Durability—Kafka uses distributed commit log which means messages persist on disk as fast as possible, hence it is durable.
- Performance—Kafka has high-throughput for both publishing and subscribing messages. It maintains stable performance even many TB of messages are stored.

Kafka is very fast and guarantees zero downtime and zero data loss. Kafka is a unified platform for handling all the real-time data feeds. Kafka supports low-latency message delivery and gives guarantee for fault tolerance in the presence of machine failures. It has the ability to handle a large number of diverse consumers. Kafka performs 2 million writes per second. Kafka persists all data to the disk, which essentially means that all the "writes" go to the page cache of the OS (RAM). This makes it efficient to transfer data from page cache to a network socket. The macrolevel Kafka architecture is given below.

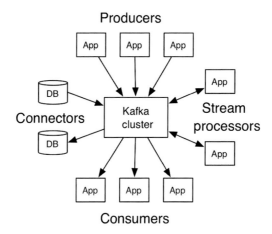

Kafka has four core APIs:

- The producer API allows an application to publish a stream of records to one or more Kafka topics.
- The consumer API allows an application to subscribe to one or more topics and process the stream of records produced to them.
- The streams API allows an application to act as a stream processor, consuming an input stream from one or more topics, and producing an output stream to one or more output topics, effectively transforming the input streams to output streams.
- The connector API allows building and running reusable producers or consumers that connect Kafka topics to existing applications or data systems. For example, a connector to a relational database might capture every change to a table.

This is a messaging broker often used in place of traditional brokers in the Hadoop environment because it is designed for higher throughput and provides replication and greater fault tolerance. Kafka works in combination with Apache Storm, Apache HBase, and Apache Spark for real-time analysis and rendering of streaming data. Kafka can message geospatial data from a fleet of long-haul trucks or sensor data from heating and cooling equipment in office buildings. In short, Kafka brokers massive message streams for low-latency analysis in enterprise Apache Hadoop. Apache Kafka supports a wide range of use cases as a general-purpose messaging system for scenarios where high-throughput, reliable delivery, and horizontal scalability are important. Apache Storm and Apache HBase both work very well in combination with Kafka.

5.4.6 Zookeeper

The various contributing modules of a distributed application can run on networked multiple nodes in a compute cluster to complete a particular task in a fast and efficient manner. Generally complicated tasks, which would take hours to complete an application by running in a single node, can be done in minutes by distributing the application modules across the cluster nodes. The time to complete the task can be further reduced by automated provisioning of servers, configuration management, application deployment, and delivery. There can be multiple clients for a distributed application as shown in the figure below.

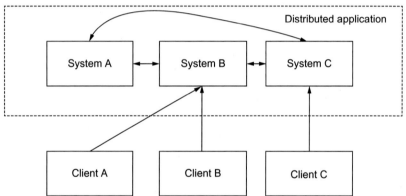

The prominent challenges of distributed applications include:
- Race condition—It is all about two or more machines performing a particular task try to access shared resources.
- Deadlock—Two or more operations waiting for each other to complete indefinitely.
- Inconsistency—Partial failure of data.

Apache ZooKeeper is one such solution for effective and efficient coordination between multiple services. ZooKeeper is a new solution used by a cluster of nodes to coordinate among themselves and maintain shared data with robust synchronization techniques. The common services provided by ZooKeeper are as follows:
- Naming service—Identifying the nodes in a cluster by name.
- Configuration management—Latest and up-to-date configuration information of the system for a joining node.

- Cluster management—Joining/leaving of a node in a cluster and node status at the real time.
- Leader election—Electing a node as the leader for coordination purpose.
- Locking and synchronization service—Locking the data while modifying it. This mechanism helps in automatic fail recovery while connecting other distributed applications like Apache HBase.
- Highly reliable data registry—The availability of data even when one or a few nodes are down.

The race condition and deadlock issues are handled using *fail-safe synchronization approach*. Another main drawback is the data inconsistency and the ZooKeeper service resolves it with *atomicity*.

6. HOW DOES IT WORK?

We are heading toward the days of distributed deployment and centralized management. That is, if there are several database systems working together for accomplishing a special job, then there is a need for centralized monitoring, measurement, and management of the cluster. There are a number of horizontal services such as name service, group services, synchronization services, configuration management, and more. In addition, with the continued growth of Hadoop ecosystem tools, there are new types of services emerging for simplifying and streamlining cross-cluster services. The workaround is that each of these Hadoop projects can just embed Zoo-Keeper without having to build the above-mentioned horizontal services from scratch into each project.

ZooKeeper provides an infrastructure for cross-node synchronization and can be used by applications to ensure that tasks across the cluster are serialized or synchronized. It does this by maintaining status type information in memory on ZooKeeper servers. A ZooKeeper server is a machine that keeps a copy of the state of the entire system and the server can persist the information in local log files. A very large Hadoop cluster can be supported by multiple ZooKeeper servers. In this case, a master server synchronizes the top-level servers. Each client machine communicates with one of the Zoo-Keeper servers to retrieve and update its synchronization information.

In summary, ZooKeeper is a distributed and open-source coordination service for distributed applications and services. It exposes a simple set of primitives that distributed applications can build upon to implement higher-level services for synchronization, configuration maintenance, and groups and naming. Coordination services are especially prone to errors such as race conditions and deadlock. The motivation behind ZooKeeper is to relieve distributed applications the responsibility of implementing multiple coordination services. The below figure says it in a clear and crisp manner.

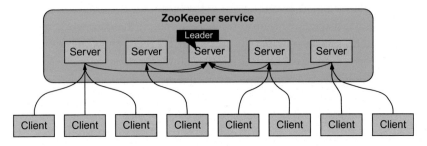

The servers that make up the ZooKeeper service must all know about each other. They maintain an in-memory image of the state, along with transaction logs and snapshots in a persistent store. As long as a majority of the servers are available, the ZooKeeper service will be available. Clients connect to a single ZooKeeper server. The client maintains a TCP connection through which it sends requests, gets responses, gets watch events, and sends heart beats. If the TCP connection to the server breaks, the client will connect to a different server.

ZooKeeper stamps each update with a number that reflects the order of all ZooKeeper transactions. Subsequent operations can use the order to implement higher-level abstractions, such as synchronization primitives. ZooKeeper components show the high-level components of the Zoo-Keeper service. With the exception of the request processor, each of the servers that make up the ZooKeeper service replicates its own copy of each of the components.

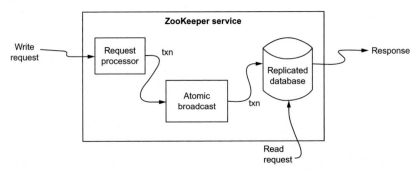

The replicated database is an in-memory database containing the entire data tree. Updates are logged to disk for recoverability, and writes are serialized to disk before they are applied to the in-memory database. Every Zoo-Keeper server services clients. Clients connect to exactly one server to submit requests. Read requests are serviced from the local replica of each server database. Requests that change the state of the service, write requests, are processed by an agreement protocol.

As part of the agreement protocol, all write requests from clients are forwarded to a single server, called the *leader*. The rest of the ZooKeeper servers, called *followers*, receive message proposals from the leader and agree upon message delivery. The messaging layer takes care of replacing leaders on failures and syncing followers with leaders. ZooKeeper uses a custom atomic messaging protocol. Since the messaging layer is atomic, ZooKeeper can guarantee that the local replicas never diverge. When the leader receives a write request, it calculates what the state of the system is when the write is to be applied and transforms this into a transaction that captures this new state.

7. CONCLUSION

The ecosystem elements described above are all open-source Apache Hadoop projects. There are numerous commercial solutions that use or support the open-source Hadoop projects. Hadoop is powerful because it is extensible and enables an easy integration with other components or third-party libraries. Its surging popularity is due to its ability to store, analyze, and access large amounts of data, quickly and cost effectively across clusters of commodity hardware.

Hadoop helps organizations make decisions based on comprehensive analysis of multiple variables and datasets, rather than a small sampling of data or anecdotal incidents. The ability to process large sets of distributed and disparate data gives Hadoop users a more comprehensive and compact view of their customers, operations, opportunities, risks, etc. This chapter has given sufficient information on the various existing and emerging technologies and tools in the ever-expanding Hadoop ecosystem.

FURTHER READING

[1] https://dzone.com/big-data-analytics-tutorials-tools-news.
[2] http://hadoop.apache.org/.
[3] https://mapr.com/products/apache-hadoop/.
[4] https://www.sas.com/en_us/insights/big-data/hadoop.html.
[5] https://www.ibm.com/analytics/us/en/technology/hadoop/.
[6] http://www.pentaho.com/solutions/hadoop.
[7] https://www.cloudera.com/products/open-source/apache-hadoop.html.
[8] https://hortonworks.com/apache/hadoop/.
[9] https://spark.apache.org/.
[10] https://databricks.com/spark/about.

[11] https://azure.microsoft.com/en-in/services/hdinsight/apache-spark/.
[12] https://www.mongodb.com/products/spark-connector.
[13] https://kafka.apache.org/.
[14] https://www.confluent.io/.

ABOUT THE AUTHOR

I have been working as the chief architect in the Site Reliability Engineering (SRE) Center of Excellence, Reliance Infocomm Ltd. (RIL), Bangalore. I previously worked as a cloud infrastructure architect in the IBM Global Cloud Center of Excellence (CoE), IBM India Bangalore for 4 years. Prior to that, I had a long stint as TOGAF-certified enterprise architecture (EA) consultant in Wipro Consulting Services (WCS) Division. I also worked as a lead architect in the corporate research (CR) division of Robert Bosch, Bangalore. In total, I have gained more than 17 years of IT industry experience and 8 years of research experience. I finished the CSIR-sponsored PhD degree in Anna University, Chennai and continued the UGC-sponsored postdoctoral research in the department of Computer Science and Automation, Indian Institute of Science, Bangalore. Thereafter, I was granted a couple of international research fellowships (JSPS and JST) to work as a research scientist for 3.5 years in two leading Japanese universities. Regarding the publications, I have published more than 30 research papers in peer-reviewed journals such as IEEE, ACM, Springer-Verlag, Inderscience, etc. I have authored 7 books thus far and I focus on some of the emerging technologies such as IoT, Cognitive Analytics, Blockchain, Digital Twin, Docker Containerization, Data Science, Microservices Architecture, fog/edge computing, etc. I have contributed 25 book chapters thus far for various technology books edited by highly acclaimed and accomplished professors and professionals.

CHAPTER EIGHT

Biological Big Data Analytics

Mohammad Samadi Gharajeh
Young Researchers and Elite Club, Tabriz Branch, Islamic Azad University, Tabriz, Iran

Contents

1. Introduction 322
2. What Is Big Data Analytics 323
   2.1 Specifications of Big Data Analytics 323
   2.2 Research Trends of Big Data Analytics 325
   2.3 Architectures for Big Data Analytics 327
   2.4 Big Data Technologies 328
   2.5 Cloud-Based Data Analytics Services 330
3. Machine Learning for Big Data Analytics in Plants 330
   3.1 Big Data Technology for the Plant Community 331
   3.2 Problems in Machine Learning for Biology 332
   3.3 Machine Learning-Based Big Data Analytics in Plants 334
   3.4 The Interface Between Machine Learning and Big Data 334
4. Big Data Analytics in Bioinformatics 335
   4.1 Types of Big Data in Bioinformatics 335
   4.2 Big Data Problems in Bioinformatics 336
   4.3 Tools for Big Data Analytics in Bioinformatics 338
5. Big Data Analytics in Healthcare 339
   5.1 Big Data in Medical Image Processing 340
   5.2 Medical Signal Analytics 340
   5.3 Big Data Applications in Genomics 342
   5.4 The MapReduce Framework for Clinical Big Data Analysis 343
6. Healthcare Information System Use Cases 343
   6.1 Advantages of the HIS Compared to Paper-Based Medical Record 344
   6.2 An RFID-Based Smart Hospital 345
   6.3 A Healthcare System Based on GIS Technology 347
   6.4 Cloud-Based WBAN for Pervasive Healthcare 348
   6.5 Collaborative Sharing of Healthcare Data in Multiclouds 349
7. Conclusions and Future Works 350
Glossary 351
References 352
Further Reading 354
About the Author 355

Advances in Computers, Volume 109
ISSN 0065-2458
https://doi.org/10.1016/bs.adcom.2017.08.002

Abstract

Big data analytics uses efficient analytic techniques to discover hidden patterns, correlations, and other insights from big data. It brings significant cost advantages, enhances the performance of decision making, and creates new products to meet customers' needs. This method has various applications in plants, bioinformatics, healthcare, etc. It can be improved with various techniques such as machine learning, intelligent tools, and network analysis. This chapter describes applications of big data analytics in biological systems. These applications can be conducted in systems biology by using cloud-based databases (e.g., NoSQL). The chapter explains the improvement of big data technology in plants community with machine learning. Furthermore, it presents various tools to apply big data analytics in bioinformatics systems. Medical signal and genomics are two major fields in healthcare environments that would be improved by this type of analytical method. Finally, the chapter discusses on several use cases of healthcare information system.

1. INTRODUCTION

Biology is the study of life in nature. It considers life organisms in one cell (e.g., bacteria) or multiple cells (e.g., animals and plants). Biological science consists of various ranges from the study of molecular mechanisms in cells to the behaviors and classification of organisms. Moreover, it indicates how species are evolved in cells to illustrate the interaction between ecosystems. Biology can be overlapped by other sciences such as medicine, biochemistry, and psychology to enhance their performance in the defined operations. Biological systems are composed of various organisms to work together in conducting certain tasks. Some of the most important developments in the science of human health and environmental sustainability can be improved with these systems. Various scientific disciplines such as computer science, bioinformatics, and physics can anticipate the ways which biological systems change over time under different conditions. These systems present efficient solutions to the most healthcare systems and environmental usages. The performance of biology systems in plants, bioinformatics, and healthcare can be improved by using big data analytics [1] and biological databases.

In this chapter, I will describe several applications of big data analytics in biological systems. Section 2 presents general features, research trends, and architectures of big data analytics. Section 3 discusses on the use of machine learning in plants via big data analytics. Section 4 considers data types and tools for big data analytics in bioinformatics. Big data analytics in healthcare as well as use cases of the healthcare information systems are studied in Sections 5 and 6, respectively. Finally, the chapter is concluded in Section 7.

2. WHAT IS BIG DATA ANALYTICS

Big data analytics discusses about challenges of the data which are vast, unstructured, and fast. It applies some of the powerful analytic techniques to operate on big data. Nowadays, the most institutions, organizations, and governments generate various types of the unprecedented and complexity data. Organizations usually use the meaningful information and competitive benefits which are gathered from massive amounts of data. One of the major challenges in this area is to extract the meaningful information from such data sources, readily and quickly. Consequently, big data can be realized by the analytics tools to improve business efficiency and make sharing purposes. In recent years, analytics tools are applied to conduct the volume, velocity, and variety of big data. Note that they are not very expensive so that some of them are available as open source. Hadoop is one of the most commonly used analytics frameworks, which consists of hardware and open-source software. It obtains massive amounts of data to distribute them onto cheap disks as well as provides various powerful tools to analyze the data efficiently. All of the technologies and tools mentioned above need to use integration of the relevant internal and external sources of data. In fact, they are essential components of a big data strategy [2].

This section represents various aspects of big data analytics including specifications of big data analytics, research trends of big data analytics, architectures for big data analytics, big data technologies, and cloud-based data analytics services.

2.1 Specifications of Big Data Analytics

The integration of big data and analytics gives more benefits to organizations. As shown in Fig. 1, main aspects of this integration can include enhancing the analytic tool results, handling the big data, the embraceable economics of analytics, learning from messy data, a special asset to merit leverage, and revealing the business changes. Enhancing the analytic tool

Fig. 1 Main aspects in the integration of big data and analytics.

results are indicated by the most analytics tools to optimize data mining or statistical analysis for large datasets. That is, these tools use the larger and more accurate samples of big data to produce the statistical data analysis. Handling the big data can be conducted by analytic tools and databases to perform big queries and parse tables in record time. Efficiency of the analytics tools has been enhanced by vendor tools and platforms in recent years. The embraceable economics of analytics are considered in big data analytics to reduce the costs of processing bandwidth and data storage. Since big data are used in both small business and big business, the analytics tools and platforms for big data have a high effect on our life. Learning from messy data is possible because this type of data is really big. The most modern tools and techniques for big data analytics are applied for raw source data by using transactional schema, nonstandard data, and poor quality data. This ability leads lots of details to be discovered and anticipated noticeable. A special asset to merit leverage can be found in big data of many small–to–midsize businesses. In fact, it is the real view of big data analytics. There are new technologies and best practices that can be used to work on the data having dozens of terabytes. These features can be integrated together to discover new insights exist in the business. Revealing the business changes is another achievement in the integration of big data and analytics, which can be obtained based on large data samples. Recent economic recession and recovery lead to change the average business beyond all recognition. Consequently, business people can share a wholesale recognition to analyze the new conditions of the business [3].

Fig. 2 indicates the most likely beneficiaries of big data analytics: customers, business intelligence, and specific analytic applications. Big data analytics gives various benefits to anything including customers. These benefits

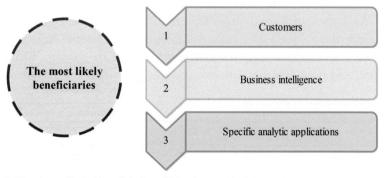

Fig. 2 The most likely beneficiaries of big data analytics.

contain better-targeted marketing, customer-based segmentation, and the recognition of sales and market opportunities. This type of analytics can help to develop definitions of the various customer-based behaviors. Business intelligence is another beneficiary of big data analytics. It can help to result more numerous and accurate business insights, understand business change, manage a better planning and forecasting, and identify the root causes of costs. Finally, specific analytic applications benefit from this analytics by considering analytic applications for the detection of fraud, the quantification of risks, or market sentiment trending [3].

2.2 Research Trends of Big Data Analytics

There are a number of actual research trends of big data analytics. As illustrated in Fig. 3, they include data source heterogeneity and incongruence, filtering-out uncorrelated data, strongly unstructured nature of data sources, high scalability, combining the benefits of RDBMS and NoSQL [4] database

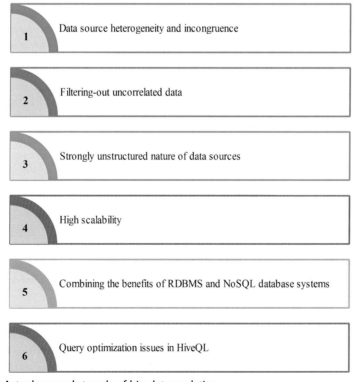

Fig. 3 Actual research trends of big data analytics.

systems, and query optimization issues in HiveQL [5]. These trends are described in the following.

Data sources—which store a large amount of big data for the target analytics processes such as scientific data repositories, sensor databases, and social networks—are strongly heterogeneous and incongruent. This process is considered in the typical integration problems that come from active literature on the data and schema integration concepts. Besides, it includes deep consequences on the various analytics types to can be designed in data services.

The uncorrelated data in various types of the data analytics are emerged frequently when data services uses large amount of data. Main reason is that these services deal with the enormous size of big data repositories. Therefore, filtering-out uncorrelated data are a wanted strategy in these conditions. It makes an essential role in the context of analytics when dealing with big data. Consequently, this strategy affects on quality of the ultimate analytics which should be designed for some purposes.

The input big data should be transformed in a suitable and structured form because of designing the meaningful analytics. Data sources that sustain big data repositories—in social network data, biological experiment data, etc.—are unstructured strongly. This issue is in contrast with the less-problematic and unstructured data which are popular in traditional database tools (e.g., XML data).

High scalability of big data analytics is considered as one of the main requirements for a data analytics system. It can be useful in cloud computing framework because cloud services deal with various users and stations that are distributed on different positions. Therefore, scalability is a feature that leads to powerful computational infrastructure (e.g., cloud systems) to be ensured considerably.

Flexibility is one of the more relevant features that should be obtained by big data analytics systems. It indicates that a large collection of analytical scenarios are covered over the same big data partition. It is essential to combine the benefits of traditional RDBMS database systems and NoSQL database systems to obtain this critical feature. Flexibility will aim to conduct data via horizontal data partitions.

Query optimization aspects of HiveQL cause to arise several open issues. From among these issues, the following five ones are considerable: (i) moving toward more expressive, sophisticated aggregations (e.g., OLAP-like rather than SQL-like); (ii) covering the advanced SQL statements (e.g., nested queries); (iii) incorporating the data compression paradigms to get a higher performance; (iv) inventing the novel cost-based optimizations (e.g., based on table or column statistics); and (v) integration with third-part tools.

2.3 Architectures for Big Data Analytics

Big data analytical systems are designed with several architectures. Many of these architectures share common computational models. MapReduce, fault tolerant graph, and streaming graph are three major architectures [6] that are discussed in the following.

MapReduce is a data-parallel architecture, which is originally developed by Google. The parallelism feature is obtained by multiple machines or nodes that perform the same duties on different data. Apache Hadoop is one of the main databases which are designed by the open-source implementation of MapReduce. This architecture applies a daemon that runs on the master and worker nodes at anytime. The master node manages the configuration and control responsibilities throughout execution process of the problem. In contrast, worker nodes perform actual computation on data. Furthermore, the master node splits the actual data, assigns them to worker nodes, and puts them into the global memory as the (key, value) pairs. Fig. 4 illustrates main elements of a basic MapReduce architecture, where {$W1$, $W2$, ..., Wn} are the worker nodes.

Since the different implementations of MapReduce process data in batch mode, they are not applicable in dealing with complex computational dependencies exist among data. In order to surpass this constraint, a fault tolerant graph-based architecture, called GraphLab, is suggested by Low et al. [7] in the first time. This architecture divides the computation among nodes

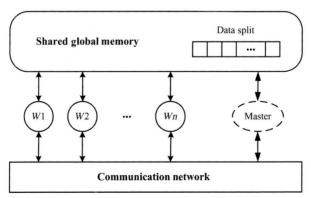

Fig. 4 The basic architecture of MapReduce [6].

in a heterogeneous way so that each of them performs some of the particular tasks. The data model is categorized into two units: a graph with computing nodes and a shared global memory. Note that the computation process is performed in execution cycles through a synchronous procedure.

While the graph-based architecture includes high disk read/write overhead, it is not more efficient for stream data. Consequently, several packages (e.g., Spark Streaming) are proposed to perform analytics on stream data in the MapReduce architecture. They transform stream data to batches internally to conduct the processing phase. Stream frameworks need an in-memory processing to control a high bandwidth. This architecture does not apply a global shared memory compared to the previous architectures in order to exchange stream data by using peer-to-peer communications, directly.

2.4 Big Data Technologies

Fig. 5 shows some of the popular big data technologies including Apache Flume, Apache Hive, MongoDB, Apache Cassandra, and Apache Hadoop. These technologies can be used to work with and manipulate big data on information systems (e.g., cloud systems) [2].

Apache Flume is a distributed and reliable system to collect, aggregate, and move large amounts of log data from many different data sources toward a centralized data store. It deploys one or more agents in a way that each agent contains within its own instance of the Java virtual machine (JVM). Every agent consists of three pluggable components including sources, sinks, and channels. It takes any data streams incoming from one or more data sources. Afterward, data will be passed toward a sink that is commonly a distributed file system (e.g., Hadoop).

Apache Hive is a big data technology that is developed by Facebook. It turns Hadoop into a data warehouse that will be accomplished via a SQL query process. Since this technology uses the SQL dialect, HIVEQL is considered as a declarative language. In Hive tool, we can describe the required results and, then, the tool will specify how to build a data flow for obtaining such results.

MongoDB is a document-oriented and open-source NoSQL database, which is one of the most well-known NoSQL databases. In compared to traditional relational databases, MongoDB applies binary format of the JSON-like documents and dynamic schemas. The query system of this

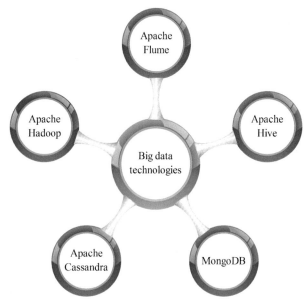

Fig. 5 Some of the well-known big data technologies.

database uses fields, range queries, and regular expression search to return particular fields and query-based compass search. Moreover, it may apply the user-defined complex JavaScript functions.

Apache Cassandra is an open-source NoSQL database that can provide big data requirements to users. It is a high-performance and high-scalable distributed database management system to design real-time big data applications as well as drive important systems for the modern and successful businesses. Cassandra contains a built-for-scale architecture to handle petabytes of information and thousands of concurrent users/operations per second. These processes are performed to manage much smaller amounts of data and user traffic.

Apache Hadoop software is a powerful framework to enable the distributed processing purposes of large datasets across multiple clusters of computers. It is designed to scale up from single servers to thousands of server machines. This target is considered to provide local computation and storage by each server. A single query can find and collect results from all cluster members that is especially more appropriate for Google's search engine. Besides, Hadoop can offer various mechanisms for storage, manipulation, and information retrieval on big data.

Table 1 Characteristics of the Cloud-Based Data Analytics Services

Service Model	Features	Target Users
Data analytics software as a service	An individual and complete data mining application or task	• End users • Analytics managers • Data analysts
Data analytics platform as a service	Developing the high-level applications as well as hiding the cloud infrastructure and data storages	• Data mining application developers • Data scientists
Data analytics infrastructure as a service	Developing, configuring, and running the data analysis frameworks or applications	• Data mining programmers • Data management developers • Data mining researchers

2.5 Cloud-Based Data Analytics Services

Since cloud-based systems deal with big data, they can be used to conduct data analytics services. Researchers and developers have attempted to design and implement big data analytics solutions in the cloud by using the software as a service (SaaS), platform as a service (PaaS), and infrastructure as a service (IaaS) models. Table 1 represents service model, features, and target users of the cloud-based data analytics services. Data analytics SaaS uses a single and powerful data mining application or task (e.g., data sources). It offers a ready-to-use knowledge discovery tool and a well-defined data mining mechanism as the internet service to end users, analytics managers, and data analysts. Data analytics PaaS utilizes a data analysis framework to develop high-level applications and hide the cloud infrastructure and data storage. Data mining application developers and data scientists can build their own analytics applications or extend the existing applications without considering any underlying infrastructure or distributed computing services. Data analytics IaaS applies a set of virtualized resources—that is offered to a programmer or data mining researcher—to develop, configure, and run data analysis frameworks or applications. Data mining programmers, data management developers, and data mining researchers can use this service type to conduct their own applications.

3. MACHINE LEARNING FOR BIG DATA ANALYTICS IN PLANTS

Big data are used in the modern biological sciences, frequently. Velocity, volume, and variety are the three popular features of big data that can be utilized to catalyze the development of innovative technical and analytical strategies. Operational costs of the data generation are not a major concern for genome-wide research. Note that computational performance of the big data analysis becomes a bottleneck. The plant science community is looking for the novel solutions to the three challenges of big data including scalable infrastructure for parallel computation, intelligent data-mining analytics, and management schemes for large-scale datasets. Machine learning is a modern emerging field of the computer science, statistics, artificial intelligence, and information theory, which can be applied by data scientists to exploit the useful information of big data in the planets [8,9]. This section indicates various aspects of the machine learning for big data analytics in plants: big data technology for the plant community, problems in machine learning for biology, machine learning-based big data analytics in plants, and the interface between machine learning and big data. The following subsections will discuss on characteristics of these aspects.

3.1 Big Data Technology for the Plant Community

Big data technology, typically, refers to three viewpoints of the technical innovation and super-large datasets: automated parallel computation, data management schemes, and data mining. Fig. 6 describes main components

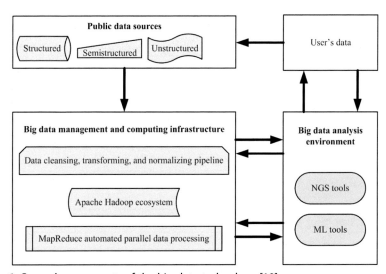

Fig. 6 General components of the big data technology [10].

of the big data technology. The following constructions are essential to build big data infrastructure for the plant science community:

- A centralized big data computing infrastructure that is placed on the top of high-performance computer clusters. Apache Hadoop ecosystem consists of Hadoop operation commands, the MapReduce programming model, the Hadoop distributed file system, and the various utilities for disparate forms of the structured, semistructured, and unstructured datasets.
- A big data storage domain for integrating the plant genome databases and public datasets, as well as the automated pipelines for transforming the query data in super large volume datasets.
- An analysis environment that is offered as a use-on-demand cloud service. It is conducted by integration of the popular bioinformatics software and machine learning-based applications in order to support the Hadoop computing infrastructure. Table 2 expresses learning task, algorithms, and methods of some machine-learning algorithms that can be used in the planet community [10].

3.2 Problems in Machine Learning for Biology

Fig. 7 entitles major problems in the machine learning for biology: missing or incomplete data, missing benchmark tools, integrating data of multiple

Table 2 Some of the Representative Machine-Learning Algorithms [10]

Learning Tasks	Algorithms and Methods
Classification	Linear discriminant analysis, quadratic discriminant analysis, logistic regression, support vector machine (SVM), neural nets
Regression	Least squares, linear models, nearest neighbor methods, regression trees, local regression, splines, wavelet smoothing, Bayesian models
Clustering	K-means clustering, spectral clustering, hierarchical clustering, association rules, local multidimensional scaling
Feature selection	Best subset selection, forward selection, least angle regression
Dimensionality reduction	Principal component analysis (PCA), factor analysis, kernel PCA, partial least squares
Ensemble learning	Boosting methods, bagging, additive regression
Network analysis	Gaussian graphical models, Bayes networks
Density estimation	Kernel density estimation

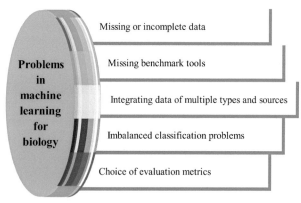

Fig. 7 Main problems in the machine learning for biology.

types and sources, imbalanced classification problems, and choice of evaluation metrics.

Missing or incomplete data are typically occurred in machine learning. For instance, some of the technical or biological reasons cause a subset of genes on microarrays to miss expression values. There are various types of the mechanisms in missing data such as missing completely at random, missing at random, and not missing at random [11]. Missing benchmark tools are another problem of the machine learning in this area. Benchmarking is an evaluation method to improve machine-learning algorithms. An appropriate benchmarking analysis should apply a set of reliable and unbiased data, some of the quantitative and qualitative evaluation parameters, and a comprehensive comparison plan [12]. Generally, there are several types and sources of data for describing the different viewpoints of the biological system under study. One of the main challenging problems in the biology is to specify how to combine various data sources for building an integrated model [13]. The imbalance of subclasses in the data is a prevalent problem in the classification analysis of biology. This problem is especially considered as a prevalent term in high-dimensional datasets. The classifier can be misled by an extreme class imbalance through overlearning the majority class and performing poorly in the prediction of the minority class [14]. Most used evaluation metrics involve threshold-based metrics (e.g., accuracy and precision), ordering-based metrics (e.g., AUC of the ROC plot), and probability-based metrics (e.g., the root-mean-square error). Choice of evaluation metrics can be encountered in the biology because of some problems in the above metrics [15].

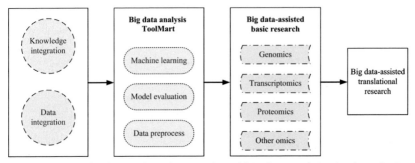

Fig. 8 A platform for the machine learning-based big data analytics in plants [10].

3.3 Machine Learning-Based Big Data Analytics in Plants

Hadoop-based machine-learning tools, big data management, and computing infrastructure are integrated and interfaced together to offer crucial services to the plant science community. This process is conducted in aspects of both translational research and big data-assisted basic research. Fig. 8 illustrates a platform that consists of GeneMart, DataMart, and ToolMart. GeneMart is defined as a knowledge base of plant genes. It integrates and categorizes a large amount of the sources which are composed of information from public genome databases. DataMart maintains public omics data in accordance with the experimental design and data types. Datasets of the DataMart are represented based on the common rules for the objectives of compiling training, validating datasets for the machine-learning training, optimization, and evaluation. ToolMart utilizes software applications for some purposes including data preprocessing, model validation, and machine-learning packages. A proper machine-learning model can be suggested according to data types and learning problems. Note that users can carry out model training, model optimization, and model evaluation via a visualized interface [10].

3.4 The Interface Between Machine Learning and Big Data

Many open-source machine-learning packages are implemented by using the MapReduce function due to the requirements of big data scale analysis. Apache Hadoop ecosystem interfaces these tools, public genome databases, and high-throughput data in the plant community. It conducts these objectives as a centralized big data analytical platform in order to help the plant science community. The practical steps that big data bioinformaticians need to take are carried out in a long term through two phases: (i) upgrading the algorithms of current software by using the MapReduce programming framework and (ii) developing the novel metadata schemes to conduct the annotational and experimental information related to a plant gene. Experimental information may need a composition of computer and human efforts with a system of the unified labeling rules. This process is done to make a group of genes and datasets based on functional descriptions and experimental designs, respectively. This procedure is practically used to compile the training example sets in a machine-learning analysis in a way that the genes and data attributes can be correspondingly obtained from the GeneMart and DataMart with regarding the same category labels. Furthermore, a data preprocessing pipeline as well as some unified algorithms and rules should be extended to cleanse, transform, and normalize big data of the planets [10].

4. BIG DATA ANALYTICS IN BIOINFORMATICS

There is a huge volume of data in bioinformatics research. Big data sources are no longer constrained to specific physical experiments or search-engine data stores. Data volume is rising in bioinformatics research because all procedures and availability of high-throughput devices are digitized at lower costs. Decreasing the computing costs and enhancing the analytics throughput support this trend in rising data volume, which are resulted by growing big data technologies. Biologists no longer apply traditional bioinformatics laboratories to figure out a novel biomarker for a disease. Moreover, various research groups depend on huge and ceaseless growing genomic data. Data size of bioinformatics information has been increased in the recent years, dramatically. Total storage size of the information is doubling every year. Therefore, it is needed to apply big data analytics to manage big data in bioinformatics [16]. This section describes various aspects of big data analytics in bioinformatics including types of big data in bioinformatics, big data problems in bioinformatics, and tools for big data analytics in bioinformatics.

4.1 Types of Big Data in Bioinformatics

There are generally five data types that are massive in size and most used in bioinformatics research: (i) gene expression data, (ii) DNA, RNA, and protein sequence data, (iii) protein–protein interaction (PPI) data, (iv) pathway data, and (v) gene ontology. Table 3 represents main features of these types. In gene expression analysis, the expression levels of thousands of genes are experimented and evaluated over various situations (e.g., separate developmental stages of the treatments and/or diseases). The expression levels for

Table 3 Main Features of Data Types in Bioinformatics Research

Data Type	Feature
Gene expression data	Analysis of the expression levels of thousands of genes with the aid of microarray-based gene expression profiling
DNA, RNA, and protein sequence data	The use of various analytical methods to identify the characteristics, functions, structures, and evolution
PPI data	Analysis of the PPI networks to give protein functions
Pathway data	Understanding molecular basis of a disease and identification of the genes and proteins
Gene ontology	Providing the dynamic, structured, and species-independent gene ontologies by using controlled vocabularies

analysis are recorded by using microarray-based gene expression profiling. Gene-sample, gene-time, and gene-sample-time are three types of microarray data. In sequence analysis, DNA, RNA, or peptide sequences are operated by using several analytical methods. Their characteristics, functions, structures, and evolution are understood by attending this process. DNA sequencing can be applied for some purposes such as the study of genomes and proteins, evolutionary biology, identification of micro species, and forensic identification. PPIs offer essential information according to all the biological processes. Hence, protein functions can be properly given by forming and analyzing the PPI networks. Anomalous PPIs are the fundamentals of various diseases (e.g., Alzheimer's disease and cancer). Pathway analysis is used to understand molecular basis of a disease. It identifies the genes and proteins which are related to the etiology of a disease. Moreover, this type of analysis estimates drug targets and manages the targeted literature searches. Gene ontology offers dynamic, structured, and species-independent gene ontologies for the three objectives of associated biological processes, cellular components, and molecular functions. It utilizes the controlled vocabularies for facilitating the query data at different levels [6].

4.2 Big Data Problems in Bioinformatics

Main categorizes of the big data problems in bioinformatics include seven terms as entitled by Fig. 9 [6]. They are described in the following with more details.

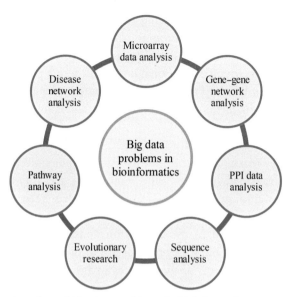

Fig. 9 Major categorizes of big data problems in bioinformatics.

Microarray data analysis: Because of reducing the operational costs and widespread use of microarray experiments, the size and number of microarray datasets are increasing quickly. Beside, microarray experiments are carried out for gene-sample-time space to store the noticeable changes over time or various stages of a disease. Fast construction of the coexpression and regulatory networks needs big data technologies with the aid of voluminous microarray.

Gene–gene network analysis: Gene coexpression network analysis predicts the relation among different gene–gene networks achieved from gene expression analysis. Differential coexpression analysis looks for those of the changes which are created by the gene complexes over time or various stages of a disease. This process leads to find the correlation between gene complexes and traits of interest. Consequently, gene coexpression network analysis is defined as a complex and highly iterative problem that needs large-scale data analytics systems.

PPI data analysis: PPI complexes and their changes contain high information about various diseases. PPI networks are considered in different domains of life sciences in cooperation with the production of voluminous data. The volume, velocity, and variety of data cause PPI complex analytics to be a genuine big data problem. Hence, an efficient and scalable architecture should be provided to give the fast and accurate PPI complex generation, validation, and rank aggregation.

Sequence analysis: RNA sequencing technology is addressed as a strong successor to the microarray technology. This successor is resulted because of its more accurate and quantitative gene expression measurements. RNA sequence data involve additional information that needs considerable machine-learning models. Therefore, big data technologies can be applied to indicate mutations, allele-specific expressions, and exogenous RNA contents (e.g., viruses).

Evolutionary research: The recent advances in molecular biological technologies have led to a noticeable source for big data generation. Various projects at microbial level (e.g., genome sequencing and microarrays) have generated huge amounts of data. Bioinformatics has been considered as an important platform for the analysis and achievement of this essential information. Functional trend of the adaptation and evolution by microbial research is an important big data problem in bioinformatics.

Pathway analysis: Pathway analysis makes an association between genetic products and phenotypes of interest. This correlation is done to estimate gene function, identify biomarkers and traits, and also make a category of the patients and samples. Association analysis on huge volumes of the data can be performed by increasing genetic, genomic, metabolomics, and proteomic data as well as by offering the big data technologies.

Disease network analysis: Large disease networks are provided by formulating various species (e.g., human). Complexity and data of these networks increase over time as well as new networks are added by using different sources in their own format. Consequently, sophisticated networks of molecular phenotypes cause to estimative genes or mechanisms for the disease-associated traits. Some of the efficient integration methods should be applied to evaluate multiple, heterogeneous omics databases.

4.3 Tools for Big Data Analytics in Bioinformatics

Fig. 10 identifies some of the important tools for big data analytics in bioinformatics. Large number of software tools (e.g., caCORRECT) is programmed for microarray data analysis to perform various analyses on microarray data. However, not all of these tools are implemented to manage large-scale data. Tools for gene–gene network analysis (e.g., FastGCN) are designed for the gene expression datasets having a large size. PPI complex that looks for to discover any problem is a highly time consuming process. The standalone implementations for the both of supervised and unsupervised PPI complex finding programs need days or even weeks of

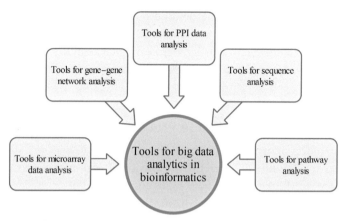

Fig. 10 Some of the main tools for big data analytics in bioinformatics.

time in order to locate the complexes of a huge dataset. Therefore, tools for PPI data analysis (e.g., NeMo) should be applied to facilitate the PPI complex finding purposes. Tools for sequence analysis (e.g., BioPig) have been designed by using the Hadoop MapReduce platform in order to manage data analytics process on large-scale sequence data. Finally, a good number of tools for pathway analysis (e.g., GO-Elite) are offered to support the pathway analysis purposes [6].

5. BIG DATA ANALYTICS IN HEALTHCARE

Because big data analytics has been expanded by researchers in the last decades, it aids to enhance the evolution of healthcare practices and research. This type of analytics has offered powerful tools to accumulate, manage, and analyze large amounts of the healthcare data. The process of care delivery and disease exploration can be improved by using data analytics techniques. However, some of the fundamental problems inherent within the big data paradigm cause to cease the adoption rate and research development within this area. Image processing, signal processing, and genomics are three major areas that have been already discussed for big data analytics in medicine. Diagnosis, therapy assessment, and planning of healthcare systems apply medical images as an important source. Medical signals contain the volume and velocity obstacles during continuous, high-resolution storage from a large number of the monitors which are connected to each patient. The development of high-throughput sequencing technology decreases the cost to sequence the human genome. Therefore these areas can be improved by big data analytics to conduct the defined purposes well [17]. This section discusses on big data analytics in healthcare by using four major aspects: big data in medical image processing, medical signal analytics, big data applications in genomics, and the MapReduce framework for clinical big data analysis.

5.1 Big Data in Medical Image Processing

Medical imaging offers fundamental information about anatomy and organ function as well as aims to detect diseases states. It can be used to conduct some of the medical requirements such as organ delineation, identifying tumors in lungs, spinal deformity diagnosis, and artery stenosis detection. Therefore, image processing techniques (e.g., enhancement, segmentation, and denoising) and machine-learning methods can be applied to increase the performance of medical imaging. Since the size and dimensionality of medical data enhance, the dependencies between medical data and design of the efficient, and accurate methods need novel computer-aided techniques. The use of computer-aided medical diagnostics and decision support systems in clinical environments is required because the number of healthcare organizations and the number of patients grow noticeably. Efficiency of the healthcare processes (e.g., diagnosis, prognosis, and screening) can be enhanced by using computational intelligence. The integration process of computer analysis and appropriate care can potentially aid clinicians in order to improve diagnostic accuracy. Furthermore, the accuracy can be improved as well as the time taken for a diagnosis can be reduced by the integration of medical images and other types of the electronic health record. Table 4 describes some of the challenges and possible solutions in medical image analysis [17].

5.2 Medical Signal Analytics

Physiological and telemetry signal monitoring devices are ubiquitous tools. Continuous data that are generated from these monitoring devices have not been generally maintained for more than a defined period of time. However, the activities toward using the telemetry and continuous physiological monitoring are increased in the last years because of improving the patient care and management [18,19]. Streaming data analytics in the healthcare systems can be specified as a systematic utilization of ceaseless waveform—signal varying against time—and associated medical record information. This trend is conducted via the applied analytical disciplines (e.g., statistical, contextual,

Table 4 Some of the Challenges in Medical Image Analysis [17]

Challenge	Possible Solutions
Preprocessing	Employing multimodal data
Compression	Reduction of the volume of data while storing the important data
Parallelization/real-time realization	The development of the scalable/parallel methods and frameworks
Registration/mapping	Arranging the consecutive slices/frames from one scan or corresponding images
Sharing/security/ anonymization	Protecting the integrity, privacy, and confidentiality of data
Segmentation	Delineation of anatomical structure (e.g., vessels and bones)
Data integration/mining	Discovering the dependencies/patterns among multimodal data
Validation	Evaluating the efficiency or accuracy of the system/ method

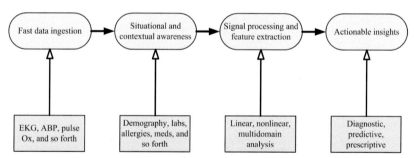

Fig. 11 Generalized analytic workflow of the clinical setting [17].

and predictive) to make appropriate decision making for patient care. Fig. 11 depicts the generalized analytics workflow of the real-time streaming waveforms in clinical settings. A platform is needed for the streaming data acquisition and ingestion that include an essential bandwidth to manage multiple waveforms at different fidelities. One of the important components for offering the conditional and contextual awareness for the analytics engine is to make an integration of these dynamic waveform data with static data gathered from the electronic health record. The system can be more robust by using the data analytics techniques as well as these rich data aim to balance the sensitivity and specificity of the estimative analytics [17].

5.3 Big Data Applications in Genomics

Big data applications in genomics are varied over a wide variety of topics. Pathway analysis and the reconstruction of networks are the main categorizes of these applications. In pathway analysis, functional effects of genes are differentially expressed based on an experiment or gene set of special interests. Analytical and experimental practices cause to generate various errors and batch effects. Moreover, the interpretation of operational effects utilizes the continuous increases in available genomic data and associates the annotation of various genes [20]. In the reconstruction of networks, the signals—which are measured by high-throughput techniques—are evaluated for reconstructing the fundamental regulatory networks. Applications developed for this category can be grouped into two broad categories consisting of the reconstruction of metabolic networks and gene regulatory networks [21]. Reconstruction of metabolic networks extends an understanding of organism-specific metabolism by using the integration of the genomics, transcriptomics, and proteomics high-throughput sequencing models. Table 5 represents brief descriptions of the popular methods and toolkits with their applications.

Table 5 Summary of the Popular Methods and Toolkits for Big Data Applications in Genomics [17]

Toolkit	Category	Selected Applications
Onto-Express	Pathway analysis	Breast cancer
GoMiner	Pathway analysis	Pancreatic cancer
ClueGo	Pathway analysis	Colorectal tumors
GSEA	Pathway analysis	Diabetes
Pathway express	Pathway analysis	Leukemia
Recon 2	Reconstruction of metabolic networks	Drug target prediction studies
Boolean methods	Reconstruction of gene regulatory networks	Cardiac differentiation
ODE models	Reconstruction of gene regulatory networks	Cardiac development

5.4 The MapReduce Framework for Clinical Big Data Analysis

MapReduce is an emerging framework for data-intensive applications that is presented by Google. It uses main ideas of the functional programming so that the programmer will define Map and Reduce tasks for processing the large sets of distributed data. Various types of clinical big data, challenges, and consequences can be implemented by the MapReduce framework, which are associated with the applications of big data analytics in a healthcare facility [22]. Some of the algorithms that are programmed by this framework are mentioned in the following.

A MapReduce-based algorithm [23] is presented for common adverse drug event (ADE) detection. It is tested in the mining spontaneous ADE reports from the US FDA. Investigation of the possibility of using the MapReduce framework in increasing the speed of biomedical data mining tasks was the main objective of this algorithm. A biometrics prototype system [24] is implemented for searching the cloud-scale biometric data and matching a collection of synthetic human iris images. A biometric-capture mobile phone application is programmed to make a secure access to the cloud [25]. It uses the biometric capture and recognition through a standard web session. The connection between a mobile user and server of the cloud is established by the Hadoop platform. A parallel version of the random forest algorithm [26] for regression and genetic similarity learning tasks is implemented for large-scale population genetic association studies. It is used to a genome-wide association study on Alzheimer disease. In this algorithm, the quantitative feature consists of a high-dimensional neuroimaging phenotype that describes the longitudinal changes of the human brain structure.

6. HEALTHCARE INFORMATION SYSTEM USE CASES

Information system (IS) can be used in various areas including healthcare, industry, education, etc., to facilitate performance of the most defined tasks. The IS framework for healthcare may lead to various behaviors in both positive and negative directions. For instance, effective use of this system will yield to higher net benefits and, thereby, more intensive use of the system. In contrast, insufficient use of the system will lead to lower net benefits and, thereby, acting as a disincentive use of the system. In comparison with existing healthcare IS frameworks, DeLone and McLean's IS model presents evident, special dimensions of the IS effectiveness or success as well as correlations between them. However, this model does not have

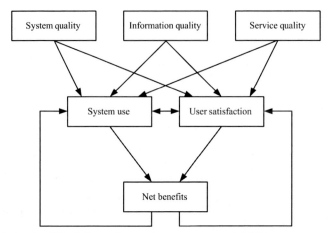

Fig. 12 Elements of the information system success model [27].

the organizational factors which are relevant to IS evaluation. Elements and correlations of the model are illustrated in Fig. 12 [28].

The remainders of this section present various use cases of healthcare information system including advantages of the hospital information system (HIS) compared to paper-based medical record, an RFID-based smart hospital, a healthcare system based on GIS technology, cloud-based wireless body area networks (WBANs) for pervasive healthcare, and collaborative sharing of healthcare data in multiclouds.

6.1 Advantages of the HIS Compared to Paper-Based Medical Record

After an electronic medical record or a HIS is introduced by researchers, most hospitals use this system to keep and update their paper-based medical records. This uses case reports information of a HIS in a hospital where the paper-based medical records are scanned and then eliminated from the handwriting archives. Patient data in the electronic medical records of the HIS are either saves in a database as the searchable text and numbers or as document images. The former model—namely regular electronic data—is composed of the chronological, text-based medical record integrated with the lab data which were stored in the textual radiology and numerical form reports. Table 6 expresses group and type of some documents that are used in the electronic medical records. The latter form consists of a structure into two categories including upon admittance and consultation. In this model, the documents in the format of old paper-based medical records are scanned

Table 6 Some Documents of the Electronic Medical Records [29]

#	Group	Type
A	Summaries	• Biographical data • Critical information (e.g., implants) • Discharge reports from other hospitals • Index of the consultations and admissions • Discharge reports • Instructions for patient upon discharge • Nurse's summaries
B	Textual medical record	• Continuous textual medical record • Referrals within the hospital
C	Lab results including tissue and body fluids	• Clinical biochemical • Other types (e.g., microbiology)
D	Organ function	Cardiovascular, senses, locomotor, etc.
E	Radiology and other imaging info	Radiological investigations, CT, MRI, and other types
F	Treatment and observation	Patient chart summary, anesthesia forms, etc.
G	Nurses' documentation	Nurse's notes and admission essential reports
H	Other health personnel	Occupational therapist, physical therapist, and others
I	Correspondence	Admission request forms, referrals, etc.
J	Certificates/notifications	Public certificates, forms, and notifications

and stored into the system as digital images in TIFF format. These images, namely scanned multiple documents, contain all sheets of the paper-based record, which corresponds to a whole document group. Upon patient discharge is done, various paper sheets are scanned, dated, and also labeled by document type singularly, namely scanned single documents. Ultimately, all of the patient data are stored in the system as regular electronic data, scanned multiple documents, and scanned single documents [29].

6.2 An RFID-Based Smart Hospital

Radio-frequency identification (RFID) is one of the modern technologies that can be used in healthcare systems to conduct different medical scenarios and requirements. It can be applied to build modern hospital of the future to improve the patients' care, optimize the workflows, decrease the operating costs, aid in avoiding severe mistakes, and reduce costly thefts. As entitled in Fig. 13, an RFID-based smart hospital is commonly designed to manage seven applications [30].

Patient identification: Patient identification error causes an unsuitable dosage of medication to patient to may be occurred and the invasive procedure to may be done. Therefore, several RFID-based patient identification projects can be implemented to cut these clinical errors, to improve the patient care and security, and to improve administration and productivity. New York's Jacobi Medical Center [31], Birmingham Heartlands Hospital [32], and German Saarbrücken Clinic Winterberg [33] are some of the RFID-based patient identification and tracking pilot projects.

Blood tracking: Mistransfusion errors (e.g., blood transfusion of the incorrect type or blood given to the wrong patient) are some of the unacceptably frequent and serious problems in the hospitals. In the smart hospital presented in [34], every blood bag arriving at the hospital takes a self-adhesive RFID label. This chip contains a memory to store a unique identification number and information on the contained blood type. The data of the patient and blood bag must match before the blood can be applied in the laboratory.

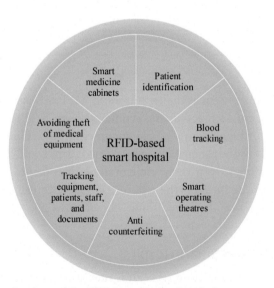

Fig. 13 Main applications of the RFID-based smart hospital.

Smart operating theaters: The patients take a RFID-tagged wristband that contains associated information and a digital picture of them. The digital picture enables the clinical team to readily confirm they have an authorized patient as well as the electronic record makes a certain about they perform the correct process.

Anticounterfeiting: Drug counterfeiting is one of the increasing problems in hospitals. This problem can be solved by incorporating RFID into the package of prescription drugs. A unique number (i.e., electronic product code) is assigned to every pallet, case, and package of drugs. Moreover, this number is used to record information of all transactions involving the product.

Tracking equipment, patients, staff, and documents: A smart hospital can be equipped with several RFID readers at strategic places such as entrances of operating theaters, recovery rooms, and main doors. Therefore, all of the medical histories and important documents will be tagged to locate them via the use of an assets tracking application. This equipment would be lead to help in decreasing the losses rate of medical files.

Avoiding theft of medical equipment: Theft of medical equipment is one of the main problems in hospitals because there are a great number of expensive medical equipments. RFID tags can be embedded into the medical equipment of a smart hospital to track and trace any stealing purposes. Since technical staff of the hospital is always aware about any material existing within the buildings, this process can reduces risks of the thefts.

Smart medicine cabinets: A smart medicine box can be built to allow the patients to achieve their own medicine management. Accordingly, a computer system connected to the internet and embedded in the cabinet can aid the patients through various ways. This advantage would be conducted by using the RFID tags which are pasted onto blister packs and RFID sensor coils.

6.3 A Healthcare System Based on GIS Technology

A framework for GIS-based healthcare system is proposed in [35] that synthesizes two GIS-based accessibility measures. It uses some of the methods to examine spatial accessibilities to initial healthcare in the Chicago 10-county region. While the availability of physicians are accounted by their surrounded requirements, the floating catchment area—namely FCA method—determines the service area of physicians by the defined threshold travel time. A close-up physician more affordable than a remote one can be specified by a gravity-based technique. Moreover, the possibility of a physician's availability will be discounted by a gravity-based potential.

The two GIS-based accessibility measures to the Chicago 10-county region are conducted based on two important factors including travel-time threshold in the two-step FCA method and travel friction coefficient in the gravity-based method. The ranges of these factors are defined reasonably as well as sensitivity evaluation is controlled by experimental values within the ranges. Table 7 represents standard deviations for the two available measures through different choices of the factors. It is worth to noting that the weighted mean of any available measure always equals to the physician-to-population ratio in the entire study field.

6.4 Cloud-Based WBAN for Pervasive Healthcare

WBANs can be applied for the noticeable deployment of pervasive healthcare applications. There are various technical challenges and issues in the integration process of WBANs and mobile cloud computing (MCC). A cloud-enabled WBAN architecture and its applications are presented in [36] that are applicable for pervasive healthcare systems. It uses some of the powerful methodologies to transmit vital sign data to the cloud server by using various terms including energy-efficient routing, cloud resource allocation, data security mechanisms, and semantic interactions.

Home, outdoor environment, and hospital are three locations that can be developed by different MCC services. For the patient at home, real-time

Table 7 Sensitivity Evaluation of the Available Measures Within a GIS-Based Healthcare System [35]

Two-Step Floating Area Method		Gravity-Based Method	
Threshold Travel Time	Standard Deviation	Travel Friction Coefficient	Standard Deviation
20	0.002570	2.2	0.000999
25	0.001550	2.0	0.000934
30	0.001240	1.8	0.000863
35	0.001110	1.6	0.000787
40	0.001040	1.4	0.000705
45	0.000953	1.2	0.000619
50	0.000873	1.0	0.000527

location information can be achieved by using various wireless location techniques such as time difference of arrival and time of arrival as well as future patient activity can be predicted with the aid of data fusion technology. A smart phone is applied to obtain and then transmit the patient's physiological information to the cloud server for the outdoor patients. When an accident happens in the outdoor environment, family member, a doctor, or a nurse can be immediately informed to rescue the patients via GPS location. Besides, for the patient at hospital, doctors or nurses can readily get the location and physiological information of the patient via smart terminals (e.g., personal digital assistants).

Fig. 14 depicts elements of a typical proxy reencryption scenario. A medical data owner uploads the encrypted information to the medical cloud server with a shared public key. Afterward, another cloud-based user requests it as well as a certain authority supports a reencryption process either within the medical cloud system or inside the authority itself. The information can be directly downloaded from the medical cloud server or via the authority.

6.5 Collaborative Sharing of Healthcare Data in Multiclouds

Interorganizational sharing and collaborative use of big data become very important in healthcare systems. A cloud system can precisely offer an environment to match the requirements of collaborating healthcare workers.

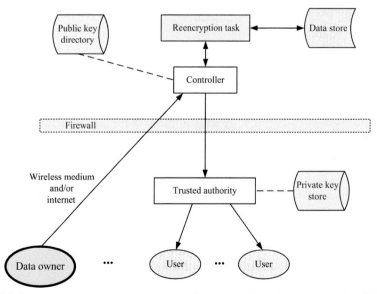

Fig. 14 A scenario presenting the process of proxy medical data reencryption [36].

In contrast, there are a considerable number of the security and privacy challenges into the cloud servers that cause much more problems and loss to the stored data. An architecture is implemented in [37] for interorganizational data sharing in order to offer a significant security and privacy for the patients' information stores in semitrusted cloud environments. It provides an attribute-based encryption for two important crucial features including selective access authorization and cryptographic secret sharing. Consequently, data can be dispersed across multiple clouds as well as the adversarial capabilities of curious cloud providers can be reduced noticeably.

In this architecture, powerful single locations of the security failure (e.g., the ABE key authority) should be decentralized precisely and the security duties within cooperation services are separated probably. More interfaces for various cloud providers are supported by the architecture as well as the streaming data and load balancing are provided accordingly. In addition, it appends the P2P interfaces for various distributed Hash tables and BitTorrent. Moreover, the authors have gathered an overview of security and privacy measures in healthcare systems that are represented in Table 8.

7. CONCLUSIONS AND FUTURE WORKS

This chapter described various definitions, representations, and experiences on biological big data analytics. It conducted these objectives through several sections including what is big data analytics, machine learning for big data analytics in plants, big data analytics in bioinformatics, big data analytics in healthcare, and healthcare information system use cases. Specifications, research trends, architectures, technologies, and cloud-based services of big data analytics were represented in the first glance. Afterward, machine learning for big data analytics in plants were conducted by using big data technology for the plant community, problems in machine learning for biology, machine learning-based big data analytics in plants, and the interface between machine learning and big data. Types, problems, and tools of the big data analytics in bioinformatics were described to discuss adequate information about big data analytics in bioinformatics. Furthermore, big data analytics in healthcare were studied by using several tools including big data in medical image processing, medical signal analytics, big data applications in genomics, and the MapReduce framework for clinical big data analysis. Finally, some of the use cases in healthcare information system were indicated based on advantages of the healthcare information systems,

Table 8 Main Security and Privacy Measures for Healthcare Data [37]

Measure	Objective	Location
Client and service authentication	Avoid unauthorized participation in the system	All participants
Network security and TLS	Avoid communication eavesdropping or modification	All participants
Federated identity management	Enhance usability and reduce management overhead	Health centers
Access control to cloud data	Avoid unauthorized retrieval of medical records from the clouds	Cloud providers
Access control to medical record	Protect access to sensitive information in medical records	Health centers
Attribute-based encryption	Protect access to sensitive information in medical records	Authors and health centers
Digital signatures and Hash-based message authentication codes	Avoid unauthorized modification of medical records and shares	Authors and health centers
Secret sharing of medical records	Enhance availability of data as well as avoid the spying on data	Health centers
Cryptographic Hash function for external identifiers	Anonymity, unlinkability between medical records and patient identifiers	Health centers
Data replication and ciphertext policy	Avoid data loss	Cloud providers

RFID-based smart hospitals, GIS-based healthcare system, cloud-based networks for pervasive healthcare, and cloud-based collaborative sharing of healthcare information.

In the future work, the effects of artificial intelligence [38,39], knowledge-based systems [40,41], and decision support systems [42] on analytical process would be studied completely in order to enhance the performance of big data analytics and reduce some of the existing problems/challenges.

GLOSSARY

Biological databases Biological databases contain variety of biological big data that cater to the needs of biological research. They proportionally improve biological knowledge to support the efficient data sharing, integration, and analysis. Available biological databases can be grouped into three categories: archival repositories, curate resources, and data integration warehouses. All of these databases provide a range of querying and mining tools to enhance the performance of knowledge discovery. Some of the databases are limited to lifetime of the underlying research project because they are intended as long-term, consistently maintained community resources, and intentionally temporary in nature.

Big data analytics Big data analytics enhances the speed and efficiency of big data on various systems. It uses the advanced analytic techniques for the very-large and diverse datasets which contain different types (e.g., structured/unstructured and streaming/batch) and different sizes from terabytes to zettabytes. Analysts, researchers, and business managers can analyze big data to make appropriate decisions by using the inaccessible or unusable data. Big data analytics utilizes various techniques such as text analytics, machine learning, and data mining to achieve high-performance results. It targets cost reduction, faster decision making, and producing new services.

Biological systems Biological systems improve the study of biology and its mechanisms in the key elements of living systems (such as DNA, RNA, proteins, and cells). They consist of novel methodological approaches to analyze the knowledge and, then, translate this knowledge into a thorough understanding. This process leads to anticipate various behaviors of the complex and dynamic networks which conduct living specimen. These systems use mathematical modeling with the aid of experimental results to obtain the defined objectives. Moreover, they combine available methods to extend novel techniques for the characterization in the behaviors of biological networks.

Bioinformatics Bioinformatics is the application of computational techniques for analyzing the information concerned to biomolecules on a large scale. It is defined as a discipline in molecular biology to contain a wide range of scientific subjects from structural biology and genomics to gene expression studies. Bioinformatics enables researchers to access the existing information for submitting new entries as they are produced. Furthermore, it develops the tools and resources which aid in the analysis of data. These tools and resources evaluate the data and interpret the given results in a biologically meaningful procedure.

REFERENCES

[1] M.S. Gharajeh, A learning analytics approach for job scheduling on cloud servers, in: A. Pena-Ayala (Ed.), Learning Analytics: Fundaments, Applications, and Trends, Springer, Berlin, 2017, pp. 269–302.
[2] J. Zakir, T. Seymour, K. Berg, Big Data Analytics, IIS 16 (2015) 81–90.
[3] P. Russom, Big Data Analytics, TDWI best practices report, vol. 19, 2011, pp. 1–38. fourth quarter. Available from: https://vivomente.com/wp-content/uploads/2016/04/big-data-analytics-white-paper.pdf/.
[4] M.S. Gharajeh, Security issues and privacy challenges of NoSQL databases, in: G.C. Deka (Ed.), NoSQL: Database for Storage and Retrieval of Data in Cloud, Chapman and Hall/CRC, Boca Raton, 2017, pp. 271–290.
[5] A. Cuzzocrea, I.-Y. Song, K.C. Davis, Analytics over large-scale multidimensional data: the big data revolution! Proceedings of the ACM 14th International Workshop on Data Warehousing and OLAP (DOLAP '11), Glasgow, Scotland, UK, 2011, pp. 101–104.
[6] H. Kashyap, H.A. Ahmed, N. Hoque, S. Roy, D.K. Bhattacharyya, Big data analytics in bioinformatics: a machine learning perspective, arXiv (2015). Available from: https://arxiv.org/pdf/1506.05101/.

[7] Y. Low, J.E. Gonzalez, A. Kyrola, D. Bickson, C.E. Guestrin, J. Hellerstein, Graphlab: a new framework for parallel machine learning, arXiv (2014). Available from: https://arxiv.org/ftp/arxiv/papers/1408/1408.2041.pdf/.

[8] J.J. Berman, Principles of big data: preparing, sharing, and analyzing complex information, Newnes, Oxford, United Kingdom (2013). Available from: https://books.google.com/books?id=gEho0DI8a2kC&printsec=frontcover/.

[9] E.K. Brauer, D.K. Singh, S.C. Popescu, Next-generation plant science: putting big data to work, Genome Biol. 15 (2014) 301–303.

[10] C. Ma, H.H. Zhang, X. Wang, Machine learning for big data analytics in plants, Trends Plant Sci. 19 (2014) 798–808.

[11] R.J.A. Little, D.B. Rubin, Statistical Analysis With Missing Data, John Wiley & Sons, Hoboken, NJ, 2014.

[12] M.R. Aniba, O. Poch, J.D. Thompson, Issues in bioinformatics benchmarking: the case study of multiple sequence alignment, Nucleic Acids Res. 38 (2010) 7353–7363.

[13] M. Linn, The knowledge integration perspective on learning and instruction, in: R. Sawyer (Ed.), The Cambridge Handbook of the Learning Sciences, Cambridge Handbooks in Psychology, Cambridge University Press, Cambridge, 2005, pp. 243–264.

[14] R. Blagus, L. Lusa, Class prediction for high-dimensional class-imbalanced data, BMC Bioinf. 11 (2010) 523–539.

[15] N. Japkowicz, M. Shah, Evaluating Learning Algorithms: A Classification Perspective, Cambridge University Press, Cambridge, England, 2011.

[16] R.J. Robison, How big is the human genome? Precision Medicine (2014), Available from: https://medium.com/precision-medicine/how-big-is-the-human-genome-e90caa3409b0/.

[17] A. Belle, R. Thiagarajan, S.M. Soroushmehr, F. Navidi, D.A. Beard, K. Najarian, Big data analytics in healthcare, Biomed. Res. Int. 2015 (2015) 370194. 1–16.

[18] P. Hu, S.M. Galvagno, A. Sen, R. Dutton, S. Jordan, D. Floccare, C. Handley, S. Shackelford, J. Pasley, C. Mackenzie, Identification of dynamic prehospital changes with continuous vital signs acquisition, Air Med. J. 33 (2014) 27–33.

[19] D. Apiletti, E. Baralis, G. Bruno, T. Cerquitelli, Real-time analysis of physiological data to support medical applications, IEEE Trans. Inf. Technol. Biomed. 13 (2009) 313–321.

[20] J. Lovén, D.A. Orlando, A.A. Sigova, C.Y. Lin, P.B. Rahl, C.B. Burge, D.L. Levens, T.I. Lee, R.A. Young, Revisiting global gene expression analysis, Cell 151 (2012) 476–482.

[21] G. Karlebach, R. Shamir, Modelling and analysis of gene regulatory networks, Nat. Rev. Mol. Cell Biol. 9 (2008) 770–780.

[22] E.A. Mohammed, B.H. Far, C. Naugler, Applications of the MapReduce programming framework to clinical big data analysis: current landscape and future trends, BioData Min. 7 (2014) 1–23.

[23] W. Wang, K. Haerian, H. Salmasian, R. Harpaz, H. Chase, C. Friedman, in: A drug-adverse event extraction algorithm to support pharmacovigilance knowledge mining from PubMed citations, Proceedings of the AMIA Annual Symposium Proceedings, Bethesda, Maryland, USA, 2011, pp. 1464–1470.

[24] N.S. Raghava, Iris recognition on hadoop: a biometrics system implementation on cloud computing, Proceedings of the IEEE International Conference on Cloud Computing and Intelligence Systems (CCIS), Beijing, China, 15–17 September 2011, 2011, pp. 482–485.

[25] F. Omri, R. Hamila, S. Foufou, M. Jarraya, Cloud-ready biometric system for mobile security access, in: Networked Digital Technologies, International Conference on Networked Digital Technologies, Macau, China, 2012, pp. 192–200.

[26] R. Díaz-Uriarte, S.A. De Andres, Gene selection and classification of microarray data using random forest, BMC Bioinf. 7 (2006) 3–15.

[27] W.H. Delone, E.R. Mclean, Measuring e-commerce success: applying the DeLone & McLean information systems success model, Int. J. Electron. Commer. 9 (2004) 31–47.

[28] M.M. Yusof, J. Kuljis, A. Papazafeiropoulou, L.K. Stergioulas, An evaluation framework for health information systems: human, organization and technology-fit factors (HOT-fit), Int. J. Med. Inform. 77 (2008) 386–398.

[29] H. Lærum, T.H. Karlsen, A. Faxvaag, Use of and attitudes to a hospital information system by medical secretaries, nurses and physicians deprived of the paper-based medical record: a case report, BMC Med. Inform. Decis. Mak. 4 (2004) 18–27.

[30] P. Fuhrer, D. Guinard, in: Building a smart hospital using RFID technologies: use cases and implementation, Proceedings of the 1st European Conference on eHealth (ECEH06), October 12–13, 2006, Fribourg, Switzerland, 2006, pp. 1–14.

[31] R. Wessel, RFID Bands at the Jacobi Medical Center, Retrieved March 26, 2006, from, http://www.rfidgazette.org/2005/12/rfidbandsatt.html, 2005.

[32] B. Heartlands, RFID-Tags Patients to Avoid Litigation, Retrieved March 26, 2006, from, http://www.bjhc.co.uk/news/1/2005/n502016.htm, 2005.

[33] J. Best, RFID comes to European Hospitals, Retrieved March 26, 2006, from, http://networks.silicon.com/lans/0,39024663,39129743,00.htm, 2005.

[34] R. Wessel, German Clinic Uses RFID to Track Blood, Retrieved March 25, 2006, from, http://www.rfidjournal.com/article/articleview/2169/1/1/.

[35] W. Luo, F. Wang, Measures of spatial accessibility to health care in a GIS environment: synthesis and a case study in the Chicago region, Environ. Plann. B. Plann. Des. 30 (2003) 865–884.

[36] J. Wan, C. Zou, S. Ullah, C.-F. Lai, M. Zhou, X. Wang, Cloud-enabled wireless body area networks for pervasive healthcare, IEEE Netw. 27 (2013) 56–61.

[37] B. Fabian, T. Ermakova, P. Junghanns, Collaborative and secure sharing of healthcare data in multi-clouds, Inf. Sys. 48 (2015) 132–150.

FURTHER READING

[38] H. Chen, R.H.L. Chiang, V.C. Storey, Business intelligence and analytics: from big data to big impact, MIS Q. 36 (2012) 1165–1188.

[39] M.T. Jones, Artificial Intelligence: A Systems Approach: A Systems Approach, Jones & Bartlett Learning, Massachusetts, 2015.

[40] R. Akerkar, P. Sajja, Knowledge-Based Systems, Jones & Bartlett Publishers, Massachusetts, 2010.

[41] C. Esposito, M. Ficco, F. Palmieri, A. Castiglione, A knowledge-based platform for big data analytics based on publish/subscribe services and stream processing, Knowl. Based Syst. 79 (2015) 3–17.

[42] H. Demirkan, D. Delen, Leveraging the capabilities of service-oriented decision support systems: putting analytics and big data in cloud, Decis. Support. Syst. 55 (2013) 412–421.

ABOUT THE AUTHOR

Mohammad Samadi Gharajeh received ASc in Computer Software from University of Applied Sciences and Technology, Iran in 2005. He received BSc in Software Engineering of Computer from University of Applied Sciences and Technology, Iran in 2009. Afterward, he received MSc in Architectural Engineering of Computer from Islamic Azad University of Tabriz, Iran in 2013. His research interests include control and learning systems, wireless sensor networks, ad hoc networks, cloud computing, robotic and autonomous systems, and reversible circuit design. He is an Editorial Board member and a Reviewer in some of the international scientific journals and is a Lecturer of university now.

CHAPTER NINE

NoSQL Polyglot Persistence

Ganesh Chandra Deka
Directorate General of Training, Ministry of Skill Development & Entrepreneurship, Government of India, New Delhi, India

Contents

1. Introduction	358
1.1 Data Integration	358
2. Programming Paradigm	360
2.1 Polyglot Programming	363
3. Polyglot Persistence	365
4. Polyglot Persistence and NoSQL	366
4.1 NoSQL Databases Access Techniques	368
5. Big Data and Polyglot Persistence	370
5.1 Polyglot Processing	373
5.2 Polyglot-Persistent Databases	373
6. Polyglot Persistence in e-Commerce	374
6.1 REST and e-Commerce Polyglot Persistence	377
7. Polyglot Persistence in Healthcare	378
8. Research Trends in Polyglot Persistence	380
9. Conclusion	382
Glossary	383
References	384
Further Reading	389
About the Author	389

Abstract

Polyglot persistence facilitates use of most suitable database technology based on the requirement of an application. There are lots of use cases as well as huge potential for Polyglot Persistence in *e-Commerce* web portals, Search Engines, and *Healthcare Information Ecosystem* applications. But, designing and implementation of an application in a Polyglot environment is not a straightforward task, since adding more data storage technologies increases complexity in programming. Further knowledge of multiple programming languages, maintainability of the Polyglot application, and tool support for Polyglot persistence are the challenges of Polyglot Persistence. This chapter discusses the various aspects of Polyglot persistence with a focus on NoSQL database.

Advances in Computers, Volume 109
ISSN 0065-2458
https://doi.org/10.1016/bs.adcom.2017.08.003

1. INTRODUCTION

Polyglot persistence facilitates use of most suitable database technology based on the requirement of an application. There are lots of use cases as well as huge potential for Polyglot Persistence in *e-Commerce* Web portals, Search Engines, and *Healthcare Information Ecosystem* applications. The survey by research agency has revealed that the market value of retail e-Commerce sales worldwide will be 4058 billion US dollars by 2020. Another survey by 451 Research commissioned by Oracle revealed that the market value of DBaaS will reach $19.0 billion by 2020. Polyglot persistence application development strategy will facilitate taking the benefits of Virtualization Technology and Cloud Computing. This chapter discusses the Polyglot Persistence as a data integration technology with a focus on NoSQL database.

1.1 Data Integration

Eighty percent of data generated by various organizations as well as by individuals are unstructured. These data are in the form of audio recordings, PDFs, and texts. Accesses to the data are usually in the form of a query, reporting system/dashboard. The update of data can happen in bulk, real-time, or near real-time.

The need for data integration emerges as the data generated by multiple systems from complex environments are to be integrated. Data integration facilitates uniform access to data available in multiple data sources such as relational, nonrelational, and hierarchical database formats [1]. Data integration is a form of data preprocessing.

"An Average Fortune 1000 company has around 48 applications and 14 databases for a total of 62 potential systems which can be integrated" [2].

Data integration merges data from multiple data sources to provide:
- uniform query interface to all sources;
- access to queries; eventually updates too; and
- distributed over LAN, WAN even the internet.

Features of data integration are [3]:
- enables processing of data from a wide variety of sources;
- facilitates processing of unstructured data from social media, email, web pages, etc.;
- syntactic and semantic checks to make sure that the data conform to business rules and policies;

- deduplication and removal of incorrectly/improperly formatted data; and
- support for metadata.

The main technologies for data integration are (Fig. 1):

    (a) extract, transform, load (ETL);

    (b) enterprise application integration (EAI); and

    (c) enterprise information integration (EII) aka data virtualization.

(a) The sources of data used in ***ETL*** include:

- Flat files
- Excel Sheet data
- Application data, such as CRM/ERP or Mainframe application data

Most difficult part of the ETL is the "Transform" component. Transform process involves:

- Data cleansing/removal of duplicates
- Resolving issues relating to data inconsistency
- Apply rules to consistently convert data to appropriate form

(b) Features of ***EAI***:

- Integrates applications and enterprise data sources for easy sharing of data.
- Integrates applications and data sources within an enterprise to solve a local problem.

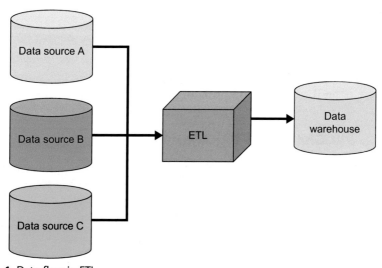

Fig. 1 Data flow in ETL.

Integration is done without significant changes in the applications and data sources. **EAI** currently subsumed into service-oriented architecture software stacks. EAI products are built in combination with the following technologies:

- J2EE (**J**ava **2** Platform **E**nterprise **E**dition);
- XML (e**X**tensible **M**arkup **L**anguage) for enterprise-wide content and data distribution;
- message queuing; and
- business process orchestration engine.

(c) **EII** aka data virtualization

- uses RDF (resource description framework) for information exchange;
- provides real-time Read and Write access;
- transform data for business analysis and data interchange; and
- manage data placement for performance, concurrency, and availability.

EAI and ETL are not competing for technologies. EAI can be source of input or act as a service for ETL. An **EAI-to-ETL** interface could be used to give an ETL product access to EAI application data. Such an interface eliminates the need for ETL vendors to develop point-to-point (P2P) adapters for these applications data sources.

EAI are focused on real-time processing, the **EAI-to-ETL** interface can also act as a real-time event source for ETL applications requiring low-latency data.

The ETL, EAI, and EII have their relative advantages and disadvantages, but normally the combination is used. Table 1 summarizes the features of ETL, EAI, and EII [4].

Polyglot persistence uses multiple database technology for storage and retrieval of data. Hence, Polyglot persistence is basically a data integration technique for an application from various data sources. Rest of the chapter discusses about data integration using Polyglot persistence.

2. PROGRAMMING PARADIGM

Before discussing about the Polyglot persistence, the concept of Polyglot programming is discussed in this section to build the foundation of Polyglot persistence in the mind of the readers.

Following are the various programming paradigms:

- **(a)** Functional programming
- **(b)** Object-oriented programming
- **(c)** Aspect-oriented programming
- **(d)** Hybrid functional programming

Table 1 Features of ETL, EAI, and EII

EIA	ETL	EII
• Integrates applications by using standard interfaces • Employs real-time operational business transaction processing through remote procedure call (RPC), remote method invocation (RMI), common object request broker architecture (CORBA), or standard messaging infrastructures • Can be used for transportation of data **between applications** and route real-time event data to other data integration applications such as an ETL process	• Backbone technology for data consolidation • In ETL, detailed profiling and analysis of the data sources and their relationships to the targets are required • Utilized for creating physical data warehouses • ETL tools can consolidate data into EAI targets and web services	• Leverages extended and enhanced data warehousing environment to address specific business needs • Similar to ETL processes, a detailed profiling and analysis of the data sources and their **relationship**s to the targets are required

(a) **Functional programming** is the oldest programming paradigm. Functional programming also known as *applicative programming* and *value-oriented programming*. Basically, functional programming is a declarative type of programming. In functional programming, everything is a function returning values instead of modifying data. Examples are **Lisp** (historically, LIPS stands for List Processing is *a programming* language for Artificial Intelligence), **ML** (meta language), **Haskell, Scala**, **F#**, **Erlang**, etc. Haskell eliminate the side effects on variables, making it easier for writing thread-safe code.

Functional programming languages are good for concurrency, but poor in developing user interfaces (UIs). This problem can be overcome by a Polyglot environment. The part of the software that needs to be highly concurrent can be written using a functional programming language, while the UIs part can be developed by using suitable general purpose languages [5].

(b) **Object-oriented programming** is a way of programming that is dominated by creating object (new data types) to solve a problem.

An object contains both data and methods that manipulate data. For example in **C#** everything is an **object**. Similar example of object-oriented programming language includes **Objective-C**, **Java**, **C++**, **Python**, and **Smalltalk**.

Out of all the object-oriented programming language *Java* and *C++* are the mostly used. However, both of them are related with certain advantage and disadvantages.

The following features available in **C++** are *not available* in Java:

- Pointer
- Multiple inheritance
- Operator overloading
- GOTO statement
- Structure and union data structure

(c) Aspect-oriented programming (**AOP**) is similar to event handling. The "events" are defined outside the code spanning over many unrelated modules. In AOP crosscutting (spanning multiple software modules) functionalities are implemented in **Aspects** instead of fusing them into core modules. **Aspects** are an additional unit of modularity, which can be reused for reducing code. An "**Aspect weaver**" takes the **aspects** and the core modules and composes the final system.

The **AOP** programming language **AspectJ** provides additional keywords and constructs to write **Aspects**. Aspects operate with reference to features in the standard Java code. An aspect can have ordinary **Java** code, i.e.

- Member variables and methods
- Can implement interfaces
- There is also aspect inheritance

The **AspectJ** compiler produces standard Java byte code executable on any Java **Virtual Machine**. The runtime library "aspectjrt.jar" is required to run any AspectJ program. AspectJ and C# also derive a set of features from **C** and **C++** *except* pointers.

Another important feature of **AspectJ** (using the JDBC library) and **C#** (using the ADO.NET library) is *Database interaction*. However, **C#** has more features for supporting functional programming than **AspectJ** [6].

C# is suitable for *mixing* OOP and declarative style using Microsoft **LINQ** (language-integrated query) services, especially for querying database and XML files [6] (Table 2).

(d) F# and **Scala** are *Hybrid Functional* programming languages, i.e., a mixture of object-oriented programming and functional programming [7]. The object-oriented programming reduces the gap between the

Table 2 Features of Programming Paradigms

Features	Functional Programming	Object-Oriented Programming	Aspect-Oriented Programming
Fundamental working block	Function	Object	Aspects
Variable	Not used	Used	Used
Working level	Function	Object	Code

problem domain and the implementation in software, while functional programming brings robustness to computing power for concurrent applications [8].

The term Polyglot (knowing/using several languages) programming refers to developing software/computer applications using a mix of programming languages. The idea behind this technique is to take advantage of different languages suitable for tackling different problems.

Web-based applications are polyglot in nature using multiple programming languages like:

- JavaScript, Flash, and HTML;
- UI such as Java, C#, and Ruby in the **Middle Tier**; and
- **SQL/NoSQL** as **Database** in a single program.

There are more than **256** programming languages, as listed at [9].

Some of the programming languages such as **HTML** are called Non-computational languages. Often a **Computational** programming language is embedded in the **Noncomputational** language specifically in internet-based applications. According to Ref. [10], "The times of writing an application in a single general purpose language are over."

Modularity paradigms in application development are defined as poly-paradigm programming (PPP).

2.1 Polyglot Programming

Polyglot programming is using several programming languages in software/application [11]. Table 3 shows some of the use cases of ployglot programming [12].

The degree of polyglotism determines the use of Polyglot programming. The degree of Polyglot programming is determined by:

- integration of the programming language used;
- organization of code;

Table 3 Polyglot Programming Examples

Sl. No.	Name of Programming Languages	Polyglot Use Case/Application
1	*Java, XML, JavaScript,* and *HTML*	JSP (JavaServer Pages)
2	SQL and HTML	PHP (Hypertext Preprocessor) pages
3	Ruby, JavaScript, and F#, in addition to Python, C#, C++, and Visual Basic	Microsoft Corporation
4	Python, JavaScript, Java, Ruby, Groovy, and Scala	Sun Microsystems
5	Ruby, Groovy, and Python	JSR 223 Servlet (Java Specification Request)
6	Java libraries in other languages	***Jython*** for Python, ***JRuby*** for Ruby Java virtual machine (JVM) is created to run on as many platforms as possible to make Java platform independent

Note: F# is an Open Source Cross Platform programming language [http://fsharp.org/].

Table 4 Applications Using Multiple Programming Languages [12]

Sl. No.	Application	Programming Languages Used
1	Civilization 4 (a computer game)	C++ and Python
2	World of Warcraft (a computer game)	C++ and Lua
3	Web application	Cascading style sheets (CSS), JavaScript, and HTML

- processes of program execution of programming language used; and
- data to be processed/manipulated.

The architecture where Polyglot programming can be utilized is [12]:

- Service-oriented architecture (SOA);
- Managed runtimes; and
- C integration;

The disadvantage of Polyglot programming includes (Table 4):

- knowledge required to use different languages;
- maintainability of different languages; and
- tools supported by different languages.

Hence, it is observed from the above discussion that a limited number of languages are possible to use in an application. Google restricts the use of programming languages to *JavaScript*, *Python*, *Java*, and *C++* [13]. Use case of *Polyglot Programming* is a web application's **UI** implemented in an OOP language **Ruby**, while **services** are implemented in a functional programming language like **Erlang**.

3. POLYGLOT PERSISTENCE

Polyglot persistence is a hybrid approach enabling usage of multiple databases in a single application/software. A Software that is capable of using more than one type of data storage is referred to as Polyglot-persistent software.

It is beneficial to use a variety of data models suitable for different parts of huge software. For instance, use relational database to store/handle structured, tabular data; a document database such as MongoDB for unstructured, object-like data; a key–value store such as Riak for a hash table; and a graph database like Neo4j for highly linked referential data [14].

Fig. 2 shows a typical Polyglot persistence use case.

An example of Polyglot-persistent software is e–Commerce applications. An e–Commerce application deals with multiple types of data, such as shopping cart, inventory, customers' orders, etc. To store all these data of an e–Commerce application in one database will require lots of data conversion work to make the data compatible originating from various sources [15]. Furthermore, implementing business transactions that maintain consistency across multiple services is another challenge. Since, transactions are short-lived session involving comparatively less data, hence NoSQL databases

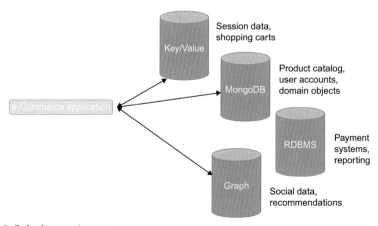

Fig. 2 Polyglot persistence.

are suitable. However, once the transaction is over the buyer Account must be Debited and sellers Account must be created. Here ACID transaction is used for which SQL database (RDBMS) are the best fit [16].

A similar example of Polyglot persistence is microservices-based applications. These applications often use a mixture of SQL and NoSQL databases [16].

Polyglot-persistent application deployed on-premise relying on local database or using multiple databases in the cloud using Database-as-a-Service (DBaaS) offers of cloud service providers. However, implementing Polyglot persistence in a distributed data management environment is a big challenge. Furthermore, since NoSQL does not enforce data integrity, hence data integration becomes the application's concern for using NoSQL database for e-Commerce applications. However, some of NoSQL databases are ACID compliant at the cost of price and complexity. Hence DBaaS will be the affordable and convenient option for developing applications/ software using multiple databases.

4. POLYGLOT PERSISTENCE AND NoSQL

NoSQL databases are designed to handle exponentially growing data intensive applications. Most of the NoSQL databases are open source, however proprietary option is also available with added advantages/features.

NoSQL provide poor transactional integrity, flexible indexing, and querying. The Polyglot persistence uses a *mediator* such as a **REST API** to issue queries for NoSQL. The mediator acts as a broker between **Application** and backend **Database**. This section discusses about the features of mostly used NoSQL database with a focus on Polyglot persistence for interoperability.

Since the language used to develop the source code of the NoSQL determines lots of its underpinning properties, let us start with the programming language used for developing the NoSQL.

As shown in Table 5, the **Document** NoSQL database is developed by using both **C++** and **Erlang** by the developer from various software vendors.

However, the **Key–value** NoSQL database is found to be more generic since the **C**, **C++**, **Java**, and **Erlang** programming language were used to write the source code of the database.

The most interesting observation is that, the majority of NoSQL database was developed by using compatible programming languages (C is the basis for **C++**, **C#**, and **Java**). Although it is not possible to run the C on Java platform, there are facilities for invoking code written in C.

Table 5 Programming Languages Used in 19 NoSQL and Search Engines

Sl. No.	Name of NoSQL	Category	Programming Language Used
1	Bigtable	Column	C
2	**MySQL**	MySQL based	
3	Redis	Key–value	
Total 3			
1	ClearDB	MySQL based	$C^{\#}$
2	Xeround	MySQL based	$C^{\#}$/C++
Total 2			
1	Hypertable	Key–value	C++
2	Xeround	MySQL based	
3	MongoDB	Document	
4	MarkLogic	Search Engine	
Total 4			
1	CouchDB	Document	**Erlang**
2	Amazon DynamoDB	Key–value	
3	Dynomite	Key–value	
4	SimpleDB	Document	
Total 4			
1	Voldemort	Key–value	**Java**
2	Cassandra	Column	
3	HBase	Column	
4	Infinispan	Data grid cloud	
5	Elasticsearch	Search Engine	
6	PNUTS	Column	**Java/JVM**
Total 6			

Java was used to develop the source code of four NoSQL database, one Search Engine and one data grid cloud. Java runs on any platform by running a JVM. Groovy, JavaScript, Python, Ruby, Perl, and PHP can all run on Java platforms.

It is observed from Table 5 that, two NoSQL databases, i.e., ClearDB and Xeround, were developed using C#, a generic programming language having high interoperability with C/C++.

Erlang is used to develop four NoSQL databases. **Erlang** has a very interesting built-in primitive for message passing between processes, without worrying about *race condition*s and *deadlock*s because of the functional nature of this programming language.

4.1 NoSQL Databases Access Techniques

The following are some of NoSQL databases and the application for which they were specifically designed (Table 6).

The NoSQL databases use various techniques to access data. Some NoSQL databases support SQL but most NoSQL databases uses programmatic querying through their unique custom APIs. For instance, PostgreSQL RDBMS provides XML, JSON, and Key–Value functionalities.

There are Data Mappers which store whole object graphs into Key–Value stores, since Key–value stores often allows limited querying. Key–value store is the main technology beneath the relational database management systems (RDBMS). In Key–value NoSQL database challenge of having fast indices is solved using the B-tree approach [17].

Document stores generally allow more complex queries. Many NoSQL databases have **plug-in** to Hadoop for Map/Reduce style of "querying."

Apache HBase allows random, real-time read/write access to big data.

Column families are used for optimization and partitioning purposes.

Table 6 NoSQL Databases Applications

Name of NoSQL	Type of Data/Application Suitable for
MongoDB	Product catalog
Redis	Shopping cart
Amazon DynamoDB	Social profile information
Neo4j	Social graph
HBase	Inbox and public feed messages
MySQL	Payment and account information
Cassandra	Audit and activity log, a query-based model

MarkLogic uses single C++ core to solve the data management problem across many data types from Text to RDF and XML to Binary. *Neo4j Doc Manager* enables Polyglot persistence for Neo4j and MongoDB [18].

The D2R declarative language describes mappings between relational database schemas and the OWL (web ontology language). The mappings are used by the D2R processor to export data from a relational database into RDF [19].

Another use case of Polyglot persistence is Force.com integrating and optimizing varieties of data persistence technologies to deliver transparent Polyglot persistence. With Force.com, coding a single API is sufficient for a given situation irrespective of the degree of complexity of the problem/application. Force.com has both RESTful API and web services (SOAP-based) APIs as well as *Salesforce Object Query Language* (SOQL) for constructing powerful database queries for Polyglot persistence [20].

Kundera is a JPA 2.1 (*Java Persistence API*) compliant object data store mapping library for NoSQL databases. *Kundera* currently supports Cassandra, HBase, MongoDBand MySQL (a relational database) [21].

Amazon DynamoDB *DBaaS* supports both document and Key–value NoSQL data store models for easy development of web, Gaming, and IoT (Internet of Things) in particular.

In an identical Polyglot persistence scenario, a *MongoDB* (document database) can be used as the principal storage (master database) and *Redis* (an in-memory database) as a caching layer to accelerate the writes.

Similarly, **LAMP** (Linux, Apache (open source web server software), MySQL (also MongoDB), and Perl/PHP/Python) an open source solution stack of software that enables rapid development of dynamic websites and web applications [22,23]. Once all the components are installed in a server, the final requirement is **Installing Dependencies** depending upon the programming language and OS (operating system) used. For example, **mod_wsgi** to be installed for **Apache** and **Python** to work together, similarly **mysqliclass** is used as interface between MySQL and PHP [24].

The following are the advantages and disadvantages of Polyglot persistence [25] (Table 7).

A Polyglot application stack might include *Redis* as a caching layer, *MongoDB* for collecting logs, *Postgres* for metadata, and *Elasticsearch* for indexing and search. This use case will require the technical knowledge of all the three databases and one search engine associated with the system if maintained by an organization [26]. The cost effective solution for such type of the disadvantages of Polyglot persistence is migration to cloud base DBaaS.

Table 7 Polyglot Persistence Advantages and Disadvantages

Advantages	Disadvantages
Faster response	Difficulties of integrating different databases
Easy to scale up application	Expertise of different databases
Power of multiple databases at the same time	Management of resources of various databases. Another disadvantage is the fragmented environments caused by Polyglot persistence by running different databases side by side. The fragmentation increases both the operational and development complexity

Table 8 Traditional Data vs Big Data

Parameter	Traditional Data	Big Data
Size	Gigabytes to Petabytes	Petabytes to Exabyte
Data source	Centralized	Distributed
Data type	Structured	Semistructured and unstructured as well as high percentage are multistructured
Data model	Stable	Flat schemas
Data interrelationship	Known complex interrelationship	Limited interrelated complexities among the data

5. BIG DATA AND POLYGLOT PERSISTENCE

Big data have hardly any structure at all. Big data are extracted from the source in its raw form. The Big data are stored in a massively distributed file system. According to Gartner survey, 73% organizations have either invested or have plan to invest in big data in the coming next 2 years [27].

Table 8 shows the features of big data and conventional data.

According to [28], challenges of big data analytics are:

(a) Challenges of dealing with **Volume**, **Velocity**, and **Variety** of big data
- scalable big and fast data infrastructures;
- coping with diversity in the data management landscape; and
- end-to-end processing and understanding of data.

Following are the roles of NoSQL database for dealing with big data at rest as well as in motion.

- Analyze huge quantity of data in less time.
- Simplify use of big data.
- Optimum utilization of existing resources for handling of big data.
- Utilize existing skills with value addition.

The essential features of NoSQL for dealing with big data are in Table 9.

Some of the big data analytics use cases are:

- Real-time intelligence
- Data discovery
- Business reporting

Hence, Polyglot persistence is the best option for big data analytics (real time, predictive, cognitive, and contextual) for creating a hybrid data processing environment combining the functionalities of RDBMS and NoSQL database.

Hadoop Ecosystem is one of the use cases of polyglot persistence in big data analytics. The user of Hadoop includes:

- Amazon/A9
- Facebook
- Google
- New York Times
- Veoh (an internet television company)
- Yahoo!

Table 9 NoSQL Databases and Big Data

NoSQL Type	Horizontal Scalability	Flexibility in Data Variety	Appropriate Big Data Types	Suitable for
Key–value stores	High	High	Yes	Stores and retrieves opaque data items and blobs identified by Key–value
Column stores	High	Moderate	Partially	Stores huge amount of data, but only retrieve a subset of these fields in queries
Document stores	Variable (high)	High	Likely	Application needs to filter records based on nonkey fields or update individual fields in records
Graph databases	Variable	High	Maybe	Complex relationships between entities

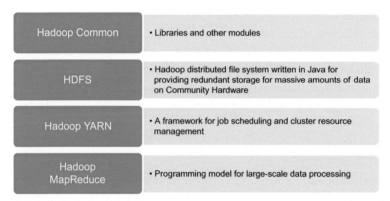

Fig. 3 Hadoop Ecosystem.

Fig. 3 shows the components of Hadoop Ecosystem which works in integration for big data processing.

Other tools of Hadoop Ecosystem:

- Hive-Developed at Facebook is a Data warehousing system for Hadoop, Hive facilitating data summarization, Adhoc queries, and the analysis of large datasets stored in Hadoop compatible file systems. Hive also provides a mechanism for querying of data using a SQL like query language called HiveQL. HiveQL supports table partitioning, clustering, and complex data types.
- Pig–Pig is a data-flow language for expressing Map/Reduce programs for analyzing large HDFS distributed datasets. Pig provides relational (SQL) operators such as **JOIN, Group By**, etc. Pig is also having easy to **plug in** Java functions.
- Cascading pipe and filter processing model.
- HBase database model built on top of Hadoop.
- Flume designed for large-scale data movement.

The big data analytic application must has the capabilities to integrate and manage lots of tools of Hadoop ecosystem. Amazon introduced Elastic MapReduce (EMR) based on Hadoop. Amazon web services developed the following big data application:

 (i) Amazon DynamoDB NoSQL.

(ii) Amazon Redshift.

Oracle has developed an Oracle NoSQL database with a focus on big data analytics. Hadoop and Oracle databases complement each other in environments where massive servers collect data requiring sophisticated analysis.

5.1 Polyglot Processing

Polyglot processing refers to capability for processing any kind of *Data*, *Workload*, and *Workflow*. Polyglot processing combines various data processing engines over NoSQL.

Apache Storm is an open source *Distributed* *real-time* computation system developed by *BackType* which was acquired by *Twitter*. The *Storm* provides real-time computation, is *horizontally scalable*, Guarantees *no data loss* and extremely *robust* and fault-tolerant. **Apache Storm** is an open source distributed real-time computation system. **Apache Storm** is language agnostic making easy to reliably process unbounded streams of data. Following five characteristics make storm ideal for real-time data processing workloads [29]:

- **(i)** *Fast*: Capable of processing *one million of 100* byte messages per second per node.
- **(ii)** *Scalable*: Run across a cluster of machines.
- **(iii)** *Fault-tolerant*: When a worker dies, automatically restart them, in the event a node dies, the worker will be restarted on another node.
- **(iv)** *Reliable*: Guarantees that each unit of data (tuple) will be processed at least once or exactly once. Messages are only replayed when there are failures.
- **(v)** *Easy to operate*: Once deployed, storm is easy to operate.

Apache Spark is a fast and general compute engine for large-scale data processing. Apart from just "*Map*" and "*Reduce*," Apache Spark defines a large set of *operations*. Apache Spark supports *Java*, *Scala*, and *Python*. Apache Spark can process data *In-memory*, but, Hadoop MapReduce persists back to the disk after a Map or Reduce action. **Apache Spark** needs lots of memory for smooth performance. Apache Flink has unified runtime engine [30].

The **Apache Lucene** based open source search engine Elasticsearch is used in addition to another NoSQL database [31] in use in many use cases. MarkLogic, a multimodel database has an embedded Search Engine in addition to its regular features.

5.2 Polyglot-Persistent Databases

The research trend in database technology is to develop hybrid database. This hybrid databases can be thought of as a borderless data center. Hybrid databases will be having both the ACID (SQL) and BASE (NoSQL) properties. These databases are compatible with many SQL applications and providing the scalability of NoSQL applications.

The MongoDB connector for BI (business intelligence) platform allows use of MongoDB data for SQL-based BI [32] applications. Couchbase*N1QL* also extends SQL for JSON [33a,33b].

MarkLogic database achieves the goal of Polyglot persistence via the technique of multimodel storage [34].

OrientDB is a multimodel open source NoSQL database that combines the power of graphs NoSQL and flexibility of document database [35].

The term NewSQL is coined by 451 group analysts *Matt Aslett* represents "A DBMS that delivers the scalability and flexibility promised by NoSQL while retaining the support for SQL queries and/or ACID, or to improve performance for appropriate workloads" [36]. Therefore, the NewSQL is likely to be the alternative to NoSQL/SQL databases for new OLTP applications [37]. The examples of popular NewSQL are:

(i) VoltDB;

(ii) Clustrix; and

(iii) NuoDB are two commercial projects that are also classified as NewSQL. NewSQL databases support the relational model and uses SQL as the query language [38].

Another very important development in Polyglot processing is the concept of hybrid transaction analytical processing (HTAP) architecture. HTAP uses *In-Memory* technologies to run transactions and analytics processing on the same database [39].

6. POLYGLOT PERSISTENCE IN E-COMMERCE

According to the survey conducted by "eMarketer," retail e-Commerce sales worldwide will be US$ 4058 billion by 2020 for products or services ordered using the internet via any device [40]. Furthermore, the survey forecasts that, the sales of retail e-Commerce will grow by 200% from US$ 1915 billion in 2016 to US$ 4058 billion by 2020.

This section discusses the e-Commerce web application with a focus on Polyglot persistence.

A Polyglot-persistent e-Commerce application is a hybrid system using multiple databases as per the requirement of an e-Commerce application. The benefits of using Polyglot persistence in e-Commerce application are:

• Productivity

• Maintainability

Using dynamic programming language such *Ruby* not only offers a higher level of abstraction but also frees the developers from lots of repetition.

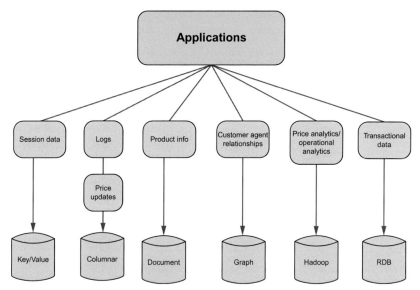

Fig. 4 Use case with various NoSQL database.

In this section, the scope of Polyglot persistence in e-Commerce will be discussed.

Fig. 4 shows a typical e-Commerce application using various types of unstructured data types. The figure shows some of the NoSQL database type suitable for storage/manipulation of data/record generated by a particular module/process of a typical e-Commerce application.

Here both SQL and NoSQL will be used by the e-Commerce application. Since, financial transaction must follow ACID, hence the transactional data must be stored in an RDBMS (SQL), whereas the other e-Commerce data, such as session data, user logs can be easily managed by NoSQL. Another important fact in the figure is the use of graph NoSQL database, a relationship intensive database to store/handle **Customer–Agent relationships**.

In this illustration, five varieties of databases including four NoSQL database type and one relational RDBMS (SQL) database type will be suitable for handling the data generated from an identical e-Commerce application.

In general, the following databases are suitable for various types of e-Commerce functionality (Table 10).

However, using multiple type of database in an application is neither technically nor financially viable. The thumb rule is use "as many as *necessary*, as few as *possible*." Currently, SQL server, Oracle, and MySQL are the three

Table 10 Database in e-Commerce

Date Source	Access Type	Suitable Database Type
User sessions	Fast access for reads and writes. Data need not be durable	**Key–Value**
Financial data	Needs durable transaction updates. Tabular structure fits the data	**RDBMS**
Point of sale	Lots of writes, infrequent reads for analytics	RDBMS (for write/reads modest) Key–Value or document (for high write/reads) or column if focus is on data analytics
Shopping cart	High availability across multiple locations. Inconsistent writes to be merged	Document/Key–Value
Recommendations	Traversing links between friends, product purchases, and product/service ratings	Graph for complex, if simple column
Product Catalog	Lots of read, infrequent writes	Document NoSQL
Reporting	SQL interfaces with reporting tools	RDBMS, column NoSQL
Analytics	Analytics on large cluster	Column
User activity/CSR logs, social media analysis	High volume of write on multiple nodes	Key–Value or document

Source: J. Serra, What is Polyglot Persistence? 2015, July 1 [Web log post]. Retrieved August 24, 2017, from http://www.jamesserra.com/archive/2015/07/what-is-polyglot-persistence/.

most preferred database options for e-Commerce websites (all having JOSN APIs to act as a mediator).

To conclude, a feasible e-Commerce solution may:

- Use of one RDBMS (SQL database) for durable financial transactions
- For various unstructured data uses
  - **(i)** Cloud-based NoSQL databases services DBaaS
  - **(ii)** Third party REST API for communication/integration between databases and applications. However, lots of NoSQL have their inbuilt REST APIs. The web service receives and sends requests through REST and serializes data in a well-understood and portable format like JSON.

6.1 REST and e-Commerce Polyglot Persistence

REST (representational state transfer)-based web services where *Everything is a Resource*. For the REST-based applications such as *Online shopping*, *Search services (engines)*, and *Dictionary services* are all *Resource*. REST ignores component implementation details, but focuses on:

- Roles of components.
- Their interactions and interpretation of data elements.

Web APIs can be used as HTTP services and can be subscribed by any client application starting from desktop to mobiles.

Following are some of the use cases of REST APIs:

- *Twitter*: REST API and streaming API with JSON. The functionalities of the REST API for Twitter include *Sampling*, *Searching*, and *Filtering* capabilities.
- *Amazon*: has a "product advertising API" in for product search, reviews, etc.
- *eBay*: Many APIs for searching, buying, and posting.
- *Flickr*: Comprehensive API set, free for noncommercial use with client code in many languages.

A similar example of Polyglot persistence is *Smartphone*App that helps driver to find a parking spot might involving the following:

- an inventory of parking lots;
- a mapping tool;
- a payment process; and
- a customer relationship management tool.

For the above-mentioned application, a *Smartphone* requires a *Graph database*, a *Geospatial database*, and a *Payment Gateway*. The graph data model is capable of integrating other data models [41] such as:

- schema and instance of relational data can be modeled with *graph*s;
- XML document is modeled as a tree and a tree is an acyclic, *undirected graph*;
- document stores consist of nested Key–value pairs, hence can be modeled as a set of trees; and
- Key–value stores as a set of nodes of the property graph.

Using MongoDB it is possible to build *location aware* applications. MarkLogic enables Polyglot persistence with a multimodel database capable of storing both the unstructured binary files such as .xslx, .pdf, and .jpg as well as the structured XML data extracted from an Excel file [34].

Key–value stores are the most useful for the case of IoT, as the data have no specific structure and come with various numbers and sizes. Data are treated as a set of bytes supplemented with a unique key. Riak® TS (Riak time series) is a database suitable for IoT and time series data [42]. Another example of IoT-specific database is DB4IoT (a database for IoT [43], http:// db4iot.com/), exclusively running on **Intel® Xeon®** processors.

There are lots of open sources time series database such as ***OpenTSDB*** suitable for storage of **IoT** generated data [44]. Amazon DynamoDB and VoltDB are the two more options [45].

Given a data stream coming from different social networks sites stored on NoSQL database (Neo4J and Mongo) an option for retrieving the information may be using the UnQL pivot query [46]. UnQL is based on SQL with an extension to query NoSQL database data. UnQL can be used for querying data stored in JavaScript object notation (JSON) format too.

The following databases support JSON in data representation:

- SQL Server 2016 and higher version [47]
- Oracle Database [1,48]
- Cassandra [49]

The **OPENJSON** rowset function (currently applies to SQL server 2016 and higher version and Azure SQL database) converts JSON text into a set of rows and columns. **OPENJSON** can be used to run SQL queries on JSON collections or to import JSON text into SQL server tables [50].

The above mentioned are few instances requiring databases to store the varieties of data. The DBaaS offers a way to simplify the management of multiple databases in the above-mentioned cases. Revenue from DBaaS offerings is expected to climb at a CAGR of 46% from $2.849 billion in 2015 to $19.0 billion in 2020 [51].

7. POLYGLOT PERSISTENCE IN HEALTHCARE

eHealth is a broad term for healthcare practice supported by ICT.

Another terminology **mHealth**, the use of mobile phones to collect and access health information. At Johns Hopkins University as of May 2015, there were over **140 mHealth** projects across the developing world (***Keynote presentation*** by ***Dr. Alain Labrique*** of Johns Hopkins University at the Catholic Relief Services 2015 ICT4Development Conference, ***Chicago, May 27, 2015***) [52].

In healthcare domain, a variety of systems with different data types from different locations is connected via network. Hence, a single ***Health Monitoring* system** can generate/collect huge volume of semistructured data

Fig. 5 Healthcare sector data sources.

such as XML, JSON, or unstructured mission-critical real-time information. Data generated by health monitoring system also includes text, numerical data, images, audio, video, sequences, time series data form IoT sensors, social media data, multidimensional arrays, etc. Fig. 5 shows the source of data in healthcare sector.

To extract meaningful information from these varieties of data, these data need to be linked together making healthcare information ecosystem a complex distributed system. Hence, building electronic health information system (EHIS) using a single data model increases the complexity of application development.

Polyglot persistence can be utilized for handling the heterogeneous data (big data) associated with EHIS. The core idea is to store structured data in a relational database, while semistructured or nonstructured data (big data) are to be stored in NoSQL database. EHIS built on "Big Data" technologies should be capable of serving the needs of clinicians in real time and in one environment.

The big data analytic application such as Hadoop can be used for analysis of the healthcare data generated by the various sources as shown in Fig. 5 for planning and decision making in healthcare.

The following are the major collector of healthcare big data.
- Pharmacies
- Laboratories

- Imaging centers (X-ray, CT scan)
- Emergency medical services (EMS)
- Hospital information systems
- Electronic medical records
- Blood banks
- Birth and death records

Big data generated by IoT will be valuable to healthcare. The sensors and wearable can be used to collect health data on patients in their homes and push all of that data into the **Cloud**. Healthcare institutions and care managers, using sophisticated tools, will monitor this massive data stream and the IoT to keep their patients health condition.

MemSQL in-memory database supports fast ingest and concurrent analytics for sensors, delivering instant actionable insight of data [53a] for various applications.

According to Das [53b], "By 2020, chronic conditions, such as cancer and diabetes, are expected to be diagnosed in minutes using cognitive systems that provide real-time 3D images by identifying typical physiological characteristics in the scans. By 2025, artificial intelligence (AI) systems are expected to be implemented in 90% of the U.S. and 60% of the global hospitals and insurance companies." Similarly, a survey by *Research and Markets* argues that the *Virtual reality healthcare* systems market will be $2.5 billion by 2020 [54].

8. RESEARCH TRENDS IN POLYGLOT PERSISTENCE

Polyglot persistence is likely to benefit traditional databases by removing unstructured and semistructured data from general purpose relational databases [36].

Neo4j has already developed "*Docker* compose" for deployment of Polyglot persistence. The *Docker Image* allows defining multiple *DockerContainer*s and link them to a single **YAML** configuration file [55].

OpenShift PaaS from *Red Hat* is *Multilanguage, Auto-Scaling, Self-service*, and *Elastic*. *OpenShift* supports MongoDB, PostgreSQL, and MySQL and multilanguage support, i.e., *Java, Perl, Python, PHP*, and *Ruby* [56].

Table 11 shows some interesting global figures of the various commodities/services relating to this chapter by 2020.

e-Agriculture focuses on the enhancement of agricultural and rural development. e-Agriculture involves the application of *Smartphones, Cloud*

Table 11 IoT, DBaaS, and Healthcare Information System Market by 2020

Services	All Figures in Billion
Cloud Shift	US$ 216 [57]
DBaaS	US$ 19.0 [36]
NoSQL	US$ 4.2 [58]
Retail e-Commerce sale	US$ 4058 [40]
IoT healthcare market	US$ 163.24 [59]
IoT healthcare market by 2021	US$ 136 [60]
Healthcare big data analytics	US$ 24.55 [59]
IoT market size	US$ 3.7 [61]
Number of IoT devices	50 billion [62]
Data generated	44 zettabyte [63]
Devices connected to the network	25 billion devices [64]
Healthcare information system	US$ 54 (market research store, [65])
IoT revenue	19 US$ Trillion [66]

Computing, ***IoT***, and ***Big Data Analytics***. The application of IoT in food factories converts the machines to cyber-physical systems with embedded intelligence and local data processing directly communicating with other machines [67].

IoT devices will be generating huge volume of unstructured ***Time Series Data***. Pushed by geographically distributed devices in milliseconds/nanoseconds deployed in healthcare, e-Agriculture and other areas. In healthcare, the telemetric and wearable IoT devices will be gathering the real-time information on health metrics for risk assessment for premium calculation of insurance company and health practitioners. Processing these huge amount of data in real-time is possible through cloud services which act as concentrators to capture and store data.

Big data analytics tool will unlock value/insight hidden in Zettabyte (1 Zettabyte $= 2^{70}$ bytes) of real-time data generated IoT devices/sensors. These data are to be stored in ***Low*** cost/***Open source*** NoSQL database such as a ***Key–value*** store or a ***Column*** family database. Graph NoSQL databases using ***relation*** are the natural choice for managing the connected IoT devices. Open source ***Column*** family database, like ***Cassandra*** or ***HBase*** can be used for storage of IoT devices generated data [68].

In compatibility, proprietary architectures, platforms, and standards are leading to vendor lock-in for the adoption of IoT in e-Agriculture. There are more than *115* protocols used by IoT devices for exchange of communication [69]. According to a McKinsey paper, interoperability will unlock more than "US\$ 4 trillion per year in potential economic impact for IoT use in 2025, out of a total impact of US\$ 11.1 trillion." Interoperability is essential for creating 40% of the potential value to be generated by IoT in various settings [70].

The IoT generated data are communicated over IP (internet protocol) to the edge servers/cloud for storage/processing. IP Version 6 (IPv6) will eventually become the default protocol for IoT for connecting the devices to the internet [52]. The networks must allow a variety of devices to connect transparently enabling high performance computing and real-time video and multimedia on demand [71]. For data interoperability, Polyglot persistence can play a big role.

Database of Things (DoT), [72], capable of storing the data in a time series, set of data from one device in relation to data of another device is highly demanded in the age of IoT.

Blockchain is the technological innovation of **Bitcoin** for the public transaction ledger in a decentralized network. Digitally recorded data are packaged in **Blocks** for storage in a linear chain. Each block in the chain contains cryptographically hashed Bitcoin transactions. Every **Block** contains a hash of the previous block for creating a chain of blocks. In lockchain distributed transaction database (ledger), all transactions are owned and monitored by everyone and controlled by none, i.e., there is no central database. Every node joins the network voluntarily and an "Administrator" of the Blockchain. Blockchain database is not stored in any single location, hence the records are public, easily verifiable, and accessible to anyone on the internet. Blockchain technology will radically improve banking, supply chain, health data, financial assets, automobiles, e-Governance, and other transaction networks [73]. Microsoft has already started the Blockchain as a Service (**BaaS**) for the customers [74].

Ethereum (https://www.ethereum.org) and other cryptocurrencies [75] in distributed computing platform are powered by Blockchain.

9. CONCLUSION

The real world applications require the data integrity offered by an RDBMS in addition to the benefits offered by NoSQL databases. None of the database meets all the requirements of a web-based application. Hence

Polyglot persistence is using a variety of data storage tools/technology for managing multiform data according to different properties and requirements will easy the situation to some extent.

Designing and implementation of a Polyglot system is not a straightforward task, since adding more data storage technologies increases complexity in programming. Further knowledge of multiple programming language, maintainability of the Polyglot application, and tool support for Polyglot persistence are the challenges of Polyglot persistence. It is observed from various studies that the number of databases used should be kept as small as possible. Selection of the data store must be based on the requirements to minimize the number of databases. The **thumb rule** for Polyglot persistence is to use "as many as *necessary*, as few as *possible*."

Some of the approaches to implement Polyglot solutions in an organization are:

- Multiple lanes
- Polyglot mapper
- Nested database
- Hadoop database.

The NewSQL databases deliver the scalability and flexibility promised by NoSQL while retaining the support for SQL queries. The NewSQL database is likely to be another option for developing applications requiring the NoSQL–SQL dual characteristics. Examples of popular NewSQL database are VoltDB. Two projects Clustrix and NuoDB, a commercial are also classified as NewSQL. Hadoop Ecosystem components **Pig** (Key–value), **GraphX** (Graph), **Drill** (SQL), **SparkSQL** (SQL), and **HBase** (Column) are the best example of Polyglot persistence use case.

GLOSSARY

Scripting Languages are used for smaller applications, but can also be used in large applications.

XML an embedded configuration language widely used in *Java* and *.NET* world. **XML** text format encodes DOM (document-object models) which is a data structure for web pages.

SAX (simple API for XML) is an alternative "lightweight" way to process XML. A SAX parser generates a stream of events as it parse the XML file.

Polyglot persistence refers continued or prolonged existence of something using several languages. The Polyglot-persistent application uses multiple data storage, i.e., RDBMS, NoSQL, BLOB, file in a single (minimum) application that best suits the requirements of an application. The parameter which plays a vital role in the development of Polyglot-persistent application are **Portability**, **Flexibility**, and strong RDBMS integration.

Nonnetworked type of Polyglot programming where the different languages are in the same file.

JSON (JavaScript object notation) is a commonly used in web to describe the data being exchanged between systems. JSON is a simple schema less language created to support web pages. JSON object enters database as a text string which is parsed to discover its fields and types.

YAML (Yet Another Markup Language) represents data in human-readable notation, almost a superset of JSON.

NewSQL systems are relational databases designed to provide real-time *OLTP* (online transaction processing) and conventional SQL-based OLAP (online analytical processing) in big data environments.

Bitcoin—A protocol that supports a decentralized, pseudoanonymous, peer-to-*peer* digital currency. The blockchain is the fundamental data structure of the Bitcoin protocol.

REFERENCES

[1] C. Boylan, Data Integration (Database Integration). Retrieved January 23, 2017, from http://www.informationbuilders.com/database-integration, n.d.

[2] K.K. Mohan, A.K. Verma, A. Srividya, R.K. Gedela, QoS Considerations for Integration-Enterprise Service Bus, Communications in Dependability and Quality Management—An International Journal 12 (1) (2009) 47–56. doi: UDC 004.421.2: 519.6, CDQM.

[3] TrustRadius, Data Integration Overview—What Is Data Integration Software? Retrieved January 24, 2017, from, https://www.trustradius.com/data-integration, 2016. 11921 North MopacExpy, Suite 330, Austin, TX 78759.

[4] R. Pethuru, Cloud Enterprise Architecture, CRC Press, 2012. http://www.crcpress.com/product/isbn/9781466502321.

[5] E. Letuchy, Facebook Chat, Retrieved January 21, 2017, from, http://www.facebook.com/note.php?note_id=14218138919&id=9445547199&index=1, 2008.

[6] Rabah, S., Li, J., Liu, M., & Lai, Y. (2010). Comparative Studies of 10 Programming Languages within 10 Diverse Criteria—a Team 7 COMP6411-S10 Term Report. arXiv preprint arXiv:1009.0305.

[7] D. Pountain, Functional Programming Comes of Age, 2006. BYTE.com (August 1994). Archived from the original on 2006-08-27. Retrieved January 24, 2017.

[8] Wikimedia Foundation Inc., Polyglot (Computing), Retrieved January 21, 2017, from: https://en.wikipedia.org/wiki/Polyglot_(computing), 2016. Available under the Creative Commons Attribution-ShareAlike License.

[9] R. Diana, The Big List of 256 Programming Languages-Check Out a List of 256 Programming Languages, From ABC to Z Shell, Retrieved January 24, 2017, from, DZone/Java Zone, 2013. https://dzone.com/articles/https://dzone.com/articles/big-list-256-programmingbig-list-256-programming.

[10] M. Wrangler, Polyglot Programming. Retrieved January 24, 2017, from http://nealford.com/memeagora/2006/12/05/Polyglot_Programming, Thought Works, a software company and a community of passionate, purpose-led individuals, n.d.

[11] B. Kullbach, A. Winter, P. Dahm, J. Ebert, in: Program Comprehension in Multi-Language Systems. Reverse Engineering, 1998, Proceedings. Fifth Working Conference on, 135–143, 1998. Available from, http://ieeexplore.ieee.org/iel4/5867/15624/00723183.pdf?tp=&arnumber=723183&isnumber=15624.

[12] H. Fjeldberg, Polyglot Programming-A business perspective (Master's thesis), Norwegian University of Science and Technology, Department of Computer and Information Science, Trondheim, Norway, 2008, Norwegian University of Science and

Technology, Department of Computer and Information Science, Trondheim, Norway, 2008, pp. 1–75. Supervisor: HallvardTrætteberg, IDI.

[13] S. Yegge, Rhino on Rails, Retrieved January 21, 2017, from, http://steve-yegge. blogspot.com/2007/06/rhino-on-rails.html, 2007.

[14] M. Neunhöffer, July 7, in: Data Modeling With Multi-Model Databases—A Case Study for Mixing Different Data Models Within the Same Data Store, 2015. Retrieved January 23, 2017, fromhttps://www.oreilly.com/ideas/data-modeling-with-multi-model-databases.

[15] MapR Technologies, Inc, Why MapR, Retrieved January 22, 2017, fromhttps://www. mapr.com/why-hadoop/why-mapr/enterprise-platform, 2016. One Platform for Big Data Applications.

[16] C. Richardson, Microservices and the Problem of Distributed Data Management, Retrieved January 23, 2017, from, DZone Integration Zone, 2016. https://dzone. com/articles/event-driven-data-management-for-microservices-par.

[17] E. De Oliveira, Why a Key–Value Store NoSQL Database Is the Best Solution for Future IoT Needs, Retrieved February 01, 2017, from, FairCom Corporation, 2016. https://www.faircom.com/insights/key-value-store-nosql-database-best-solution-future-iot-needs.

[18] W. Lyon, Neo4j Doc Manager: Polyglot Persistence for MongoDB& Neo4j, Retrieved January 25, 2017, from, Neo Technology, Inc., 2015. https://neo4j.com/blog/neo4j-doc-manager-polyglot-persistence-mongodb/.

[19] C. Bizer, D2R Map—Database to RDF Map—Chris Bizer. Retrieved January 25, 2017, from http://wifo5-03.informatik.uni-mannheim.de/bizer/d2rmap/D2Rmap. htm, FU Berlin, n.d.

[20] Salesforce.com, Inc, Multi Tenant Architecture–Developer. Force.com, Retrieved January 26, 2017, from, San Francisco, CA. https://developer.salesforce.com/page/ Multi_Tenant_Architecture, 2016.

[21] K. Prasad, Polyglot Persistence, Retrieved January 26, 2017, from, GitHub, Inc, 2016. https://github.com/impetus-opensource/Kundera/wiki/Polyglot-Persistence.

[22] Wikimedia Foundation Inc., LAMP (Software Bundle), Retrieved March 08, 2017, from, https://en.wikipedia.org/wiki/LAMP_(software_bundle), 2017.

[23] K. Payne, A Step by Step Guide to Install LAMP (Linux, Apache, MySQL, Python) on Ubuntu, Retrieved March 08, 2017, from, http://blog.udacity.com/2015/03/step-by-step-guide-install-lamp-linux-apache-mysql-python-ubuntu.html, 2015.

[24] The PHP Group, Themysqli Class, Retrieved March 08, 2017, from, http://php.net/ manual/en/class.mysqli.php, 2017.

[25] N. Dhandala, The Pros and Cons of Polyglot Persistence, Retrieved January 25, 2017, from, http://opensourceforu.com/2015/08/the-pros-and-cons-of-polyglot-persistence/, 2015. Tags: e-Commerce, Enterprise, Network, NoSQL, Persistence.

[26] S. Pimentel, The Rise of the Multimodel Database, Retrieved March 26, 2017, from, http://www.infoworld.com/article/2861579/database/the-rise-of-the-multimodel-database.html, 2015.

[27] Stamford, Gartner Survey Reveals That 73 Percent of Organizations Have Invested or Plan to Invest in Big Data in the Next Two Years, Retrieved February 26, 2017, from, Gartner, Inc, 2014. http://www.gartner.com/newsroom/id/2848718.

[28] D. Abadi, R. Agrawal, A. Ailamaki, M. Balazinska, P.A. Bernstein, M.J. Carey, S. Chaudhuri, et al., The Beckman Report on Database Research, Retrieved January 28, 2017, from, Beckman Center of the National Academies of Sciences & Engineering Irvine, CA, USA, 2013. http://beckman.cs.wisc.edu/beckman-report2013.pdf.

[29] Apache Storm, Retrieved January 22, 2017, from http://hortonworks.com/apache/ storm/, n.d.

[30] S. Edlich, N*SQL, Retrieved January 24, 2017, from, http://nosql-database.org, 2011. Owned by: Prof. Dr. Stefan Edlich Stubenrauchstr.9c 14167 Berlin Gemany.

[31] A. Fowler, NoSQL For Dummies (India ed.), Wiley India Pvt. Ltd. A Wiley Brand, New Delhi, 2015.

[32] MongoDB, Inc, MongoDB Connector for BI, Retrieved February 18, 2017, from, https://www.mongodb.com/products/bi-connector, 2017.

[33] (a) ReactiveCouchbase, 2017. ReactiveCouchbase/repository. Retrieved August 24, 2017, from https://github.com/ReactiveCouchbase/repository/blob/master/snapshots/org/reactivecouchbase; (b) R. Zicari, New Gartner Magic Quadrant for Operational Database Management Systems. Interview with Nick Heudecker, Retrieved February 18, 2017, from, http://www.odbms.org/blog/2016/11/new-gartner-magic-quadrant-for-operational-database-management-systems-interview-with-nick-heudecker/, 2016 ODBMS.org.

[34] D. Feldman, Avoiding the Franken-Beast: Polyglot Persistence Done Right.Retrieved January 24, 2017, from http://www.marklogic.com/blog/polyglot-persistence-done-right/, MarkLogic®/Agile Development, Data Management, n.d.

[35] OrientDB Ltd, Why Orientdb?Graph/Document Multi-Model Database. Retrieved January 26, 2017, from http://orientdb.com/why-orientdb/, OrientDB Open Source Project. n.d.

[36] 451 Research, NoSQL, NewSQL and Beyond: The Drivers and Use-Cases for Database Alternatives, Retrieved January 23, 2017, from, http://www.bing.com/cr?IG=548773F8E30444D19B7B1B3FA576F38F&CID=365A0BE5B3476EDE2F1A01F6B2766FB1&rd=1&h=g47O96F_f6AXs9xLPkf6FUjtDHpwEH142OcqpiMER6w&v=1&r=https%3a%2f%2f451research.com%2freport-long%3fficid%3d1389&p=DevEx, 2011. 5083.1, The 451 Group: Commercial Adoption of Open Source.

[37] M. Stonebraker, New SQL: An Alternative to NoSQL and Old SQL for New OLTP Apps, Retrieved February 16, 2017, from, http://cacm.acm.org/blogs/blog-cacm/109710-new-sql-an-alternative-to-nosql-and-old-sql-for-new-oltp-apps/fulltext, 2011.

[38] K. Grolinger, W.A. Higashino, A. Tiwari, M.A.M. Capretz, Data Management in Cloud Environments: NoSQL and NewSQL Data Stores, Retrieved February 16, 2017, from, http://journalofcloudcomputing.springeropen.com/articles/10.1186/2192-113X-2-22, 2013. Journal of Cloud Computing.

[39] M. Pezzini, D. Feinberg, N. Rayner, R. Edjlali, Hybrid Transaction/Analytical Processing Will Foster Opportunities for Dramatic Business Innovation, Retrieved February 15, 2017, from, Gartner, Inc., 2014. https://www.gartner.com/doc/2657815/hybrid-transactionanalytical-processing-foster-opportunities.

[40] Statista Inc., Global retail e-commerce sales 2014-2020, Retrieved January 26, 2017, from, https://www.statista.com/statistics/379046/worldwide-retail-e-commerce-sales/, 2014Worldwide; eMarketer; 2014 to 2016.

[41] A. Maccioni, O. Cassanon, Y. Luo, J. Castrejon-Castillo, G. Vargas-Solar NoXperanto: Crowdsourced Polyglot Persistence. [Research Report] RRLIG-047, 2014, http://basho.com/use-cases/iot-sensor-device-data/.

[42] Basho, IoT/Sensor/Device Data. Retrieved January 29, 2017, from http://basho.com/use-cases/iot-sensor-device-data/, n.d.

[43] A Database for Internet of Things (IoT) Data, Retrieved January 29, 2017, from http://db4iot.com/. n.d.

[44] The OpenTSDB Authors, OpenTSDB—A Distributed, Scalable Monitoring System, Retrieved January 29, 2017, from, http://opentsdb.net, 2016. OpenTSDB 2.3.0.

[45] VoltDB, Inc, IoT Sensors, Retrieved January 29, 2017, from, https://www.voltdb.com/solutions/verticals/iot-sensors, 2016.

[46] Dataversity, UnQL: A Standardized Query Language for NoSQL Databases, Retrieved January 22, 2017, from, http://www.dataversity.net/unql-a-standardized-query-language-for-nosql-databases/, 2013. Dataversity Education, LLC.

[47] Microsoft Developer Network, JSON Data (SQL Server), Retrieved January 26, 2017, from, https://msdn.microsoft.com/en-us/library/dn921897.aspx, 2016.

[48] XML DB Developer's Guide, 39 JSON in Oracle Database. Retrieved January 26, 2017, from http://docs.oracle.com/database/121/ADXDB/json.htm#ADXDB6246, Oracle, n.d.

[49] DataStax, Inc, Inserting JSON Data Into a Table, Retrieved January 26, 2017, from, https://docs.datastax.com/en/cql/3.3/cql/cql_using/useInsertJSON.html, 2017.

[50] Microsoft, Convert JSON Data to Rows and Columns With OPENJSON (SQL Server), Retrieved February 19, 2017, from, https://msdn.microsoft.com/en-us/library/dn921879.aspx, 2017. SQL Server 2016 and later.

[51] Oracle, Database as a Service (DBaaS): Use Cases and Adoption Patterns A Report on Research Commissioned by Oracle, Retrieved January 26, 2017, from, http://www.oracle.com/us/products/database/451-report-dbaas-3038788.pdf, 2016. 451 Research, LLC and/or its Affiliates.

[52] International Telecommunication Union, Harnessing the Internet of Things for Global Development, Retrieved February 5, 2017, from, A Contribution to The UN Broadband Commission for Sustainable Development, 2016. https://www.itu.int/en/action/broadband/Documents/Harnessing-IoT-Global-Development.pdf.

[53] (a) MemSQL Inc, Real-Time Solutions for IoT. Retrieved February 05, 2017, from http://www.memsql.com/solutions/internet-of-things/, n.d.; (b) R. Das, 2016, April 26, Five Technologies that Will Disrupt Healthcare By 2020. Retrieved August 31, 2017, from https://www.forbes.com/sites/reenitadas/2016/03/30/top-5-technologies-disrupting-healthcare-by-2020/#51b467e16826.

[54] D. Ergürel, Virtual Reality Healthcare Systems Market to Reach $2.5 Billion by 2020, Retrieved March 05, 2017, from, https://haptic.al/healthcare-virtual-reality-market-9d23a2cc9cd6#.ze5plsida, 2016. Haptical.

[55] W. Lyon, NoSQL Polyglot Persistence: Tools and Integrations With Neo4j, Retrieved January 23, 2017, from, Neo Technology, Inc, 2016. https://neo4j.com/blog/nosql-polyglot-persistence-tools-integrations/#why-polyglot-persistence.

[56] S. Gulati, Thinking Beyond RDBMS: Building Polyglot Persistence Java Applications. Retrieved January 25, 2017, from http://events.linuxfoundation.org/sites/events/files/slides/Thinking%20Beyond%20RDBMS%20–%20Building%20Polyglot%20Persistence %20Java%20Applications.pdf, The Linux Foundation®, n.d.

[57] Gartner, Gartner Says by 2020, Retrieved January 28, 2017, from, STAMFORD, Conn., 2016. http://www.gartner.com/newsroom/id/3384720.

[58] L. Person, NoSQL Market by Type (Key–Value Store, Document Databases, Column Based Stores and Graph Database) and Application (Data Storage, Metadata Store, Cache Memory, Distributed Data Depository, e-Commerce, Mobile Apps, Web Applications, Data Analytics and Social Networking)—Global Opportunity Analysis and Industry Forecast, 2013–2020, Retrieved January 28, 2017, from, Allied Market Research, 2015. https://www.alliedmarketresearch.com/NoSQL-market.

[59] Home, Healthcare Analytics Market Worth 24.55 Billion USD by 2021. Retrieved January 28, 2017, from http://www.marketsandmarkets.com/PressReleases/healthcare-data-analytics.asp, n.d.

[60] M. A. Malik, Internet of Things (IoT) Healthcare Market by Component (Implantable Sensor Devices, Wearable Sensor Devices, System and Software), Application (Patient

Monitoring, Clinical Operation and Workflow Optimization, Clinical Imaging, Fitness and Wellness Measurement)—Global Opportunity Analysis and Industry Forecast, 2014–2021. Retrieved January 28, 2017, from, https://www.alliedmarketresearch.com/iot-healthcare-market, n.d.

[61] L. Columbus, Roundup of Internet of Things Forecasts and Market Estimates, 2016, Retrieved January 28, 2017, from, Forbes, 2016. http://www.forbes.com/sites/louiscolumbus/2016/11/27/roundup-of-internet-of-things-forecasts-and-market-estimates-2016/#289c2e8e4ba5.

[62] S.Z. Aeris, 50B IoT Devices Connected by 2020—Beyond the Hype and Into Reality, Retrieved January 28, 2017, from, Syed Zaeem Hosain, Cto, Aeris, 2016. http://www.rcrwireless.com/20160628/opinion/reality-check-50b-iot-devices-connected-2020-beyond-hype-reality-tag10.

[63] A. Adshead, Data Set to Grow 10-Fold by 2020 as Internet of Things Takes Off, Retrieved January 28, 2017, from, Antony Adshead, Storage Editor–TechTarget–ComputerWeekly, 2014. http://www.computerweekly.com/news/2240217788/Data-set-to-grow-10-fold-by-2020-as-internet-of-things-takes-off.

[64] A. Staff, Terabyte Terror: It Takes Special Databases to Lasso the Internet of Things, Retrieved January 29, 2017, from, WIRED Media Group, 2016. https://arstechnica.com/information-technology/2016/06/building-databases-for-the-internet-of-data-spewing-things/.

[65] Market Research Store, Global Healthcare Information System Market Set for Rapid Growth, To Reach Around USD 54.0 Billion By 2020, Retrieved January 29, 2017, from, http://www.marketresearchstore.com/news/global-healthcare-information-system-market-190, 2015.

[66] MongoDB, Internet of Things, Retrieved February 01, 2017, from, MongoDB, Inc., 2017. https://www.mongodb.com/use-cases/internet-of-things.

[67] G.M. Siddesh, G.C. Deka, K.G. Srinivasa, L.M. Patnaik, Cyber-Physical Systems: A Computational Perspective, CRC Press, Taylor & Francis Group, Boca Raton, 2016.

[68] CITO Research, Six Essential Skills for Mastering the Internet of Connected Things. Retrieved January 29, 2017, from http://www.evolvedmedia.com/wp-content/uploads/2015/05/CITO-Research_Neo-Technology_Six-Essential-Skills-for-Mastering-the-Internet-of-Connected-Things_White-Paper_2014.pdf, Sponsored by Neo4j n.d.

[69] G. Schatz, The Complete List of Wireless IoT Network Protocols, Retrieved February 06, 2017, from, LINK LABS, 2016. https://www.link-labs.com/blog/complete-list-iot-network-protocols.

[70] J. Manyika, M. Chui, P. Bisson, J. Woetzel, R. Dobbs, J. Bughin, D. Aharon, The Internet of Things: Mapping The Value Beyond The Hype-Executive Summary, Mckinsey Global Institute, Retrieved February 5, 2017, from, McKinsey & Company, 2015. file:///C:/Users/HP%20LAPTOP/Downloads/Unlocking_the_potential_of_the_Internet_of_Things_Executive_summary.pdf.

[71] Cisco, Cisco VNI Mobile Forecast (2015–2020)-Cisco, Retrieved February 06, 2017, from, http://www.cisco.com/c/en/us/solutions/collateral/service-provider/visual-networking-index-vni/mobile-white-paper-c11-520862.html, 2016.

[72] S. Samaraweera, Database of Things (DoT): The Future of Database. Has It's Time Come? Retrieved February 05, 2017, from, Virtusa Corporation, 2015. http://blog.virtusapolaris.com/2015/10/database-of-things-dot-the-future-of-database-has-its-time-come/.

[73] S. Norton, CIO Explainer: What Is Blockchain? Retrieved February 18, 2017, from, Dow Jones & Company, Inc, 2016. http://blogs.Wsj.Com/cio/2016/02/02/cio-explainer-what-is-blockchain/.

[74] The Windows Club, Blockchain Technology & Microsoft's Blockchain as a Service Strategy, Retrieved February 19, 2017, from, http://www.thewindowsclub.com/blockchain-microsoft-plans-develop-service, 2016.
[75] O. Ogundeji, Blockchain News, Retrieved February 18, 2017, from, https://cointelegraph.com/tags/blockchain, 2017.

FURTHER READING

[76] D.G. Chandra, M.D. Borah, in: Cost benefit analysis of cloud computing in education, In Computing, Communication and Applications (ICCCA), 2012 International Conference on, IEEE, 2012, pp. 1–6.
[77] Deka, G. C. (2014). A survey of cloud database systems IT Professional, 16(2), 50–57. https://doi.org/10.1109/MITP.2013.1, Survey of 15 popular NoSQL Databases.
[78] D.G. Chandra, BASE analysis of NoSQL database, Future Gener. Comput. Syst. 52 (2015) 13–21.
[79] G.C. Deka, Handbook of Research on Securing Cloud-Based Databases With Biometric Applications, IGI Global, 2014.
[80] G.C. Deka, P.K. Das, An Overview on the Virtualization Technology. Handbook of Research on Cloud Infrastructures for Big Data Analytics, IGI Global, 2014, pp. 289–321.
[81] G.S. Tomar, N.S. Chaudhari, R.S. Bhadoria, G.C. Deka, The Human Element of Big Data: Issues, Analytics, and Performance, CRC Press, 2016.
[82] Deka, G. C., Siddesh, G. M., Srinivasa, K. G., &Patnaik, L. M. (2016). Emerging Research Surrounding Power Consumption and Performance Issues in Utility Computing. Information Science Reference.
[83] Deka, G. C., Zain, J. M., & Mahanti, P., ICT's role in e-Governance in India and Malaysia: a review, (2012) arXiv Preprint arXiv:1206.0681.

ABOUT THE AUTHOR

Ganesh Chandra Deka is currently Deputy Director (Training) at Directorate General of Training, Ministry of Skill Development and Entrepreneurship, Government of India.

His research interests include ICT in Rural Development, e-Governance, Cloud Computing, Data Mining, NoSQL Databases, and Vocational Education and Training. He has published more than 57 research papers in various conferences, workshops, and International Journals of repute including IEEE & Elsevier. So far he has organized 08 IEEE International Conference as Technical Chair in India. He is the member of editorial board and reviewer for various Journals and International conferences.

He is the co-author for four text books in fundamentals of computer science. He has edited seven *books* (three IGI Global, USA, three CRC Press, USA, one Springer) on *Bigdata*, *NoSQL*, and *Cloud Computing* in general as of now.

He is Member of *IEEE*, the *Institution of Electronics and Telecommunication Engineers*, India and Associate Member, the *Institution of Engineers*, India.

Printed in the United States
By Bookmasters